AF552497

ENDOCRINOLOGY

ENDOCRINOLOGY

By

Dr. P.R. Yadav
Lecturer
Department of Zoology
D.A.V. College
Muzaffarnagar (U.P.)

D P H

DISCOVERY PUBLISHING HOUSE
NEW DELHI-110002

Reprinted - 2019

First Published - 2004

ISBN: 978-81-7141-856-5

Endocrinology

Published by:

DISCOVERY PUBLISHING HOUSE PVT. LTD.

4383/4B, Ansari Road, Darya Ganj
New Delhi-110 002 (India)
Phone: +91-11-23279245, 23253475; 43596065
E-mail: discoverybooksindia@gmail.com
discoverypublishinghouse@gmail.com
web: www.discoverypublishinggroup.com

Printed at:
Infinity Imaging Systems
Delhi

PREFACE

The present title is directed toward the advanced undergraduate Zoology student who has some background in physiology as well as toward the graduate student seeking an introduction to the field. Because various disciplines encompossed by the field of endocrinology are changing so rapidly, it is difficult to keep any text book on this subject current. The exploration of integrative mechanisms among organisms of phylogenetically different levels of complexity has been rewarding and the results are being applied to such basic problems as metabolism, reproduction, development, ecologic adaptation, and organic evolution. The chemical co-ordination of individuals and groups of individuals is recognized as a pervasive theme—throughout the world of life, and the subject is inextricably incorporated into the contemporary biology. Our knowledge of the mechanisms whereby major functions of the body are integrated through the diffusible products of the nervous and endocrine system has assumed such proportions that endocrinology can no larger be adequately taught as part of another course

In the preparation of this book large number of books and research papers have been consulted. So no authenticity is claimed.

The author wishes to express his deepest appreciation to the many people who have contributed in one way or the other to the preparation of this title.

The author expresses his gratitute to Mr. Wasan and staff of M/s Discovery Publishing House for their whole hearted co-operation in the publication of this book.

The author tried hard to be accurate and upto date in statement and realises the impossibility of completely avoiding errors therefore, the author will greatly appreciate having his attention called to any questionable statement.

Author

CONTENTS

1

INTRODUCTION

Hormones are molecules that are synthesized and secreted by specialized cells, released into the blood, and exert biochemical effects on target cells at a distance from their site of origin. Some hormones such as thyroid-stimulating hormone act exclusively on one target tissue (in this case the thyroid gland); other hormones such as insulin and thyroid hormone *act* on many cell types, including (in this example) liver, brain, and skin. The specificity of hormone action is determined by the presence of specific hormone receptors on or in these target cells. The cellular response, however, is determined by the genetic programming of the particular cell. Thus, the same hormone may have different actions on different tissues. For example, adrenal glucocorticoids cause cytolysis of lymphocytes but induce enzymes necessary for the production of glucose in the liver.

CLASSES OF HORMONES

Hormones have diverse molecular structures ranging from single modified amino acids (*epinephrine*) through lipids (estrogen, cortisol) to proteins (glucagon, insulin, growth hormone). In correlating properties of hormones within physiologic actions, it is generally useful to categorize them as (1) peptides and proteins, (2) steroids, and (3) amines or amino acid derivatives. This listing is representative and not complete, since new molecules with hormonal activity are continually being discovered. For example, an area of investigation that is particularly active at present is the identification of new gastrointestinal hormones.

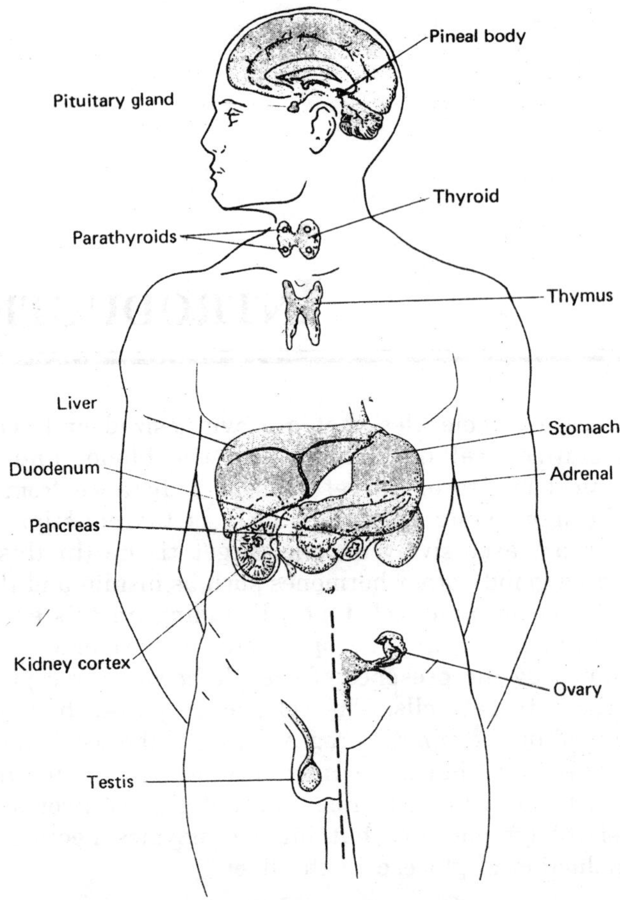

Fig. 1.1. Approximate locations of the endocrine glands of the human. Although the liver and kidneys add important materials to the blood, they are not usually considered to be endocrine glands.

Hormones as Regulatory Molecules

Hormones are one class of regulatory molecules and act in somewhat the same ways as other regulatory molecules such as the neurotransmitters, paracrine agents, and metabolites. In general, the mechanism of intercellular regulation consists of perception of a stimulus by a cell by means of a specialized cell receptor, followed by the cellular response. In the case of hormone action, the stimulating hormone becomes bound to its specific receptor on the cell, and the result is a cascade of intracellular events including integration of several stimuli and amplification of the original bound stimulus, culminating in the cellular response. The stimulus may consist of only a few molecules of hormone per target cell. Intracellular amplification generally involves either intracellular messengers such as cAMP and Ca^{2+} or the synthesis of new RNA and protein. Cellular regulation by hormones, then, follows the same general pattern of *cell activation* observed in the control of other biologic processes; examples are the development of an egg following fertilization, the contraction of muscle following electrical excitation, and the activation, following exposure to antigen, of a quiescent lymphocyte into one that actively synthesizes specific antibody.

Hormone Specificity Through Unique Receptors

Binding sites for peptide hormones and catecholamines are generally large glycoproteins on or in cellular membranes, including the plasma membrane, Golgi complex, and in some cases nuclear membrane. Hormones initially bind to receptors present on the cell surface membrane (*plasma membrane*) that is exposed to the cell environment. It is not clear whether binding sites in intracellular membranes, such as those on the Golgi complex, endoplasmic reticulum, and nuclear membrane, play a role in hormone action or are merely being transported to or from the plasma membrane as part of a general turnover of membrane proteins. In contrast to polypeptide hormone the known receptors for steroids and thyroid hormones are located chiefly within the cell. Steroid receptors are located in the cytoplasm; thyroid receptors are located in the plasma membrane, nucleus, and mitochondria. Steroid and thyroid hormones are small lipophilic molecules that are assumed to be taken up through the cell membranes; the mechanism of cellular penetration by these molecules is not known, but evidence exists for mechanisms other than simple diffusion. Once inside the cell, these hormones interacts with their intracellular receptors. In some cases, the action of these hormones may be initiated by their initial interactions with plasma membrane receptors at the cell membrane.

While receptors (as distinct from binding proteins) must have the potential to be coupled to a biologic response, the receptor itself is a separate unit involved with recognition. In

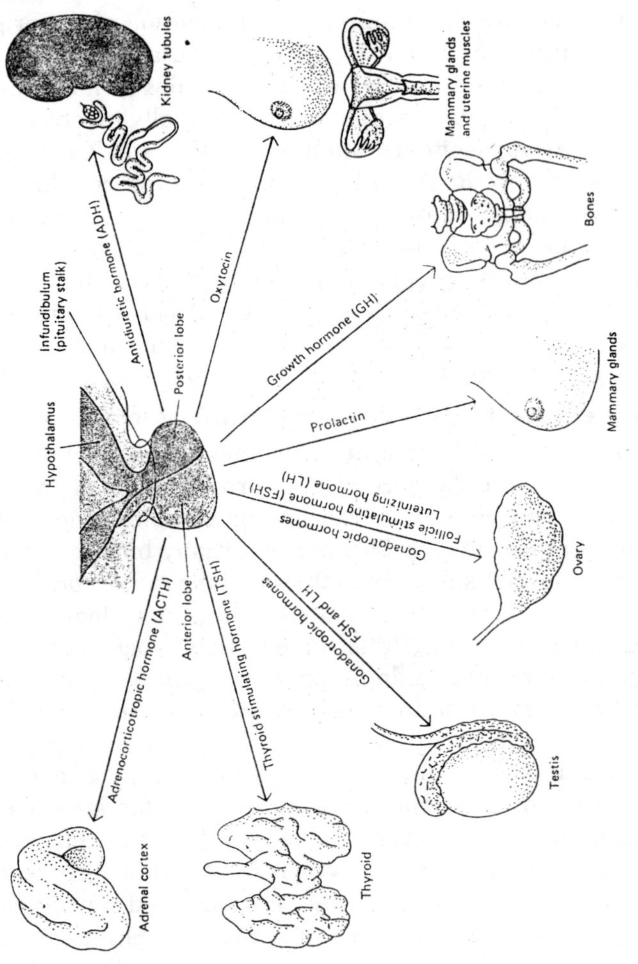

Fig. 1.2. The pituitary gland is suspended from the hypothalamus by a stalk of neural tissue.

general, hormone receptor interactions are noncovalent; they are reversible, involving hydrogen bonding and hydrophobic and electrostatic forces. Both membrane and intracellular receptors can be solubilized and isolated and their properties studied. In the case of membrane receptors, which are normally coupled to an effector protein (adenylate cyclase, Ca^{2+} channel), addition of a new receptor by cell fusion or reimplantation of a solubilized receptor to a previously unresponsive cell possessing the effector system will allow that cell to respond to that hormone. Abnormalities in receptors or absences of specific receptors are now known to underlie some pathologic states, including insulin resistance, testicular feminization, and certain types of dwarfism, diabetes insipidus, and pseudohypoparathyroidism.

Synthesis and Secretion of Hormones

Polypeptide and Amine Hormones

The cells that secrete polypeptide hormones have many common structural features related to the synthesis, packaging, and release of these molecules. Polypeptide hormones, along with other proteins destined for export from the cell, are synthesized on membrane bound ribosomes and then sequestered into the cisternae of the endoplasmic reticulum. By contrast, proteins destined to remain in the cell are synthesized on free ribosomes.

Initially, the messenger RNA (mRNA) specific for the polypeptide hormone attaches to a free ribosome, and this is followed by initiation of translation at an AUG initiation codon. The initial amino acid sequence of the forming peptide (specified by the mRNA) then causes the ribosome-mRNA complex to bind to the endoplasmic reticulum. This initial signal peptide, composed largely of hydrophobic amino acids, induces formation of a pore or channel such that the newly synthesized hormone is extruded through the membrane of the endoplasmic reticulum and into its lumen. The complete molecule, including the signal peptide, is referred to as a *prehormone*; its lifetime is very short, however, as the signal peptide is rapidly cleaved off by a trypsinlike enzyme present in the endoplasmic reticulum.

In many cases, further modification of the hormone are required before its secretion, such as the addition of sugar to the glycoprotein hormones hCG, TSH, LH, and FSH. If extra amino acids are still present following cleavage of the signal peptide, this form of the molecule is referred to as a *prohormone*. In the case of insulin, the prohormone proinsulin is a single chain, which allows folding and opposition of the 2 ends of the molecule so that interchain disulfide bonds can be correctly formed. When the middle sequence of proinsulin,

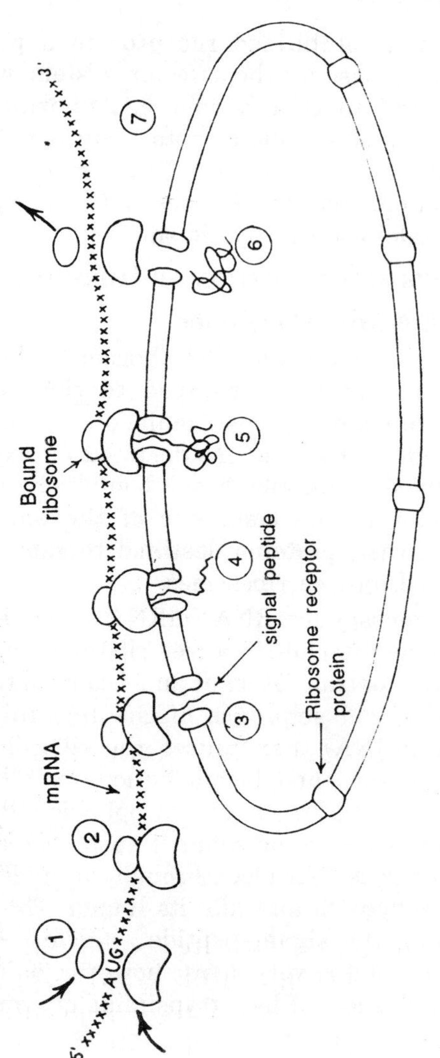

Fig. 1.3. Synthesis and sequestration of polypeptide hormones. Synthesis begins at the left as ribosomal subunits aggregate (1) and begin to read the mRNA specific for the hormones (2). Subsequent steps shown include (from left to right) synthesis of the initial signal peptide that interacts with ribosome receptor protein (3) extrusion of the nascent peptide into the lumen of the endoplasmic reticulum (4-5); release of the protein into the cisternae of the endoplasmic reticulum (6); and dissociation of ribosomal subunit from the mRNA.

the connecting or C peptide, is removed, the mature 2-chain form of insulin remains.

Following synthesis and sequestration within the lumen of the endoplasmic reticulum, the newly synthesized hormone moves to the Golgi region by vesicular transport, and is there packaged into granules or vesicles. This intracellular transport is guided by microtubules and can be blocked by drugs such as colchicine that disrupt microtubules. It is in the Golgi region

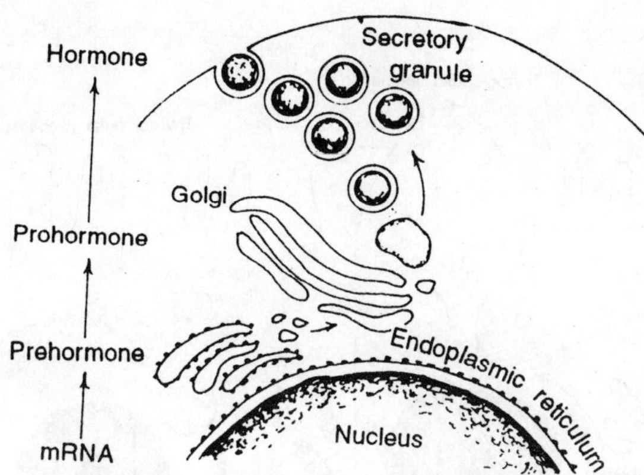

Fig. 1.4. Intracellular transport of protein from the endoplasmic reticulum to the Golgi complex, packaging of newly synthesized polypeptide hormone into secretory granules, and release of hormone by exocytosis. At left are listed the molecular forms of the hormone (or information coding for the hormone) present in that portion of the cell.

that terminal glycosylation occurs as well as conversion of prohormones to mature hormones. In some cases, substances such as proteins, amines (e.g., dopamine), ATP, and Ca^{2+} are packaged along with the hormone into secretory granules. It is in this form that the hormone is stored within the cell until its secretion is required.

Thyroid Hormones

The synthesis of thyroid hormones proceeds in part by the steps of synthesis and packaging discussed above in that thyroid hormones are first synthesized within the matrix of large glycoprotein, thyroglobulin. Newly synthesized thyroglobulin is packaged in the Golgi region into vesicles and secreted into a specialized compartment, the follicular lumen. During this process of secretion, iodide that has been transported into the thyroid is oxidized by a peroxidase located on the apical plasma membrane. Oxidized iodine combines covalently with tyrosine residues of the protein to form mono- and diiodotyrosine. Two iodotyrosine residues then undergo a coupling reaction to form iodothyronine while still within the thyroglobulin matrix.

The mature thyroglobulin in the follicular lumen thus can be considered a prohormone, since it contains a number of

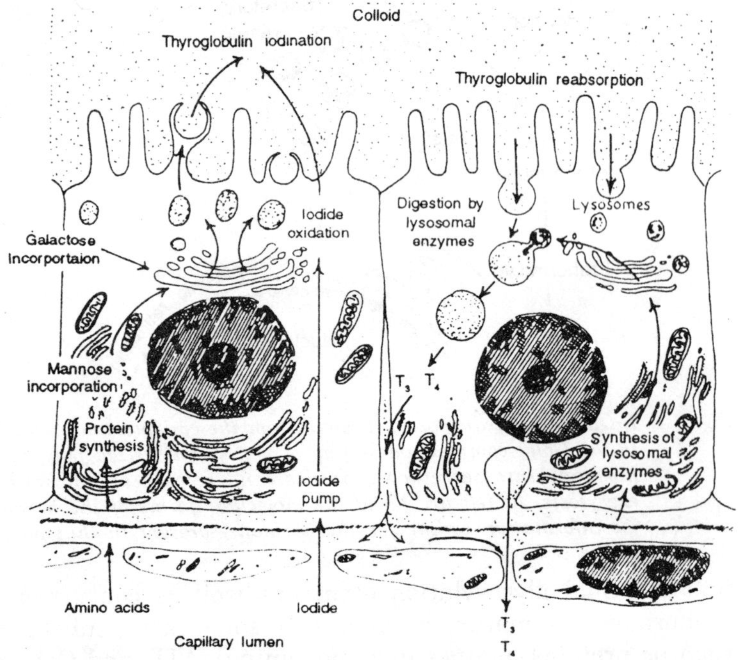

Fig. 1.5. Processes of synthesis and iodination of thyroglobulin (left) and its reabsorption and digestion (right). These events occur in the same cell.

potential thyroid hormone molecules within its structure. Conversion of the prohormone thyroglobulin to thyroid hormone requires phagocytosis or pinocytosis of colloid from the thyroid follicular lumen followed by lysosomal digestion. The released thyroid hormone then leaves the cell by an unknown avenue. The supply of thyroid hormone stored within follicular lumens may be sufficient to maintain normal secretory rates for several months.

Catecholamines

Cells secreting the catecholamine hormones epinephrine and norepinephrine appear similar to cells secreting polypeptide hormones in that they contain prominent secretory granules. The process of hormone synthesis and packaging is different, however, in that catecholamines are synthesized from the amino acid tyrosine by a series of enzymatic steps occurring in the cytoplasm. Dopamine is then taken up and

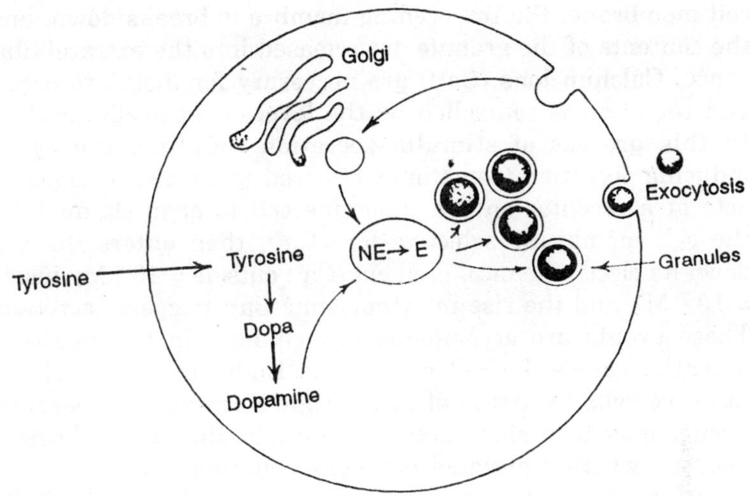

Fig. 1.6. Synthesis and packaging of norepinephrine (NE) and epinephrine (E). Tyrosine is converted in the cytoplasm to dopamine, which is concentrated within secretory granules and converted to NE and E.

concentrated within secretory granules by a transport process localized in the granule membrane. The enzyme dopamine β-hydroxylase, which is localized chiefly in the granule, catalyzes the conversion of dopamine to norepinephrine. Cells of the adrenal medulla also contain an enzyme called phenylethanolamine-N-methyltransferase (PNMT) that catalyzes the conversion of norepinephrine to epinephrine. The cytoplasmic localization of PNMT suggests that norepinephrine and epinephrine can also be transported across the granule membrane. Within the granules, catecholamines exist in a complex with ATP along with proteins of unknown function called *chromogranins*. All of the granule contents are released together when the granule fuses with the plasma membrane.

Stimulus-Secretion Coupling

Amounts of polypeptide and catecholamine hormones sufficient to maintain normal secretory rates for hours to days are stored in secretory granules. Variations in the rate of hormone secretion into the blood are acutely brought about by control of the rate at which secretory granule contents are released from the cell by exocytosis. In this process, the membrane surrounding the secretory granule fuses with the cell membrane, the intervening membrane breaks down, and the contents of the granule are released into the extracellular space. Calcium ions (Ca^{2+}) are necessary for fusion to occur, and secretion is controlled by the level of intracellular Ca^{2+}. In this process of stimulus-secretion coupling, the agent inducing secretion (sometimes referred to as a secretagogue) acts at a receptor on the endocrine cell to open channels in the cell membrane selective for Ca^{2+}; then enters the cell down its electrochemical gradient (Ca^{2+} outside = 10^{-3} M; inside = 10^{-7} M), and the rise in cytoplasmic Ca^{2+} triggers secretion. These events are accompanied by changes in the electrical potential across the cell membrane. Endocrine cells, related to nerve cells by virtue of their origin form neuroectodermal tissue, may also show action potentials similar to those of neurons when stimulated to release hormones.

Endocrine cells are now known to contain contractile proteins such as actin and myosin. There is increasing evidence that these proteins may be important both in moving secretory granules to the plasma membrane and bringing about their subsequent fusion. Along with Ca^{2+} in some secretory cells, cAMP may also play an ancillary role in control of hormone secretion.

Synthesis and Secretion of Steroid Hormones

In contrast to polypeptide hormones; appreciable amounts of steroid hormones are not stored within the cells that produce them. Steroid hormones are able to pass through membranes and leave the cell rapidly after synthesis. As a result, the rate of secretion of most steroid hormones is controlled by the rate of synthesis. Steroid synthesis begins with cholesterol, and the various steps (primarily oxidative) in synthesis take place within the cytoplasm, smooth endoplasmic reticulum, and mitochondria. Cells secreting steroids generally contain an abundant supply of these organelles plus prominent lipid, droplets containing the cholesterol precursor.

Control of the rate of steroid synthesis is by regulation of rate-limiting enzymes in the biosynthetic pathway—specifically, the initial hydroxylations and side chain cleave of cholesterol within the mitochondrion that result in the production of pregnenolone. In the case of adrenal and gonadal steroids,

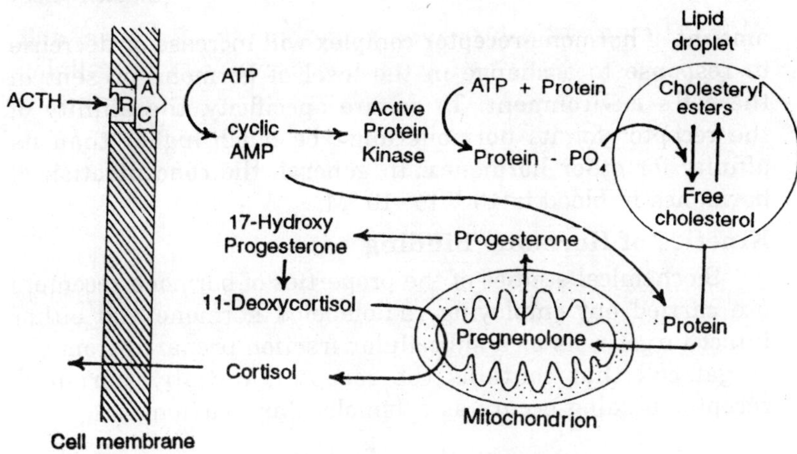

Fig. 1.7. Synthesis and secretion of the steroid hormone cortisol from the adrenal cortex in response to stimulation by ACTH.

whose synthesis is controlled by tropic hormones, the stimulating hormone interacts with a specific receptor that leads to activation of adenylate cyclase and a rise in cAMP. Cyclic AMP then brings about a change in the activity of enzymes that are rate-limiting for steroid synthesis. This process also requires the synthesis of new protein that is rapidly turning over; the role of these proteins is as yet unknown.

Ca^{2+} also plays a role in the regulation of steroid synthesis. Ca^{2+} is necessary in the adrenal cortex both for ACTH to increase the level of cAMP and for the subsequent action of cAMP. In the mitochondria of kidney tubule cells, the level of Ca^{2+} appears to be an important regulator in the production of another steroid hormone, 1,25 dihyroxycholecalciferol, which is the active form of vitamin D.

Common Mechanism of Hormone Action

Hormone Receptors

Hormone receptors on or in cells have 2 distinct roles: to distinguish a particular hormone from other hormones and regulatory molecules and to translate the hormonal signal into an appropriate cellular response. A receptor must have an affinity for its hormone that is high enough in relation to the concentration of the hormone in the blood so that the amount of hormone-receptor complex will increase or decrease in response to a change in the level of hormone present in the cell's environment. To ensure specificity, the affinity of the receptor for its hormone must be much higher than its affinity for other hormones. In general, the concentration of hormones in blood is 10^{-11} to 10^{-9}M.

Kinetics of Hormone Binding

Biochemical studies of the properties of hormone receptors are carried out employing radiolabeled hormone and either intact target cells or a subcellular fraction prepared from the target cell that contains the receptor. Initially, hormone-receptor binding occurs as a bimolecular reaction:

$$[H] + [R] \underset{k_2}{\overset{k_1}{\rightleftharpoons}} [HR] \qquad ...(1)$$

where [H] is the concentration of free hormone, [R] the concentration of free receptor, [HR] the concentration of the hormone-receptor complex, and k_1 and k_2 the rate constants for association and dissociation. When such a system is at equilibrium, the rates of the forward and backward reactions are equal, and

$$\frac{[H][R]}{[HR]} = \frac{k_2}{k_1} = K_d \qquad ...(2)$$

where K_d is the equilibrium dissociation constant, a measure of the receptor's affinity the hormone. Since the biologic response to the hormone is controlled by a signal generated in proportion to the number of hormone-receptor complexes, the equation can be rearranged as

$$\text{Biologic response } \alpha \, [HR] = [H]\,[R]\frac{1}{K_d} \qquad ...(3)$$

In addition to the concentration of the free hormone, the concentration of receptors [R] and their affinity (K_d) influence the biologic response. It is thus important to be able to determine these parameters.

For any value of K_d and with the total number of receptors [R_0] remaining constant, the number of hormone-receptor complexes, usually referred to as "bound hormone" shows a

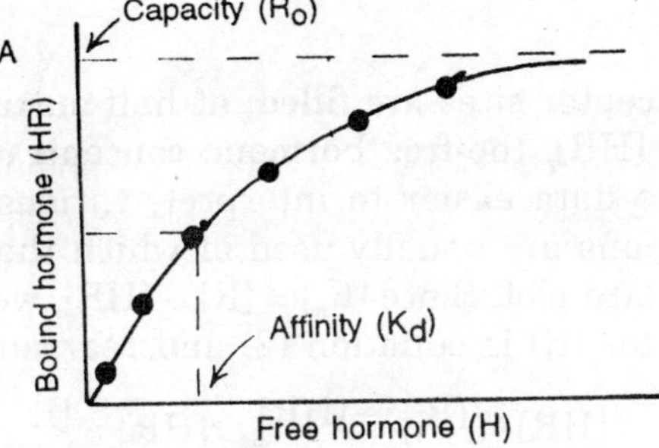

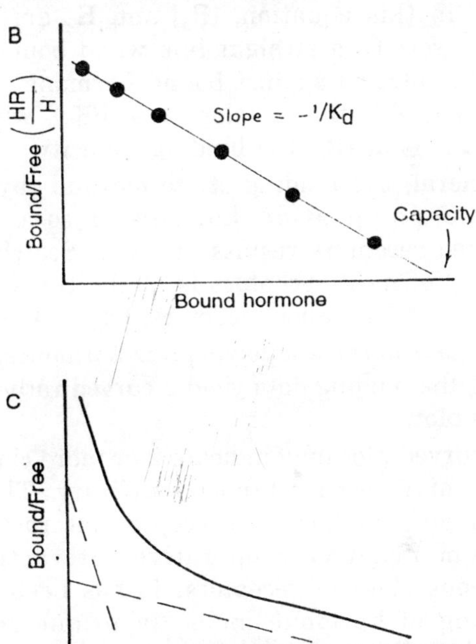

Fig. 1.8. Binding of hormone to its specific receptor A; Plot of bound versus free hormone over a range of free hormone concentration. The dashed line indicates the asymptote; the value of free hormone at which half the receptors are saturated is the K_a or affinity constant. B: Scatchard transformation of the data in (A). The linear plot is indicative of a single binding site C: Curvilinear Scatchard plot. The dashed lines indicate the contribution of 2 separate binding sits the sum of whose occupancy yields the curvilinear Scatchard plot.

predictable hyperbolic relationship with regard to the concentration of free hormone. At high concentrations of hormone, [HR) approaches a maximum equal to $[R_0]$, since all of the receptor sites are filled; at half-maximal occupancy, when [R] = [HR], the-free hormone concentration equals K_d, To make the data easier to interpret, various mathematical transformations are usually used of which the most common is the Scatchard plot. Since $[R_0] = [R] + [HR]$, we can substitute $[R_0] - [HR]$ for [R] is equation (2) and rearrange to yield

$$\frac{[HR]}{[H]} = \frac{[R_0]-[HR]}{K_d} = \frac{[HR]}{K_d} + \frac{[R_0]}{K_d} \qquad ...(4)$$

Since, in this equation, $[R_0]$ and K_d are constants, the equation describes a straight line when bound/free hormone ([HR)/[H]) is plotted against bound hormone [HR]. The slope of this line yields the K_d (slope = $- 1/K_d$), whereas the line cuts the X axis at $[R_0]$ the binding capacity.

In general, the binding of steroid and thyroid hormones (and some polypeptide hormones) to target cells or semipurified receptors results in linear Scatchard plots. The affinity (K_d) of most receptors is in the range of 10^{-10} to 10^{-8} M, and the total number receptors, $[R_0]$ = 4000-100,000/cell.

In the case of certain polypeptide hormones, however, such as insulin, the binding data yield a curved rather than a linear Scatchard plot.

This curved plot may indicate 2 or more classes of specific receptors of differing binding affinity. The curvilinear Scatchard plot for hormone-receptor interaction could also result from negative cooperative interactions among a homogeneous class of receptors. It has been proposed that the binding of hormone molecule to one receptor would decrease the affinity of the surrounding receptors for the same hormone. Evidence has been presented for negative cooperative interaction in the binding of insulin, although the issue is controversial.

Relation of Hormone-Receptor Occupancy to Biologic Response

The simplest relationship between receptor occupancy and the evoked biologic response is seen when there is a close correlation between the concentration of hormone required to occupy the receptor and for induction of the biologic response. For example, the fractional occupancy of glucocorticoid receptors in cultured hepatoma cells parallels induction of the enzyme tyrosine aminotransferase. In other cases, only a small percentage of the hormone receptors must be occupied to generate a maximal response. These additional receptors, which are occupied when hormone concentration increases above that necessary to induce the maximal biologic response, are termed "spare" or reverse receptors. All the receptors are identical, potentially functional, and contribute to hormone sensitivity. An example is the stimulation of glucose transport in fat and muscle cells by insulin; in these cells, occupancy of only 2-10% of the insulin receptors leads to a maximal increase in the rate of glucose transport. This relationship increases the sensitivity of the system and in the case of the fat cell allows a receptor with K_d of about 10^{-9} M respond to plasma concentrations of insulin that are approximately 100 times lower. Since "spare" receptors increase hormone sensitivity (equation [3]) a decrease in the number of spare receptors produced either by regulatory processes or by experimental manipulation leads to a decrease in hormone sensitivity (measured as the concentration of hormone necessary to elicit a half-maximal response). The maximal

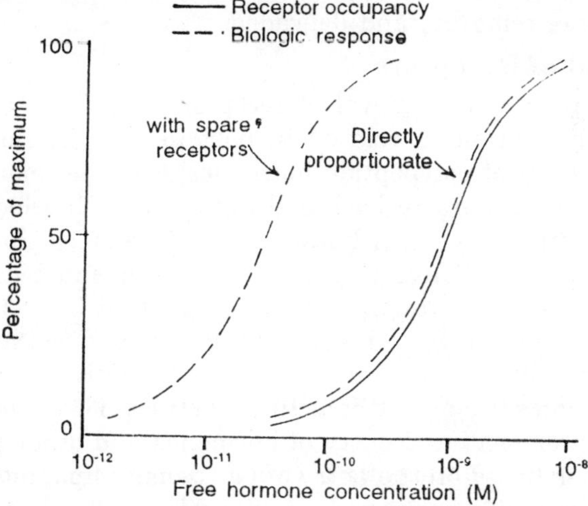

Fig. 1.9. Relationship between receptor occupancy and induced biologic response as free hormone concentration is increased. Receptor occupancy is 50% of maximum when the free hormone concentration equals the K_d of the receptor.

response of the tissue, however, is preserved if sufficient hormone is presented to the cell.

Regulation of Receptor Concentration

The number of receptors for polypeptide and catecholamine hormones is extensively regulated by the homologous hormone and in some cases by other hormones. Most commonly, there is negative regulation of the number of hormone receptors in response to change in plasma hormone levels. In response to a high plasma concentration of the hormone, the number of receptors decreases "down-regulation" or "densitization" resulting in reduced sensitivity of the target cell. An example of this is seen in obesity, where high circulating levels of insulin are associated with a decreased number of functional insulin receptors. Regulation in the opposite direction ("up-regulation") is seen in the developing ovarian follicle, where increasing local concentrations of estrogen and FSH increase the number of receptors to luteinizing hormone (LH).

Regulation of the number of polypeptide receptors in the plasma membrane is believed to be due in part to the fact that receptors, like other membrane proteins, are not static but are being continually synthesized, inserted into the plasma membrane, removed, and degraded.

Removal of Receptors

The process of removal of receptor from the cell surface plasma membrane is related in part to the phenomenon of cellular entry of polypeptide hormones, Since hormones have been isolated, characterized, and radioactively labeled to high specific activity, more is known about hormones than about their receptors. Convincing ultrastructural and biochemical evidence is now available that polypeptide hormones and growth factors such as insulin, prolactin, and nerve growth factor rapidly penetrate into target cells, with all or part of the molecules remaining structurally intact. Three models for the receptor mediated entry of hormones and other proteins. Model A is based on the entry of low-density lipoproteins and some polypeptide hormones into fibroblasts. The ligand binds to receptors located over specialized areas termed coated pits, The membrane containing the receptor and ligand is then endocytosed, forming a coated vesicle that subsequently loses its coat and fuses with lysosomes, leading to degradation of both receptor and ligand. This model, however, does not account for the uptake of hormones into cells such as the liver and pancreas, where coated vesicles are not prominent, nor does it account for the presence of intact hormone in the cytoplasm unless intact hormone is able to escape from the endocytotic vesicles. Alternative models B and C (for which little direct evidence exists) involve nonpinocytotic mechanisms. In B, occupancy of the receptor triggers the entry of the hormone-receptor complex; whereas in C, occupancy of the receptor leads to internalization of the hormone but not the receptor. This last model has some similarity to the entry of nascent polypeptides into the cisternae of the endoplasmic reticulum in that binding leads to the opening of a previously

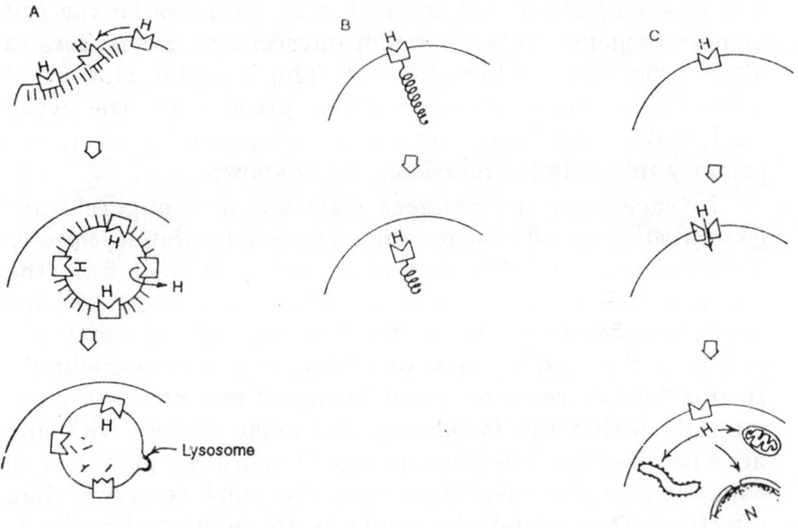

Fig. 1.10. Possible modes of internalization of polypeptide hormones (H) and receptors. In A, the hormone receptor complexes are taken up by endocytosis from coated pits, followed by fusion of the endocytotic vesicle with lysosomes. B and C are hypothetical mechanisms to allow the hormone receptor complex or hormone alone to be directly internalized. Free internalized hormone could then bind to intracellular receptors on target organelles such as endoplasmic reticulum, mitochondria, or nucleus (N).

hidden pore. Model B as well as A would lead to a depletion of membrane receptors while C would not, although there could be a time-dependent step in C before rebinding of a new hormone molecule could occur, thus accounting for down-regulation. At present, it is not clear what biologic activities of hormones initially binding to plasma (cell surface) membrane receptors are mediated either by the internalized hormone or its receptor. Other internalized proteins, however, such as nerve growth factor and diphtheria toxin, are established as having intracellular sites of actin.

Intracellular Messengers

Steroids and thyroid hormones are able either to interact with the plasma membrane or to penetrate intracellularly and combine with intracellular messengers that can directly mediate their actions. By contrast, some polypeptide hormones and catecholamines that interact with receptors on the cell membrane generally act through intracellular messengers to alter processes within the cell. The 2 major classes of intracellular messengers known at present are the cyclic nucleotides and ions. For many hormones however, the primary intracellular messenger is unknown.

Intracellular messengers ("second messengers") are generated by an effector protein such as adenylate cyclase or a Ca^{2+} channel protein that is an entity distinct from the hormone receptor itself. Both receptor and effector proteins are believed to be mobile within the plane of the membrane and are influenced by the state of fluidity of membrane lipids. In this "mobile receptor model" occupied receptors can move laterally within the membrane and make contact with and activate effectors. The phenomenon of spare receptors can be explained on the basis that there are more receptors than effectors. This model also explains the ability of multiple hormones, each acting on distinct receptors, to activate the same effector system. In fat cells, for example, at least 5 different hormones activate adenylate cyclase, whereas in liver 3 different hormones increase intracellular Ca^{2+}. It is unlikely that a single protein could contain 5 different receptor sites, whereas there could easily be a number of receptors clustered around and able to make contact with and activate a single effector protein.

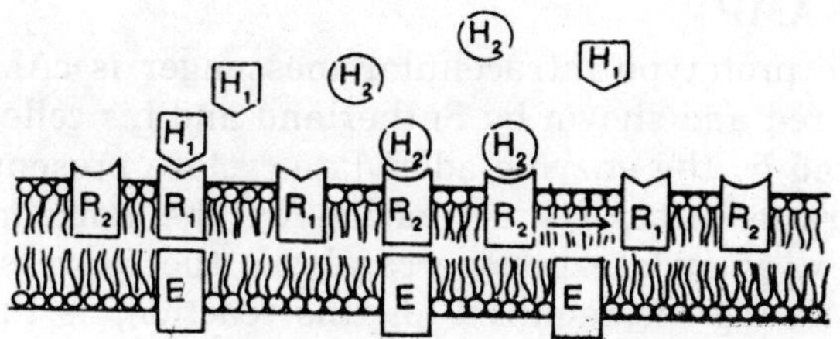

Fig. 1.11. Highly schematic model of 2 types of hormone receptors, R_1 and R_2, which selectively bind hormones H_1 and H_2 and are able to diffuse laterally and interact with a single type of effector, E, in a fluid mosaic membrane.

The most convincing evidence for the separateness of receptors and effectors has been the demonstration that when receptors to foreign hormone are added to a cell possessing adenylate cyclase, they confer responsiveness to the foreign hormone. This has been done both by fusion of 2 different types of cells—one containing receptors to the other only adenylate cyclase—and by adding an extract of solubilized receptor to the new cell. In both cases, the added proteins and lipids mix rapidly with those already present in the membrane, after which the previously unresponsive cell is able to respond to the hormone.

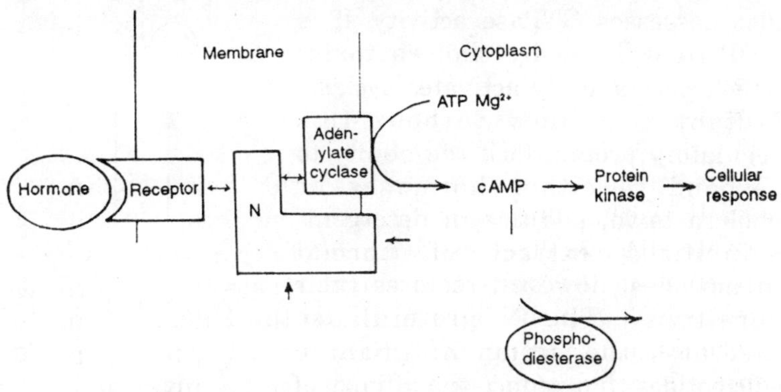

Fig. 1.12. Model of receptor-adenylate cyclase coupling.

Genetic studies of clones from cultured cells have also shown that receptor and effector proteins are specified by different genes.

Cyclic AMP

The prototype intracellular messenger is cAMP, initially discovered and shown by Sutherland and his colleagues to be produced by the enzyme adenylate cyclase present in plasma membrane. After its activation by the hormone-receptor complex, adenylate cyclase catalyzes the conversion of ATP to cAMP. Mg^{2+} is required for this reaction, as the substrate is Mg^{2+}—ATP complex. Adenylate cyclase is a distinct protein that can be solubilized with detergent and is readily separable from the receptor. Adenylate cyclase has also been found to be activated by cholera toxin and guanine nucleotides, particularly GTP and its non-hydrolyzable analogs. Recent biochemical and genetic evidence indicates that these agents act on neither the receptor nor the adenylate cyclase but on a distinct regulatory protein termed the guanine nucleotide-binding protein, or N protein. This protein was originally isolated as a GTP-binding protein and also possesses GTPase activity. It is ADP-ribosylated by cholera toxin, which permanently activates cyclase. Mutant cell lines without this regulatory protein lack the ability to respond to either hormones or cholera toxin, addition of detergent-solubilized extract of normal membranes, however, restores their sensitivity. The N protein also mediates the action of guanine nucleotides that reduce the affinity of the receptor for the hormone. While the N protein clearly serves as a transducer or coupling protein to link the receptor to the cyclase, the exact mechanism of this process is unknown. One possible mode that

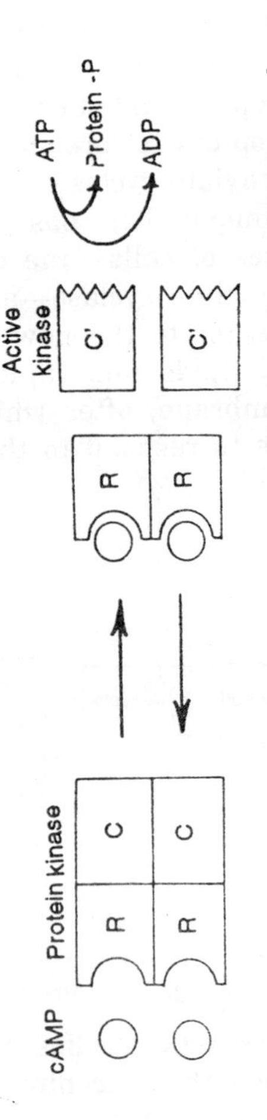

Fig. 1.13. Activation of protein kinase by cAMP. R—regulatory subunit; C—repressed catalytic subunit; C'—active catalytic subunit.

binding of the hormone to the receptor causes the regulatory protein to bind GTP a state that allows the N protein to activate the cyclase. The GTPase activity of the protein would convert the bound GTP to GDP and terminate the action of the cyclase. Another possibility recently proposed is that the receptors and N proteins normally exists as inactive oligomers. Hormone binding would lead to dissociation and allow the interaction with catalytic cyclase.

The activity of adenylate cyclase may also be modulated by other factors. Ca^{2+} inhibits adenylate cyclase in a number of tissues, whereas the Ca^{2+}-calmodulin complex is known to activate certain cyclases. The activity of adenylate cyclase is also affected in various cells by prostaglandins and adenosine.

Cyclic AMP normally exists at very low levels (10^{-8} to 10^{-6} M) in all cells. It is continually being degraded by a specific enzyme, cAMP phosphodiesterase, such that a rise is cellular cAMP induced by hormone activation of adenylate cyclase is rapidly reversed when removal of the hormone stops the production of cAMP. This phosphodiesterase is inhibited by methylxanthines such as caffeine and theophylline.

All of the presently known actions of cAMP in eukaryotes involve phosphorylation of proteins catalyzed by cAMP-activated protein kinase. Normally, this protein kinase is inactivated by binding to a regulatory subunit. When cAMP is elevated, it binds to this regulatory subunit, causing it to release the active catalytic subunit. The active kinase then catalyzes the phosphorylation from ATP of serine and threonine residues of target proteins. This phosphorylation brings about changes in the activity of these proteins. Examples are the phosphorylation of phosphorylase kinase in liver and muscle to activate glycogenolysis and of hormone-sensitive lipase in adipose tissue to activate lypolysis. A great many other structural and enzymatic proteins are phosphorylated in response to various hormones, but in most instances the physiologic effects are not yet known.

Calcium (Ca^{2+})

Ionic calcium (Ca^{2+}) is a well-established intracellular messenger important in coupling excitation to contraction in muscle, excitation to transmitter release in nerves, and secretagogue action to secretion in various exocrine and

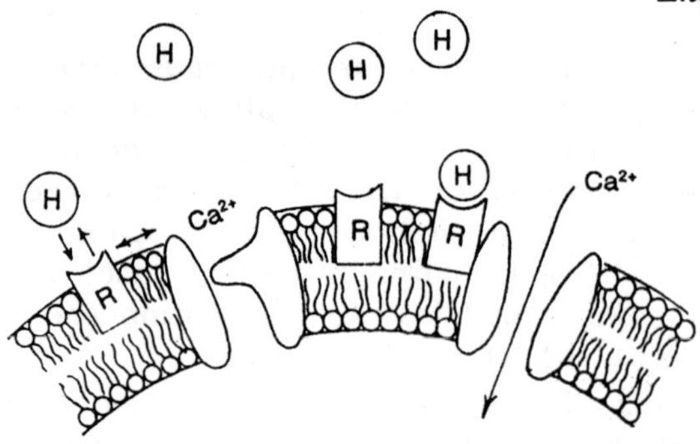

Fig. 1.14. Hormone receptors (R) that bind hormone (H) and control the opening of selective Ca^{2+} channels in the cell membrane.

endocrine glands. The importance of Ca^{2+} as an intracellular messenger in the action of a number of hormones is becoming increasingly apparent.

The free cytoplasmic Ca^{2+} concentration is highly regulated to maintain a very low intracellular concentration. In various cell types, a Na^+-Ca^{2+} counter transport mechanism has been described that uses the energy of Na^+ entering the cell down its electrochemical gradient to drive the expulsion of Ca^{2+}, as well as a Ca^{2+}-activated ATPase that directly uses ATP to transport Ca^{2+} out of the cell. Ca^{2+} is also taken up (buffered) by intracellular organelles, including mitochondria and endoplasmic reticulum. At rest, the cytoplasmic concentration of Ca^{2+} is maintained at 10^{-7} M or less. A rise in cytoplasmic Ca^{2+} can then be brought about either by the opening of Ca^{2+} channels in the membrane or by release of Ca^{2+} from intracellular organelles such as endoplasmic reticulum. Little is known about the mechanisms controlling these processes. Analogous to the activation of adenylate cyclase, multiple hormones and neurotransmitters may activate the same Ca^{2+} channels, suggesting separate receptor and Ca^{2+} channel subunits. An accelerated turnover of membrane phosphatidylinositol usually accompanies the increase in cytoplasmic Ca^{2+} and is thought to perhaps participate in the opening of Ca^{2+} channels on the release of Ca^{2+} from intracellular organelles.

Probably most actions of Ca^{2+} as an intracellular messenger result from its binding to Ca^{2+}-binding proteins. The major

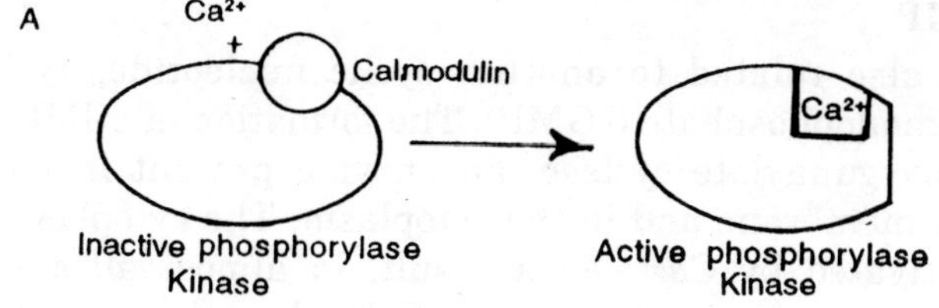

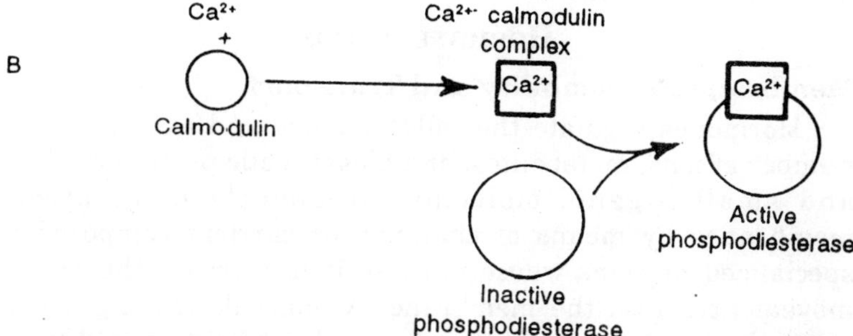

Fig. 1.15. Mechanisms of enzyme activation by Ca^{2+} acting through binding to the Ca^{2+} receptor protein calmodulin. A—Ca^{2+} combines with calmodulin present as a subunit of phosphorylase kinase to active the enzyme. B—Ca^{2+} combines with free calmodulin, and the Ca^{2+}-calmodulin complex then binds to and activates phosphodiesterase.

Ca^{2+}-binding protein involved in the action of hormones in *calmodulin*, a small, highly acidic protein that binds four Ca^{2+} per mole with an affinity (K_d) of about 10^{-6} M. Calmodulin may exist as a integral subunit of an enzyme, as is the case with phosphorylase kinase. Alternatively, the Ca^{2+}-calmodulin complex may bind to other proteins and alter their activity. An example is the activation of Ca^{2+} regulated cyclic nucleotide phosphodiesterase by the calcium-calmodulin complex in brain and other tissues. In the liver, where Ca^{2+} mediates the action of a number of hormones (i.e., epinephrine, vasopressin, and angiotensin), a number of proteins are phosphorylated following exposure of hepatocytes to these hormones. It seems likely, therefore, that many of the actions of Ca^{2+} will be found to be mediated by Ca^{2+} activated kinases.

Cyclic GMP

Ca^{2+} is also related to another cyclic nucleotide, cyclic guanosine monophosphate (cGMP). The formation of cGMP is catalyzed by gunaylate cyclase, an enzyme present in both the plasma membrane and in the cytoplasm. The cytoplasmic form is activated by Ca^{2+}. As a result, in almost all cases where there is a rise in cytoplasmic Ca^{2+}, there is a parallel rise in cGMP. While cGMP-activated protein kinases exist, there is at present no clear evidence for a role of cGMP in hormone action. It has been described as a second messenger in search of a hormone.

Hormone Action

Membrane Permeability and Transport

Hormones regulate the cellular entry and exit of a large number of ions, metabolites, and biosynthetic precursors. Ions and small organic molecules frequently move across membranes by means of channels or carriers composed of specialized proteins whose purpose is to increase the rate of movement across the membrane. A molecule moving across a membrane down its electrochemical gradient is said to be undergoing *passive transport*. Passive transport may show kinetic characteristics of diffusion or, if mediated by a carrier, saturation kinetics analogous to enzyme activity. When cellular energy is used (usually by means of an ATPase enzyme) to move the transported molecule against its electrochemical gradient, it is undergoing *active transport*. Hormones affect all of these types of transport processes. These actions are important both in regulating the biosynthetic and metabolic activities of individual cells and in regulating the movement of ions and other molecules across cells such as in the intestine, kidney tubules, and bone.

Hormones regulate the movement of molecules across membrane by at least 4 different mechanisms: (1) alteration of the affinity of the transport mechanism for the molecule being transported; (2) activation of previously inactive transport mechanisms present in the plasma membrane; (3) insertion into the plasma membrane of preexisting transport mechanisms present inside the cell; and (4) synthesis of new transport proteins.

Glucose and amino acid transport

The cellular uptake of glucose and amino acids is stimulated by hormones in a large number of target tissues. The earliest and best-studied example is the stimulation of glucose entry into fat and muscle cells by insulin. Glucose crosses the plasma membrane by means of a transport protein, the glucose carrier, that facilitates glucose entry into cells. Because the number of carriers is limited, this "facilitated diffusion" shows saturation kinetics that can be characterized by a K_m (the concentration at which transport is half-maximal) and a J_{max} (the maximal rate of transport). Insulin increases the J_{max} either by uncovering previously inactive carriers or by bringing about the insertion of new carriers from the cell interior into the plasma membrane. This action of insulin is rapid (seconds to minutes) and does not require new protein synthesis. Other hormones also increase glucose uptake by their target cells; examples include TSH acting on thyroid cells and cholecystokinin acting on pancreatic acinar cells.

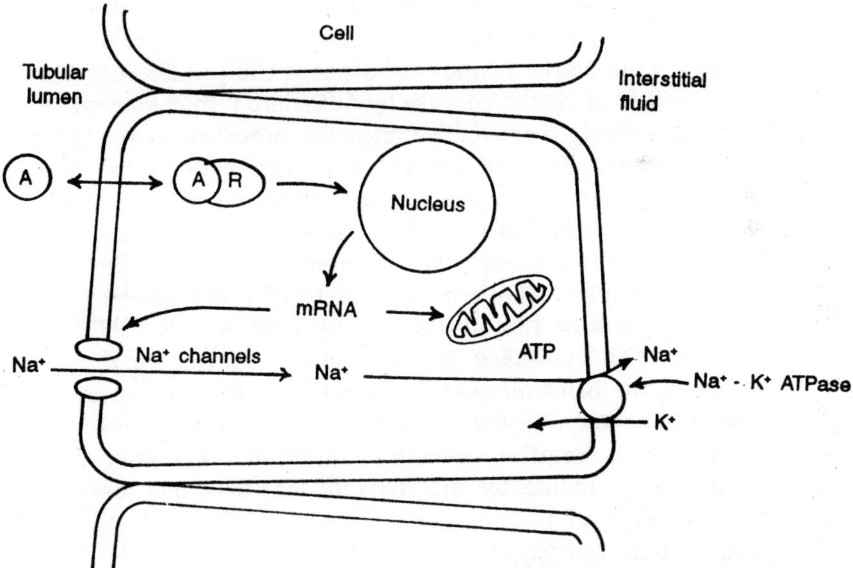

Fig. 1.16. Mechanism by which aldosterone increases the resorption of Na^+ *by cells of the distal nephron. Aldosterone (A) enters the cell and binds to a cytoplasmic receptor (R).*

The cellular uptake of amino acids also involves a number of transport proteins that mediate amino acid entry by facilitated diffusion. In some cases, the entry of amino acids is driven by Na^+, which also binds to the carrier and, by moving down its electrochemical gradient, serves to concentrate the amino acid inside the cell. Acceleration of amino acid transport is brought about by a variety of hormones, including insulin, glucagon, growth hormone, and thyroid hormone. Thyroid hormone appears to act directly on the plasma membrane independently of its effects on nuclear functions.

Other small molecules such as uridine and thymidine that are important as biosynthetic precursors enter cells by carrier-mediated facilitated diffusion. The transport of these molecules is stimulated by a number of polypeptide tropic hormones and by gonadal steroids acting on their respective target tissues.

Ions

Cellular ionic balance is regulated to a great extent by the Na^+-K^+ ATPase that actively transports Na^+ out of and K^+ into cells. This active transport balances the passive entry of Na^+ and efflux of K^+ that occurs by diffusion. Other ions either diffuse passively across the plasma membrane or their movements are coupled to Na^+ via carrier proteins. Two hormones that increase the activity of the Na^+-K^+ ATPase in a number of tissues are insulin and thyroid hormone. The effects of insulin are rapid and separate from its effects on glucose transport. The stimulation by thyroid hormone is slower, requiring the synthesis of new Na^+-K^+ ATPase molecules. The increased Na^+ transport induced by thyroid hormone is the basis for part of the action of thyroid hormones to stimulate oxygen consumption and heat production (calorigenesis), Another specialized form of ion transport is the uptake of iodide by the thyroid; which is increased by TSH.

Epithelial transport

Epithelia are specialized sheets or tubules of cells whose function is to separate 2 compartments and regulate the transport of water and other molecules. For example, in the distal tubule of the kidney, aldosterone acts to increase the reabsorption of Na^+. Na^+ enters the cell from the lumen of the tubule down its electrochemical gradient by means of a specialized transport protein and then is actively transported out of the cell into capillaries by means of Na^+-K^+ ATPase located in the basolateral membrane of the tubular cell. Aldosterone induces the synthesis of new Na' channel protein that facilitates the entry of Na^+ into the cell. It also induces the production of other proteins that increase mitochondrial ATP synthesis for the transport of Na^+ out of the cell,

In an analogous manner, vitamin D, a steroidlike substance, increases the transepithelial absorption of Ca^{2+} and PO_4^{3-} from the lumen of the small intestine, Growth hormone also increases Ca^{2+} and PO_4^{3-} absorption, while Ca^{2+} absorption is inhibited by adrenal glucocorticoids,

Other hormones regulating epithelial transport include parathyroid hormone in bone and kidney; gastrin, which increases the secretion of H^+-rich gastric juice by parietal cells of the stomach; and secretin, which increases the secretion of HCO_3^- rich pancreatic juice by the duct cells of the pancreas.

Enzyme Activation

Many effects of polypeptide and catecholamine hormones on intracellular enzymes involve alterations in the activity of preexisting enzymes rather than the synthesis of new enzymes, These effects are generally slower than those regulation membrane transport, occurring in minutes to hours, but are also independent of the synthesis of new protein, Insulin, for example, has well-established effects on the activity of various enzymes involved in the metabolism of glucose, amino acids, fatty acids, and nucleotides.

Major mechanisms by which hormones affect enzyme activity are phosphorylation and dephosphorylation. The first and best-studied example of phosphorylation is the regulation of glycogen metabolism in liver and muscle. The activity of both glycogen synthetase and phosphorylase is regulated by phosphorylation; in the case of phosphorylase, addition of phosphate activates the enzyme, whereas in the case of glycogen synthetase, it inhibits the enzyme. Thus, hormones such as epinephrine and glucagon, by increasing the activity of several kinases (including phosphorylase kinase), bring about

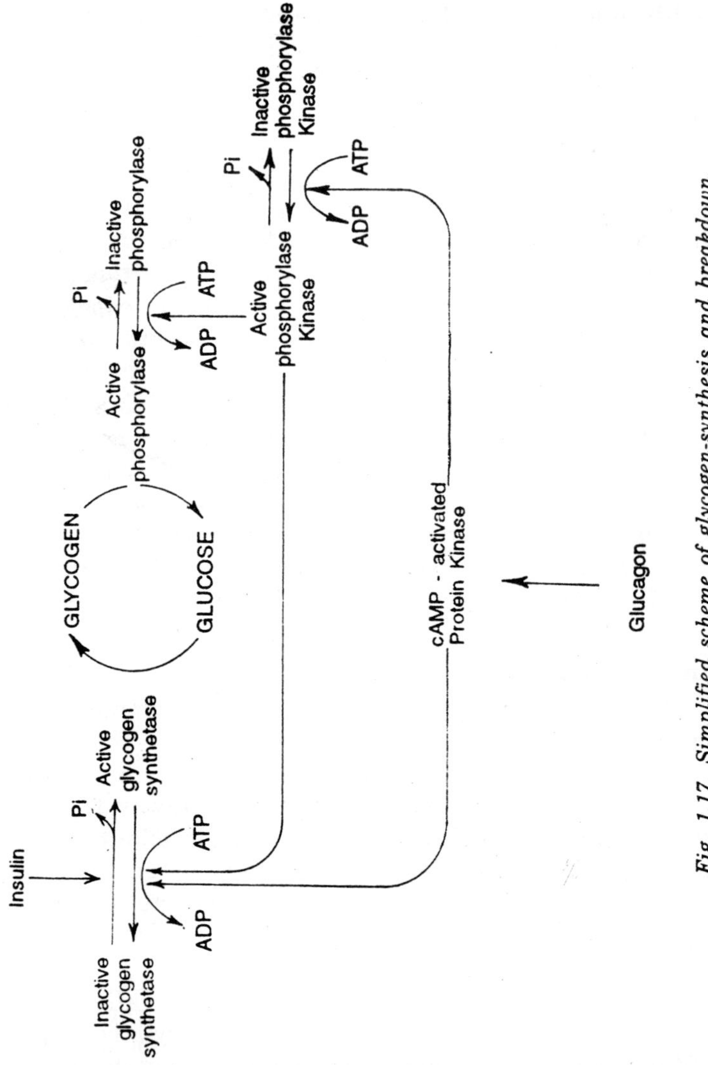

Fig. 1.17. Simplified scheme of glycogen-synthesis and breakdown.

the net conversion of glycogen to glucose. Insulin, possibly by activating a phosphatase, increases the activity of glycogen synthetase and accelerate the formation of glycogen.

Many other enzymes responsive to insulin have been shown to be regulated by phosphorylation and dephosphorylation independently of the cAMP-activated protein kinases, A well-studied example is the regulation of pyruvate dehydrogenase, a mitochondrial enzyme complex that produces acetyl-CoA necessary for lipid synthesis. Exposure of adipose tissues to insulin rapidly increases pyruvate dehydrogenase activity as a result of dephosphorylation. Insulin also activates pyruvate kinase and hydroxymethyl-glutaryl-CoA reductase and inhibits hormone-sensitive lipase—all by dephosphorylation processes. It is not known in any of these cases whether insulin activates phosphatases or inhibits kinases.

Both of the second messengers (cAMP and Ca^{2+}) can act by phosphorylation mechanisms. An example where the 2 intracellular messengers act together is the regulation of phosphorylase kinase, the enzyme that phosphorylates phosphorylase. Phosphorylase kinase is a multisubunit enzyme and can be activated independently by cAMP-activated protein kinase, which phosphorylates its α and β subunits, or by Ca^{2+} which binds to the δ subunit, calmodulin, The 2 intracellular regulators potentiate each other in that cAMP-induced phosphorylation increases the sensitivity of the enzyme to Ca^{2+}. Almost all hormones (including steroids) alter the state of phosphorylation of small number of proteins in target cells. In most cases, however, the identity and functional role of the phosphorylated protein are unknown.

Besides phosphorylation, other enzymes can be directly activated by specialized binding proteins. Calmodulin, the protein that serves as the Ca^{2+} receptor for hormones utilizing Ca^{2+} as intracellular messenger, binds to and regulates a number of enzymes, including cyclic nucleotide phosphodiesterase, certain adenylate cyclases, and the Ca^{2+} ATPase present in the cell membrane that acts to transport Ca^{2+} out of the cells.

Other ways in which enzymes activities can be regulated that may ultimately prove to be affected by hormones include aggregation of subunits, adenylation, methylation, and changes in the local concentration of ions or cofactors.

Nuclear Regulation

Certain actions of steroid, thyroid, and polypeptide hormones are brought about by increasing the number of a few specific types of mRNA. This effect on mRNA is caused by altering either the rate of synthesis and degradation of mRNA molecules or their processing from precursor RNAs and appearance in the cytoplasms. Hormone actions mediated by this mechanism are generally delayed in onset and blocked by inhibitors of RNA synthesis such as actinomycin D and cordycepin.

Steroid hormones

The mechanism of nuclear regulation has been studied most extensively for steroid hormones. Furthermore, most of the known actions of steroids appear to follow this pathway. As previously mentioned, steroid hormones bind to receptors in the cytoplasm. Specific cytoplasmic receptor proteins have been found for all classes of steroids and bind biologically active steroids in relation to the steroid's potency. These

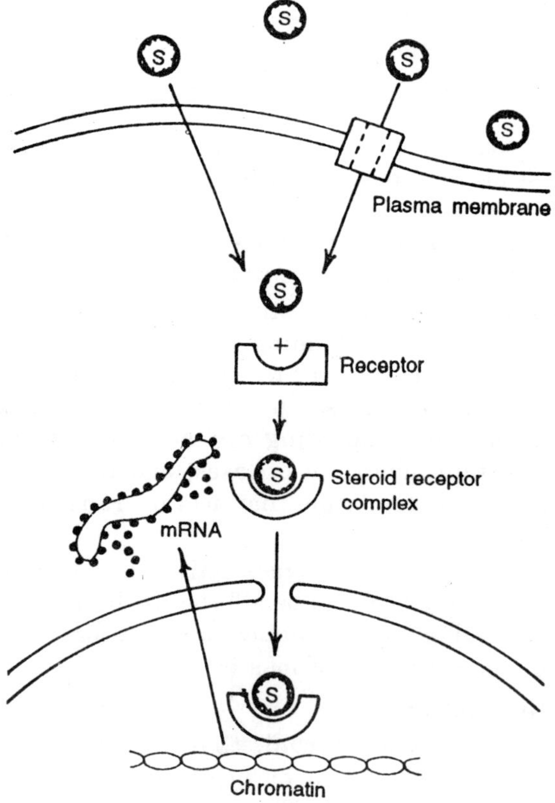

Fig. 1.18. Proposed mechanism of steroid hormone action.

steroid receptors are asymmetric proteins weighing 50-300 thousand daltons.

Following the binding of steroid to its receptor there is a reduction of the number of receptors in the cytosol and an increase in the nucleus. The steroid-receptor complex has the ability to bind to isolated nuclei, whereas neither the free steroid nor the receptor has this ability. Steroid hormone antagonists are structurally similar molecules that bind to the receptor in the cytoplasm but are unable to confer the ability to bind to the nucleus. The ability to enter the nucleus results from a change in the receptor (presumably allosteric in nature) termed "activation" that under physiologic conditions rapidly follows binding of agonist steroids. In cells responding to estrogens, activation appears to result in aggregation of 2 receptors to form a dimer.

Steroid-promoted nuclear binding is rapid and is usually maximal by 30 minutes. Nuclear binding sites termed "acceptors" are present in the chromatin fraction of nuclei. In studies of both isolated nuclei and chromatin, it has been impossible to saturate acceptor sites; it is generally concluded, therefore, that there are more nuclear acceptors than steroid receptors. Although steroid-receptor complexes bind to isolated DNA, there is no species or tissue specificity in this binding.

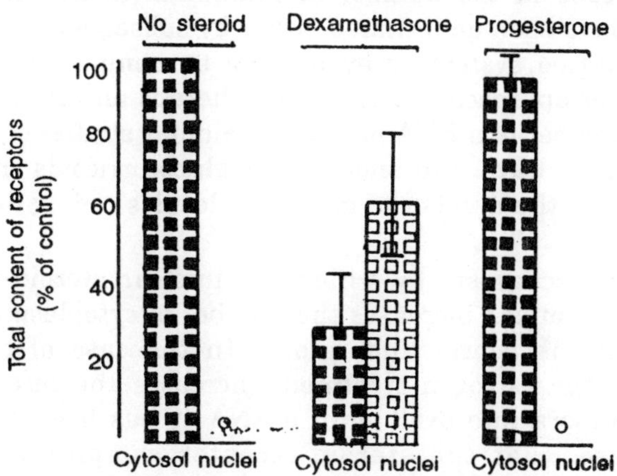

Fig. 1.19. Distribution of steroid receptors in the cell under conditions where it is exposed to no steroid, an agonist (dexamethasone) or an antagonist (progesterone).

It seems reasonable to conclude, therefore, that binding is also affected by chromatin proteins. In the case of the progesterone receptor, the receptor appears to be a dimer with one subunit that binds to DNA, while the other binds to a specific type of acidic protein present only in chromatin of target tissues. These observations suggest a model for steroid hormone action in which the protein-binding subunit localizes the receptor to a specific location in chromatin, while the DNA-binding subunit then acts as a nuclear regulator. A currently unexplained paradox is that while there are many more acceptor sites than genes being regulated, the stimulation of mRNA formation correlates well with the number of nuclear steroid receptor complexes. Following removal of the steroid from the environment of the target cells, there is a relatively rapid dissociation of steroid from the receptor and a return of the unoccupied receptor to the cytosol.

It is generally believed that the actions of steroid hormones result from increased production of specific mRNA. In only a few cases, however, has this been convincingly demonstrated. A well-studied example is the increase in the protein ovalbumin that is induced in the chick oviduct by estrogen. In parallel with the increase in ovalbumin, there is an increase in the number of ovalbumin mRNA molecules. These have been quantitated either by subsequent translation in a cell-free system or by binding to complementary DNA The latter approach indicates that there is an actual increase in the number of mRNA molecules. Similar studies of pituitary tumor cells have also shown that glucocorticoids induce an increase in the number of mRNA molecules coding for growth hormone.

The mechanism by which the nuclear steroid-receptor-acceptor complex increases the number of cytoplasmic mRNA molecules is poorly understood. In the case of the chick oviduct, the estrogen treatment increases the incorporation of precursors into ovalbumin mRNA It has been suggested, therefore, that the steroid-receptor complex alters the structure of chromatin to promote the binding of RNA polymerase to specific genes, resulting in increased production of new mRNA While newly synthesized mRNA must be processed and only a portion enters the cytoplasm, there is no evidence to date for post-transcriptional regulation of this processing mechanism by steroids.

Thyroid hormones

Thyroid hormones are now also known to regulate nuclear function. Thyroid hormones readily enter cells, but their uptake mechanism is unknown. They may either diffuse through the lipid bilayers of the cell membrane by virtue of their hydrophobicity or enter cells by means of a transport mechanism similar to that used by amino acids. After entry, thyroid hormones bind to cytoplasmic binding proteins that concentrate hormone intracellularly. In contrast to steroid hormone action, however, these cytoplasmic binding proteins do not translocate thyroid hormones. It is the free thyroid hormone in the cell that binds directly to receptors on or in the nucleus. Other receptors are present in mitochondria and may mediate certain effects of thyroid hormone on oxygen consumption.

Nuclear receptors for thyroid hormones are acidic proteins that can bind to DNA but are also influenced by the histone

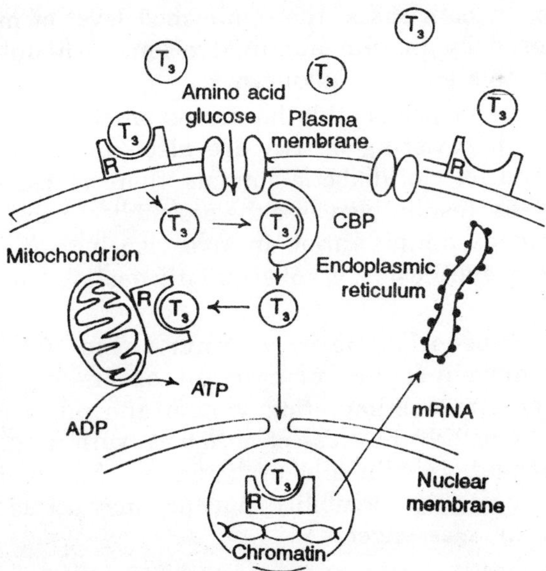

Fig. 1.20. Proposed mechanism of thyroid hormone action.

proteins present in chromatin. It is this interaction with histones that maintains the higher specificity for triiodothyronine (T_3) as compared to thyroxine (T_4). The finding of thyroid hormone receptors associated with chromatin suggests that the hormone influences nuclear functions. It is possible that the binding of thyroid hormone to these receptor proteins may increase the content of cytoplasmic mRNA. Specific mRNAs known to be increased by thyroid hormones include those coding for growth hormone, liver $\alpha 2\mu$-globulin, and Na^+-K^+ ATPase. In general, effects of thyroid hormones are slow, taking 1-2 days to reach a maximum.

Polypeptide hormones

A number of polypeptide hormones—including growth hormone, prolactin, and insulin—also have prominent effects on nucleic acid and protein synthesis. Effects of insulin related to transcription have been reported in liver, adipose tissue, and mammary gland. Administration of insulin has been shown to increase both RNA polymerase activity and chromatin template activity in the liver in vivo. More recent studies have shown that in experimental diabetes, there is decreased production of mRNA for albumin in the liver and amylase in the pancreas. In both cases, the diminished level of mRNA can be restored by insulin administration, although the mechanism of this effect is unknown.

Other effects of polypeptide hormones on the ribosomal translation of preexisting mRNA are also known. In the muscle, fat, and liver of diabetic animals, there is decreased protein synthesis. Insulin increases the assembly of polysomes. Growth hormone administration in vivo also increases the rate of protein synthesis by subsequently isolated muscle ribosomes.

It is not presently known whether these effects of polypeptide hormones are primary or secondary ones. Polypeptide hormones stimulating protein and nucleic acid synthesis all belong to the class of hormones with no known second messenger. Insulin and prolactin are known to be internalized, raising the possibility that they may act as their own intracellular messenger.

2

Hypophyseal Gland

Immunoassay techniques that measure the secretory products of the pituitary and advances in the neuroradiologic evaluation of the sella turcica have improved the diagnosis of pituitary disorders. Trans-sphenoidal microsurgery and bromocriptine now allow earlier and more effective therapy. This chapter describes the clinical evaluation of the anterior pituitary (adenohypophysis), including its anatomy, embryology, and histology; the physiologic action and secretion of its 6 major hormones; and disorders of the gland with respect to neuroradiologic findings and secretory hypo- and hyperfunction.

Embryology

The human fetal pituitary anlage is initially recognizable at 4-5 weeks of gestation, and rapid cytologic differentiation leads to a mature hypothalamic-pituitary unit at 20 weeks. The adenohypophysis originates from Rathke's pouch, an ectodermal evagination of the oropharynx, and migrates to join the neurohypophysis (an outpouching of the third ventricle). The portion of Rathke's pouch in contact with the neurohypophysis develops less extensively and forms the intermediate lobe. This lobe remains intact in some species, but in humans its cells become interspersed with those of the anterior lobe. These cells do, however, develop the capacity to synthesize and secrete adrenocorticotropic hormone (ACTH) and its related peptide hormones.

Remnants of Rathke's pouch may persist at the boundary of the neurohypophysis, resulting in small colloid cysts. In

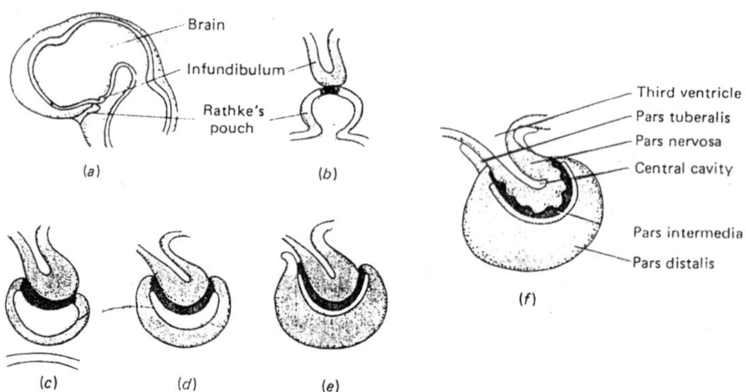

Fig. 2.1. Diagrams showing progressive stages in the embryonic development of the pituitary gland.

addition, cells may persist in the lower portion of Rathke's pouch beneath the sphenoid bone, the pharyngeal pituitary. These cells have the potential to secrete hormones and have been reported to undergo adenomatous change.

Anatomy

Gross Anatomy

The pituitary gland lies at the base of the skull in a portion of the sphenoid bone called the *sella turcica* ("Turkish saddle"). The anterior portion, the tuberculum sellae turcicae, is flanked by posterior projections of the sphenoid wings, the anterior clinoid processes; the dorsum sellae forms the posterior wall, and its upper comers project into the posterior clinoid processes. The gland is surrounded by dura, and the roof is formed by a reflection of the dura attached to the clinoid processes, the diaphragma sellae. In normal individuals the arachnoid membrane and, therefore, cerebrospinal fluid are prevented from entering the sella turcica by the diaphragma sellae. The pituitary stalk and its blood vessels pass through an opening in this diaphragm. The lateral walls of the gland are in direct apposition to the cavernous sinus. The optic chiasm lies 5-10 mm above the diaphragma sellae and anterior to the stalk. The size of the pituitary gland, of which the anterior lobe constitutes two-thirds, varies considerably. It measures approximately 15 × 10 × 6 mm and weighs 500-900 mg; it may double in size during pregnancy. Since the sella turcica tends to conform to the shape and size of the gland, there is considerable variability in the contour of this body structure.

Blood Supply

The anterior pituitary is the most richly vascularized of all mammalian tissues, receiving 0.8 mL/g/min from a portal circulation connecting the median eminence of the hypothalamus and the anterior pituitary. This arterial blood is supplied by the internal carotid arteries via middle and inferior hypophyseal arteries. The arteries form a capillary network in the median eminence of the hypothalamus that recombines in long portal veins draining down the pituitary stalk to the anterior lobe, where they break up into another capillary network and re-form into venous channels. The posterior pituitary is supplied directly from branches of the

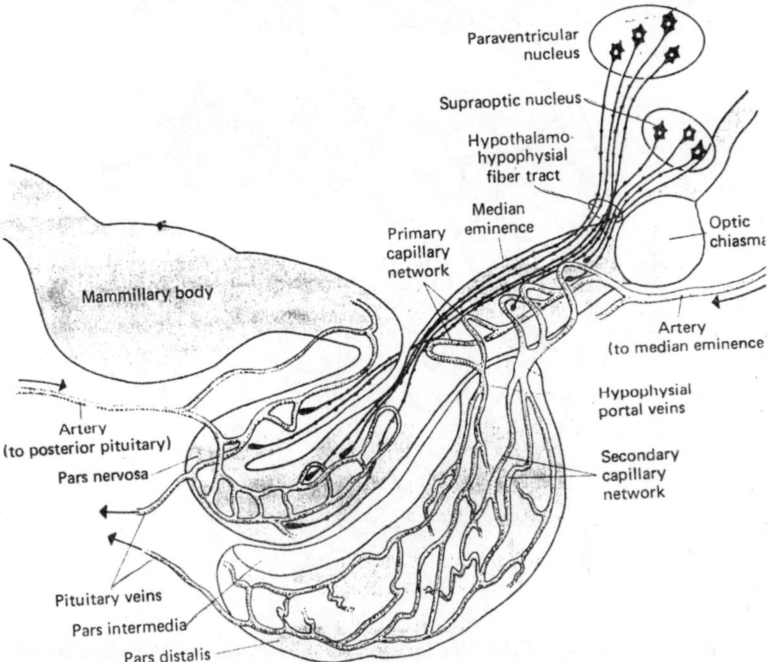

Fig. 2.2. Diagram of the anatomic connections between the hypothalamus and the pituitary gland.

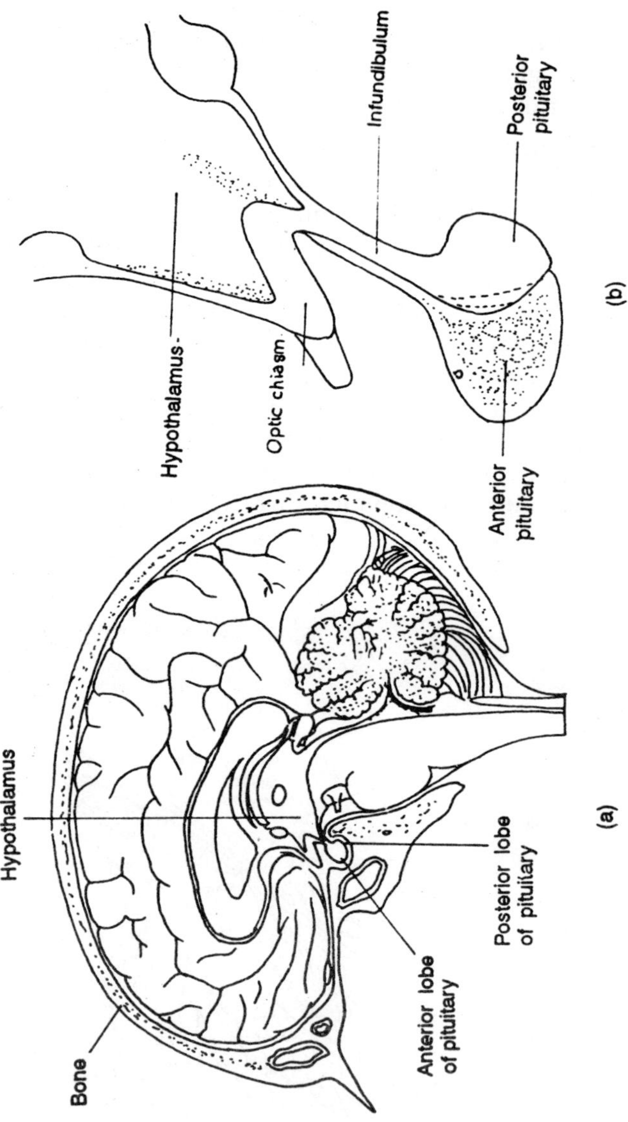

Fig. 2.3. Anatomy of the pituitary gland (a) Relation of the pituitary gland to the hypothalamus and to the remainder of the brain. (b) Schematic enlargement of the pituitary gland and its connection to the hypothalamus.

middle and inferior hypophyseal arteries. Venous drainage of the pituitary, the route through which anterior pituitary hormones reach the systemic circulation, is variable, but venous channels eventually drain via the cavernous sinus posteriorly into the superior and inferior petrosal sinuses to the jugular bulb and vein.

Recent evidence indicates that there is also retrograde blood flow between the pituitary and hypothalamus, providing a means of direct feedback between pituitary hormones and their neuro control.

Histology

The cells of the anterior pituitary were originally classified using techniques for staining intracellular granules as acidophils, basophils, and chromophobe cells. Immunocyto-chemical and electron microscopic techniques now permit classification of cells by their specific secretory products:

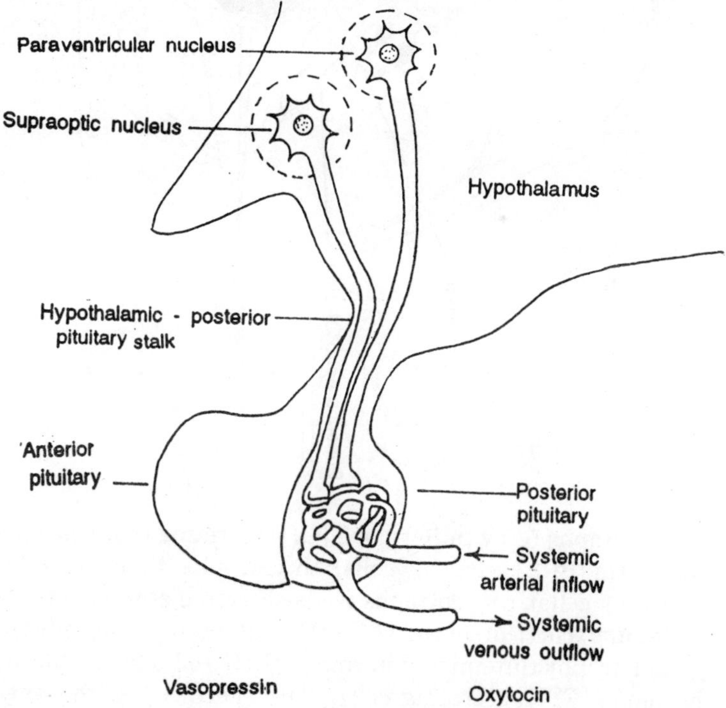

Fig. 2.4. Relationship of hypothalamus and posterior pituitary.

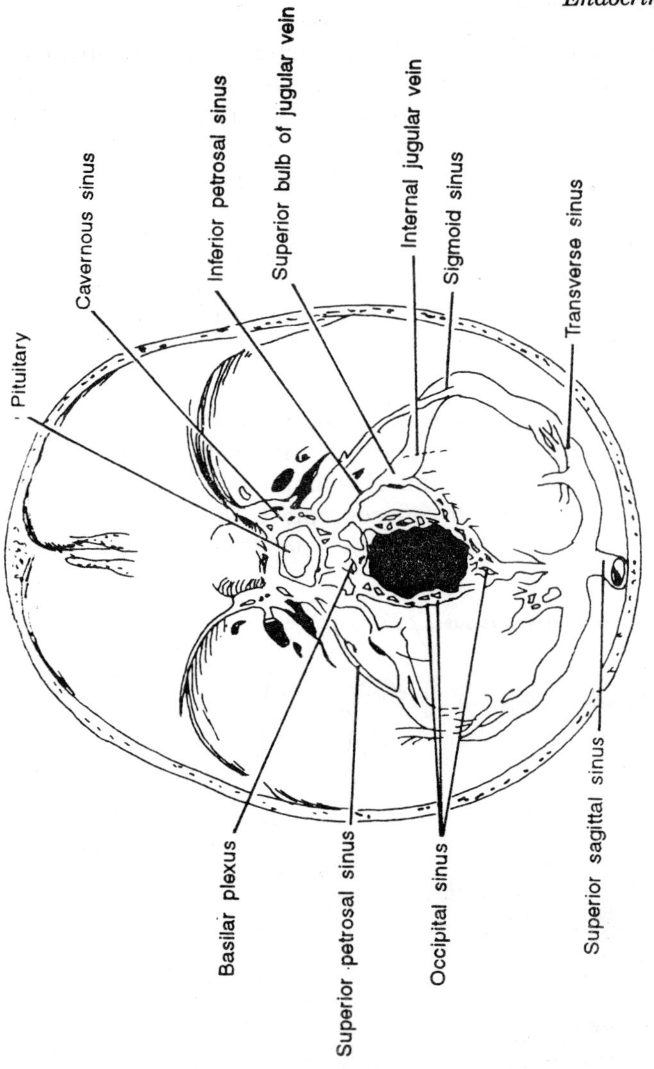

Fig. 2.5. Venous drainage of the pituitary gland—the route by which adenohypophyseal hormones reach the systemic circulation.

somatotrophs (growth hormone [GH]-secreting cells), lactotrophs (prolactin [PRL]-secreting cells), thyrotrophs (thyrotropin [TSH]-secreting cells), corticotrophs (cells secreting adrenocorticotropic hormone (corticotropin, [ACTH] and related peptides), and gonadotrophs (luteinizing hormone [LH]-and follicle-stimulating hormone [FSH]-secreting cells). The specificity of the antibody used determines the ability of immunocytochemical techniques to differentiate the various cell types.

Somatotrophs

The GH-secreting cells are identified as acidophilic in standard hematoxylin and eosin preparations and are usually located in the lateral portions of the anterior lobe. Granule size by electron microscopy is 300-400 nm in diameter. These cells account for about 4-10% of the net weight of the anterior pituitary.

Lactotrophs

A second but distinct acidophil-staining cell randomly distributed in the anterior pituitary is the PRL-secreting cell. Granule size averages approximately 550 nm on electron microscopy. These cells proliferate during pregnancy as a result of elevated estrogen levels and account for the increase in gland size.

Thyrotrophs

These TSH-secreting cells because of their glycoprotein product, are basophilic and also show a positive reaction with periodic acid-Schiff (PAS) stain. The thyrotroph granules are small (50-100 nm), and these cells are usually located in the anteromedial and anterolateral portions of the gland. During states of primary thyroid failure, the cells demonstrate marked hypertrophy, increasing overall gland size.

Corticotrophs

ACTH and its related peptides, the lipotropins (LPH), and the endorphins are secreted by basophil-staining cells that are embryologically of intermediate lobe origin and usually located in the anteromedial portion of the gland. Electron microscopy shows that these secretory granules are about 360 nm in diameter. In states of glucocorticoid excess (both exogenous and endogenous), corticotrophs undergo degranulation and a microtubular hyalinization known as Crooke's hyaline degeneration.

Gonadotrophs

Both LH and FSH originate from these basophil-staining cells, whose secretory granules are about 200 nm in a diameter. These cells are located in the lateral portion of the gland. They become hypertrophied and cause the gland to enlarge during states of primary gonadal failure such' as Klinefelter's syndrome and Turner's syndrome.

Other cell types

Despite immunocytochemical staining with antibodies directed against all of the known anterior pituitary hormones, some cells remain, unstained. These are chromophobes by conventional staining methods, but electron microscopy has identified secretory granules in many of them. It is not certain whether they represent undifferentiated primitive secretory cells or whether they produce an as yet unidentified hormone, such as adrenal androgen-stimulating hormone or ovarian growth factor.

Growth Hormone

Chemistry

Human growth hormone (HGH) has been purified and its amino acid sequence determined. It has a molecular weight of 21,500 and contains 188 amino acid residues and 2 disulfide bridge. Although the molecular weights of growth hormones vary from 21,500 to 47,800, the heavier proteins may represent polymers having a unit molecular weight of 22,000. C.H. Li has suggested that bovine growth hormone (BGH) has a molecular weight of 45,000 and is composed of two chains with one C-terminal and two N-terminal amino acids and four disulfide bridges. Tryptic digestion of BGH results in a preparation that is metabolically active in humans and irnmunochemically reactive with rabbit anti-human growth hormone. The molecular weight of purified tryptic BGH is approximately 22,000; some components of tryptic BGH have electrophoretic mobilities similar to HGH and several laboratories have reported molecular weights of 22,000 for BGH.

Growth hormones stimulate formation of specific antibodies; their immunochemical reactions with the sera produced by these antibodies have become a basis for hormonal assay. Similarity of structure among the growth hormones is indicated by the broad cross reactivity that exists between the hormones of various species and the antisera to growth hormones; for example, antiserum to rat or rabbit GH gives a precipitin reaction with GH prepared from rabbits, cattle, sheep, and pigs but no cross reaction with human or monkey GH.

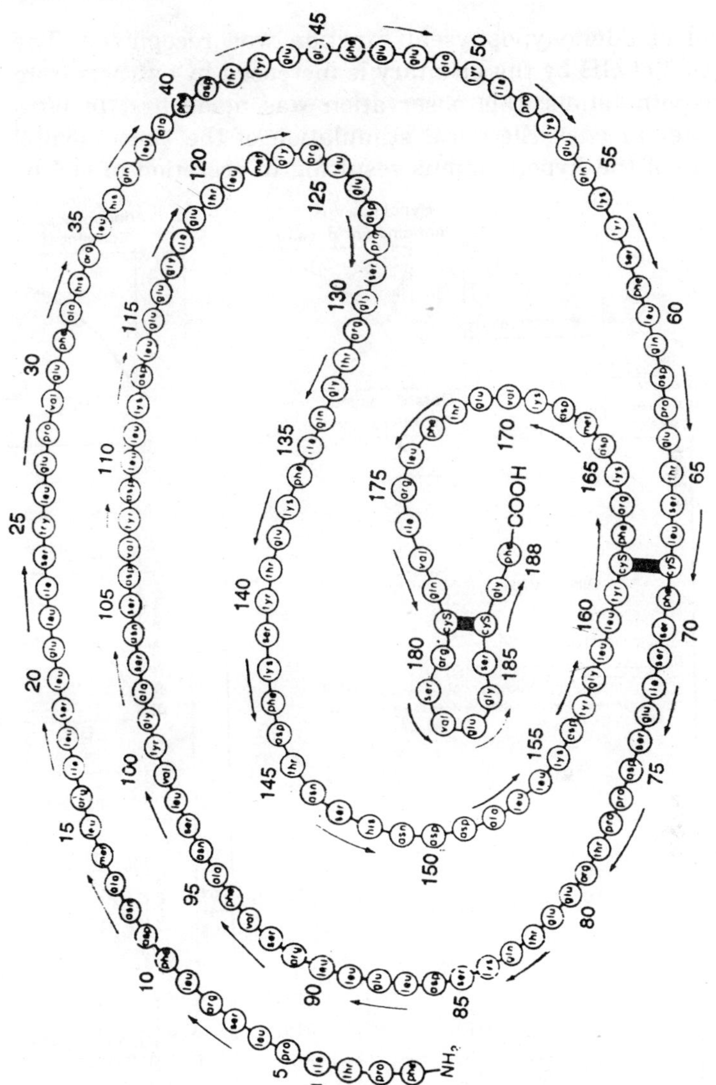

Fig. 2.6. The amino acid sequence of human growth hormone (HGH).

Regulation of secretion

The factors controlling the secretion of GH by the pituitary are unusually complex. Although it was known that lesions placed in the rat hypothalamus impaired growth and caused degranulation of acidophils in the adenohypophysis, the mechanism was not understood until hypothalamic hormonal control of adenohypophyseal function was recognized. The secretion of GH by the pituitary is increased by extracts from the hypothalamus; this observation was made first *in vitro* and later *in vivo*. Electrical stimulation of the ventromedial nucleus of the hypothalamus resulting in secretion of GH by

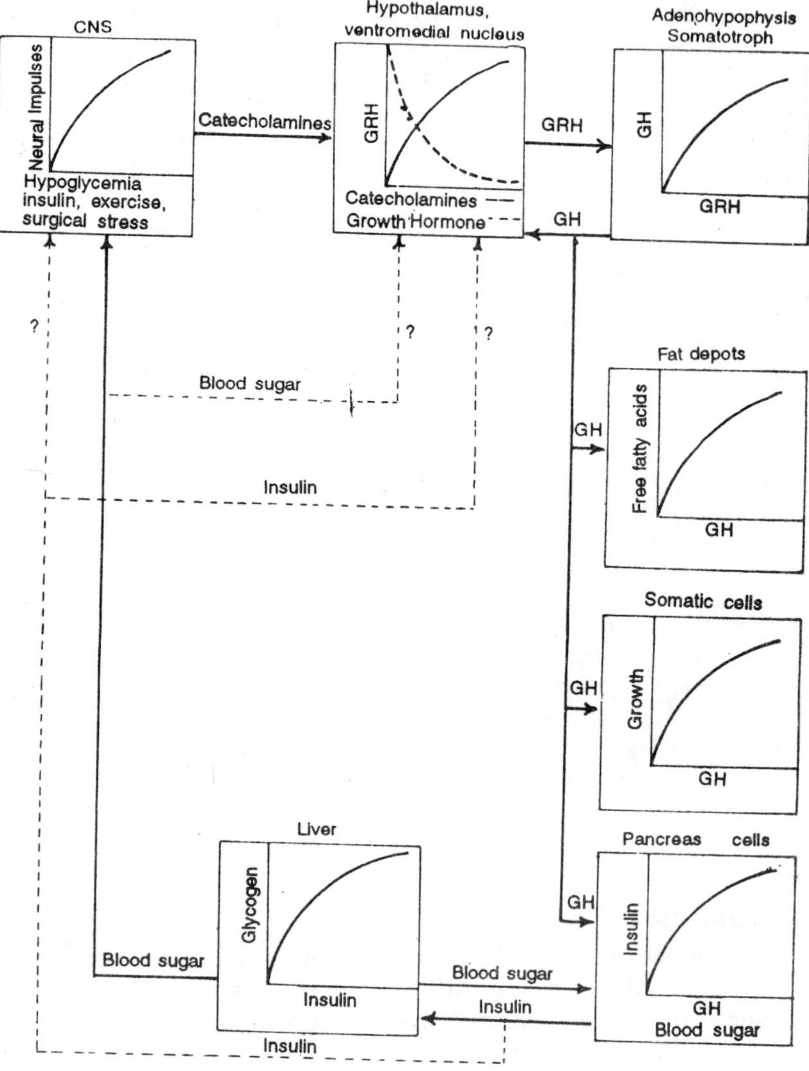

Fig. 2.7. Control of GH secretion.

the adenohypophysis has established this nucleus as one area that secretes GRH. Stravation causes a reduction of GH levels in the pituitary and GRH levels in the hypothalamus. Insulin causes a discharge of pituitary GH and administration of glucose blocks this discharge α-2-Deoxy-D-glucose stimulates GH secretion and hyperglycemia; it causes a cellular glucopenia when converted to α-2-deoxy-D-glucose-D-phosphate, which partially inhibits phosphoglucose isomerase and glucose-6-phosphate dehydrogenase. Lesions in the hypothalamus block the increase in plasma GH levels that would otherwise be induced by insulin.

Insulin may induce GH release by causing a secretion of norepinephrine or L-dopamine. Drugs such as reserpine, α-methyl-dihydroxyphenylalanine (α-methyl-DOPA), α-methyl-α-tyrosine, or tetrabenazine cause depletion of norepinephrine reserves in the central nervous system and are able to prevent insulin-induced release of GH from the pituitary. Since in these experiments extreme hypoglycemias were induced, it is likely that the effect of low blood sugar levels on secretion of GH is also blocked by depletion of catecholamine (epinephrine or norepinephrine) stores in the central nervous system. Guanethidine and tyramine (depletors of peripheral norepinephrine stores) did not block the insulin-induced discharge of GH by the pituitary. The effect of insulin or reserpine on the concentration of GRH in the hypothalamus has also been examined: insulin causes and reserpine prevents depletion of hypothalamic stores of GRH. It is likely that the catecholamines may be the neurohumors involved in causing neurosecretory cells of the hypothalamus to release neurohormones. The induced secretion of insulin may complete the positive feedback control of GH secretion; both hormones are found in inverse concentrations in the plasma and each indirectly causes release of the other. A short loop between the pituitary and hypothalamus and the changing concentrations of glucose in the plasma may form the negative-feedback control.

Effect

The initial recognition of the pituitary's relationship to growth came from the correlation of gigantism, dwarfism, and acromegaly (due to hyperproduction of GH in the adult) with pathology of that gland. Evans first showed that BGH causes growth of young rats. The stimulatory action of GH on all major organs is most obvious in adult animals in which the visceral enlargement causes protrusion beyond the normal confines. Associated with growth is a markedly positive nitrogen balance; GH causes a reduction in plasma amino nitrogen, increases amino acid transport across cell membranes, and increases protein synthesis.

Growth hormone causes mobilization of nonesterified fatty acids from fat deposits while it also inhibits glucose utilization by muscle tissues and decreases the sensitivity of hypophysectomized animals to insulin. The first of these actions is the so-called ketogenic effect and the latter two are diabetogenic actions. Greater utilization of fat is also reflected by the respiratory quotient.

The development of sensitive radioimmunochemical assay techniques for highly purified pituitary hormones has made it possible to determine blood concentrations of these hormones in humans. The plasma concentration of GH in fasting human subjects is 1 to 2 ng/ml plasma; after exercise it rises to 17 to 53 ng/ml, reaching the highest point in the first hour of a 2-hr exercise period. Plasma free fatty acid concentrations also rise, while the blood glucose level remains constant. These changes are prevented if the subject ingests glucose. By making depot fat available, GH may playa role in proving cellular fuel, since fatty acids are metabolized only after they are mobilized from depots.

Current studies on GH action point to a role in the transcription or translation steps leading to protein biosynthesis. Widnell and Tata have show an *in vivo* increase in liver *ribonucleic acid* (RNA) polymerase within 24-hr after GH treatment. Growth hormone has also been reported to stimulate *transfer RNA* (tRNA) and *messenger RNA* (mRNA) formation. Injection of GH into hypophysectomized rats can partially restore the lost ability of ribosomes to incorporate amino acids into proteins.

Thyroid-Stimulating Hormone

Chemistry

Thyroid-stimulating hormone (TSH) has not been prepared in pure form largely because of its instability and its formation of complexes with inert protein present in the preparation; TSH is also very difficult to separate from *luteinizing hormone* (LH), a contaminant in most preparations. The potency of the best preparations has varied from 20 to 200 IU/mg. TSH is a glycoprotein having a molecular weight of about 28,000. The carbonhydrate moiety, covalently' bonded to a protein, appears to be a single oligosaccharide unit containing mannose, glucosamine, galactosamine, and fucose. Human TSH shows little cross reaction with bovine TSH antibodies prepared from rabbit serum, indicating immunochemical dissimilarities between the hormones of the different species.

Regulation of secretion

Adenohypophyseal secretion of TSH is under dual control. There is evidence that thyroid hormone acts directly on the pituitary. Thyroid gland implants into the adenohypophyseal area of thyroidectomized rats retard the appearance of thyroidectomy cells among the thyrotrophs closest to the implants. Thyroxine (T_4) reduces the rate of ^{131}I release from the thyroids of hypophysectomized rabbits with implants of pituitary tissue in the anterior chamber of the eye. A dose of

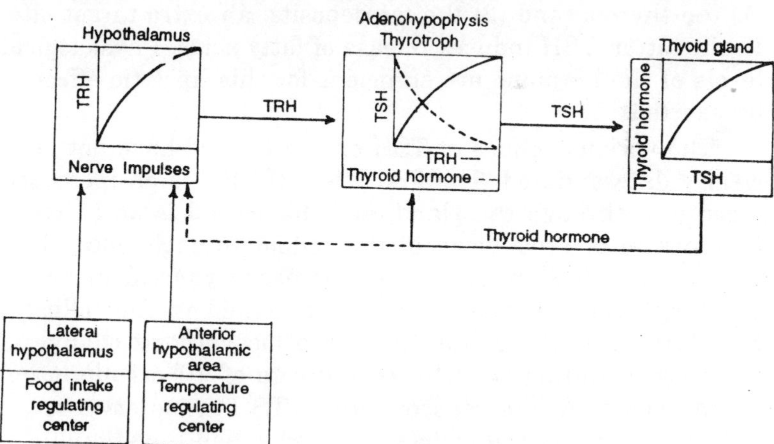

Fig. 2.8. The regulation of thyroid-stimulating hormone (TSH) secretion occurs on two levels; one involves the direct interaction of thyroid hormones with the adenohypophysis and the second involves the hypothalamic control which is modulated by environmental and core temperatures, by food intake, and perhaps by the action of thyroid hormones on the hypothalamus.

thyroxine so low that it causes no systemic effect, depresses ^{131}I rate of release from rabbit thyroid when instilled into the pituitary, but is ineffectual when injected into the hypothalamus. In organ culture of pituitaries, thyroxine reduces the secretion of TSH.

The existence of a hypothalamic feedback pathway is suggested by the isolation from the hypothalamus of neurohormone (TRH) which induces the release of TSH. Other less direct evidence is that the thyroids of goitrogen (compounds causing enlargement of the thyroid) treated rats hypertrophy more than do the glands of rats given the same goitrogen but bearing bilateral hypothalamic lesions. In the latter case, the destroyed link in the chain is assumed to be the source of a substance regulating TSH secretion by the pituitary. The role of thyroid hormones in the regulation of TRH secretion is not fully established; a number of different stimuli are probably involved and the thyroid hormone may act through these stimuli by affecting caloric homeostasis.

Effects

Thyroid-stimulating hormone acts on two different sites: (1) the thyroid, and (2) the fat deposits, an extra-target site. In the latter, TSH induces release of fatty acids. Physiological levels of the hormone are sufficient for this lipolytic effect to be exerted.

The various actions of TSH on the thyroid have not been clearly differentiated. The first effect of TSH is an increased blood flow through the gland, and the second is an increase in the rate of breakdown of the colloid thyroglobulin. This proteolytic effect results in an increased thyroid hormone discharge and a reduced amount of colloid in the follicles. Simultaneously there is an increased biosynthesis of thyroid hormone, resulting in an increased rate of ^{131}I accumulation by the gland. A late consequence of TSH stimulation is an increase in size of the follicle cells and, when the stimulus is long-acting and intense, an increased number of cells and follicles.

Long-Acting-Thyroid Stimulator

In 1956 Adams and Purves described in the plasma of hyperthyroid patients a stimulatory substance which induced the discharge of organically bound ^{131}I from rat thyroid glands. Response to a dose of TSH required 2 to 3 hr while response to serum from hyperthyroid patients required 12 to 16 hr. Long-acting-thyroid stimulator (LATS) has been extracted from this serum and has been characterized as a 7-Svedberg (7-S) ... α-globulin. It seems to be a specific α-globulin rather than an α-globulin-TSH complex. The molecule has been hydrolyzed to A and B chains with the A chain showing LATS activity. This substance is not present in the pituitary. LATS loses activity after incubation with minced thyroid tissues. Thus LATS may be an autoantibody against a thyroid component.

Exophthalmos-Producing Substance

Surgical intervention in the clinical condition of Graves disease is frequently followed by an exophthalmos (bulging of the eyes). Guinea pigs also show an exophthalmos after thyroidectomy. Although crude TSH preparations may induce exophthalmos, highly purified preparations do not. Many pituitary extracts and various subtractions have been examined; some have high concentrations of exophthalmos-producing substance and low levels of TSH activity, whereas others show the opposite condition. The consensus is that TSH and another hormone are involved; however, the exophthalmos factor has not yet been characterized.

Adrenocorticotrophin

Chemistry

Adrenocorticotrophin (ACTH) is a single-chain polypeptide consisting of 39 amino acid residues and having a molecular weight of about 4,500; the ACTH molecule is the smallest of the adenohypophyseal hormones and has been completely characterized and synthesized. The hormone causes depletion of adrenal ascorbic acid and induces steroidogenesis. Similar effects can be elicited by a polypeptide consisting of the first 20 amino acids from the N-terminal end of the ACTH molecule. The amino acid sequences of ACTH of four species (pigs, cattle, sheep, and humans) have been determined. Species differences are confined to the amino acid residues numbered 25 through 33, and have no apparent effect on biological activity.

Regulation of secretion

A number of diverse factors affect the adenohypophyseal secretion of ACTH. ACTH simultaneously inhibits the cellular secretion of CRH via a short negative feedback loop and stimulates its target cells in the adrenal cortex. Furthermore, the cortical steroids in pharmacological doses may modify the activity of the adrenocortical cells and in smaller dosage the cells that secrete CRH. Stress (any of a large variety of noxious stimuli) may provoke an increased secretion of ACTH, acting on the hypothalamus via central nervous system connections. Steroids implanted in the midbrain usually cause a reduction in adrenocortical secretion of corticosterone. This probably induces diffusely distributed steroid receptors to discharge, inhibiting the hypothalamic release of CRH. Hypothalamic extracts containing CRH have been shown to regulate ACTH secretion by the pituitary, and ACTH implanted into the hypothalamus has been shown to reduce the ACTH content of the adenohypophysis, probably due to the inhibition of neurosecretory-cell secretion of CRH. Glucocorticoids may have a direct action on the adenohypophyseal secretion of ACTH: they depress oxygen consumption of pituitary cultures and, when implanted into the Vicinity of the pituitary, prevent the adrenal hypertrophy that occurs after unilateral

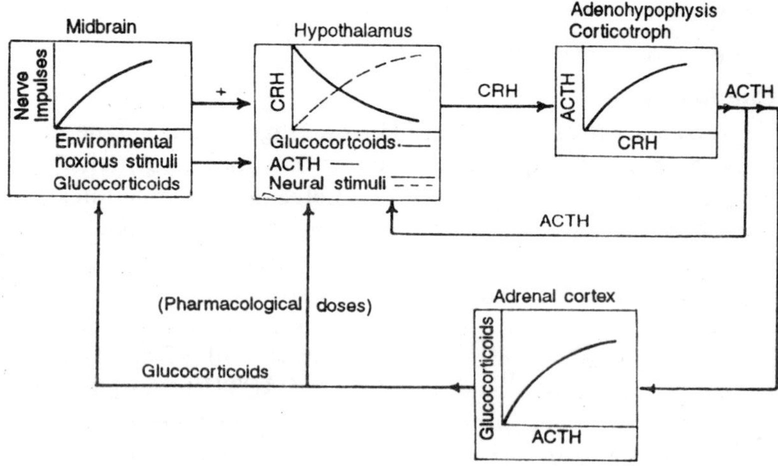

Fig. 2.9. The regulation of adenohypophyseal secretion of adrenocorticotrophin by neural and hormonal stimuli.

adrenalectomy. Despite these indications of direct action of cortical steroids on the pituitary, the consensus is that the primary site of feedback action is the hypothalamus and perhaps other nervous areas. Vasopressin has also been shown to stimulate ACTH secretion but is not believed to be physiologically important.

Effects

Adrenocorticotrophin acts upon three sites. The targets that it primarily influences are the two innermost zones of cells in the adrenal cortex (the zona fasciculata and zona reticularis). These cell layers constitute the site of glucocorticoid synthesis and secretion. The zona glomerulosa (the outer cell layer of the cortex), which secretes aldosterone, is minimally affected by ACTH. The hormone also affects fat depots, where it exerts a lipolytic action in a manner similar to GH, TSH, and catecholamines. It stimulates melanocytes, causing darkening of skin in a manner similar to *melanocyte-stimulating hormone* (MSH).

The adrenal cortex is affected by ACTH in several ways: (1) It causes a reduction in the ascorbic acid content. (2) It concurrently converts cholesterol to glucocorticoids and increases their secretion rate. (3) It raises the level of metabolic activity of adrenal tissue through an increase in oxygen consumption and glucose utilization. (4) It stimulates cell division of the inner two layers and causes the cortex to grow. (5) It increases adenylcyclase activity and causes c-AMP synthesis with enhanced steroidogenesis proportionate to the ACTH dosage. Protein-synthesis inhibitors block the response to c-AMP but not the effect of ACTH on c-AMP concentration and glycogenolysis. The increased steroidogenesis may be due to increased conversion of cholesterol to pregnenolone.

Prolactin

Chemistry

Prolactin was among the first of the pituitary hormones to be prepared in almost pure form. The molecular weight of 23,000 pure ovine prolactin, calculated on the basis of sedimentation and diffusion, agrees with the result of amino acid analysis of 205 residues.

Human prolactin has all the properties of the best preparations of HGH and the two are indistinguishable by many criteria. However, the prolactin prepared from other species is distinct from their growth hormones.

Regulation of secretion

There is evidence that the hypothalamus continually inhibits prolactin secretion by the adenohypophysis. Transplanted pituitary tissue continues to produce prolactin while the secretions of other hormones cease. Furthermore, cultured pituitary tissues release less prolactin to the medium when grown in a flask containing hypothalamic tissue than

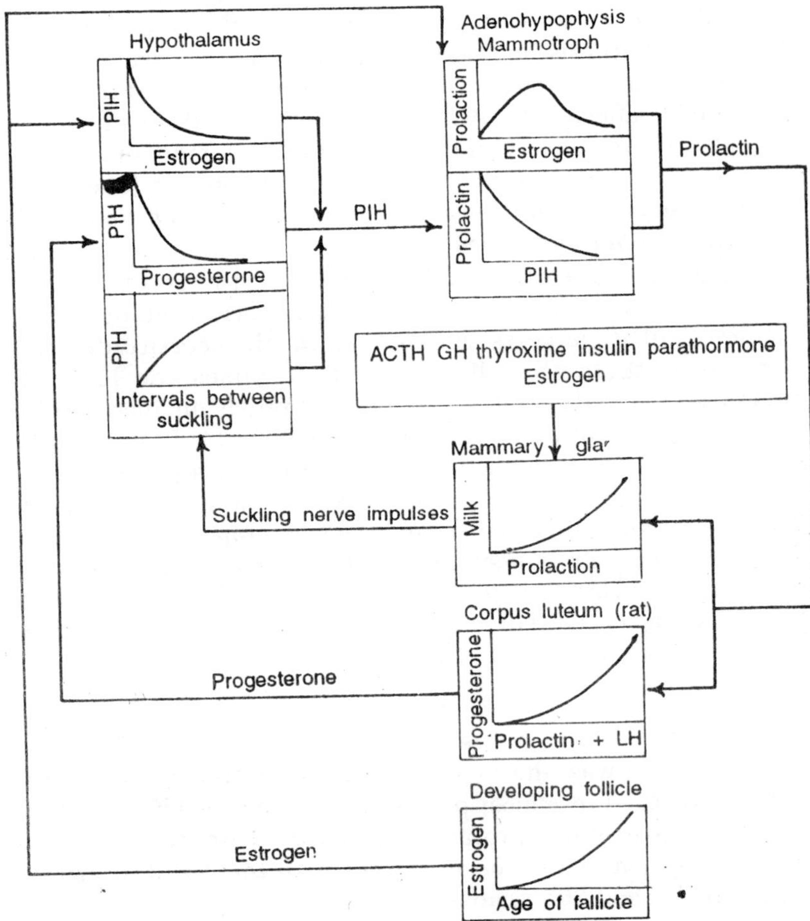

Fig. 2.10. The regulation of the secretion of prolactin.

when grown alone; under similar conditions, more GH is formed and released when hypothalamic tissue is present. The hypothalamus is the source of a continuously secreted substance that inhibits prolactin release from the adenohypophysis. This substance is called *prolactin-release-inhibiting hormone* (PIH).

Secretion of PIH is decreased in the lactating animal; termination of lactation has been associated with increased secretion of PIH. Therefore there should be an inverse relationship between amounts of prolactin and PIH secreted. Such a relationship has been observed and correlated with frequency of suckling by rat pups. As time of weaning approaches, the pups suckle less frequently and the prolactin content of the pituitary rises, reflecting the increased secretion of PIH.

The corpus luteum in the rat fails to function in the absence of prolactin, but this luteotrophic action of the hormone is seen only in rodents. Maintenance of a functional corpus luteum in rats is assumed, therefore, to be evidence of prolactin secretion. Estrogens, progesterone, and testosterone cause corpora lutea to continue functioning. Estrogen induces prolactin secretion by the pituitary; testosterone probably has a dual effect (1) causing secretion of estrogen by the ovary, and (2) blocking neurosecretory-cell secretion of PIH. The progesterone effect is an example of positive-feedback action, the pituitary production of prolactin is unaffected by large doses of the steroid.

Effects

It is apparent in nonmammalian vertebrates that prolactin has functions other than those of a lactogenic hormone. In pigeons, prolactin causes the crop sac to secrete "crop milk." In tadpoles the hormone inhibits tail resorption and urea excretion, and promotes growth after hypophysectomy. Prolactin stimulates a water drive in the land form of a newt (the red eft), and in some teleosts causes melanogenesis. In some euryhaline fish transferred to fresh water after hypophysectomy, administration of prolactin makes survival possible.

Gonadotrophins

Chemistry

Although no pure gonadotrophic hormones are available, five different preparations (three of non pituitary origin) are currently in common use. *Follicle stimulating hormone* (FSH) and LH are directly prepared from pituitary tissue. *Human menopausal gonadotrophin* (HMG), isolated from the urine of menopausal women, shows both Fishlike and LH-like activity. The serum from pregnant mares contains a gonadotrophin (PMS) which is predominantly FSH-like but with a small LH-like action. PMS is formed in the uterine endometrium and accumulates in the blood because it cannot be excreted by the kidney. *Human chorionic gonadotrophin* (HCG) appears in the urine of women about 7 days after conception and reaches a peak in 6 wk. It exerts an LH-like effect, but FSH-like action appears with high doses.

The high glycosidic content of FSH and LH permits easy aqueous extraction from pituitary tissue. Ovine LH in acid solution dissociates into two nonidentical subunits with a molecular weight of about 15,000; in neutral or slightly alkaline solution, it has a molecular weight of 28,000 to 30,000. The two polypeptide subunits have different amino acid compositions. The LH molecule is believed to be a globular nonhelical molecule stabilized by cystine disulfide bridges. FSH contains 5 percent sialic acid which is essential for its activity, since neuraminidase digests this oligosaccharide and inactivates the hormone. HMG shows both FSH-like and LH-like activities, which tend to remain in the same ratio through purification procedures. This suggests a single molecule with dual actions. HCG derived from the placenta and PMS from the uterine endometrium are usually rich in carbohydrate, containing 30 and 50 percent, respectively. The high carbohydrate content may reflect contamination of these substances. The molecular weight of HCG is about 30,000 whereas that of PMS is 23,000. The apparent tendency of PMS to aggregate in electrolyte solutions may explain its failure to pass through renal membranes.

Regulation of secretion

It is difficult to discuss one of the pituitary gonadotrophins, FSH or LH, without simultaneously considering the other, since functionally they complement each other. Spermatogenesis is possible only if both are secreted together, since FSH is needed in the intermediate stages of sperm development and *interstitial-cell-stimulating hormone* (ICSH) is needed to stimulate the androgen secretion necessary for sperm maturation. In the female, ovulation occur only if both FSH and LH are sequentially available.

The negative-feedback loop illustrates our current ideas on the regulation of gonadotrophin release in the female. The testosterone levels in the blood are thought to regulate activity of the neurosecretory cells that secrete the gonadotrophin-releasing hormones, FRH and LRH. Although the testes of some animals are rich sources of estrogens (for

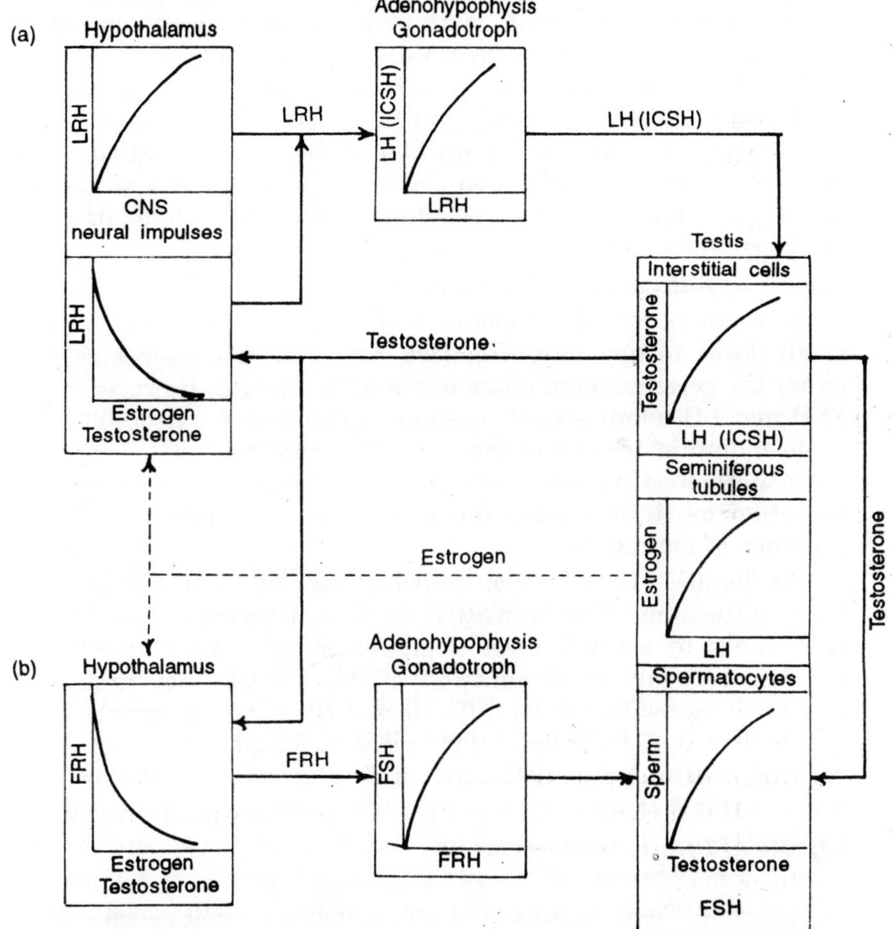

Fig. 2.11. The regulation of gonadotrophin secretion in male mammals.

example, bull, stallion), the role of estrogens in male reproductive physiology and biochemistry is unknown. There is some evidence that the sex steroids directly affect the adenohypophysis; however, these pathways are not illustrated here.

Regulation of gonadotrophin secretion in the female is much complicated than in the male. The secretion of both FSH and LH occurs at a constant basal level. In the human an intense surge of LH and FSH occurs on approximately the fourteenth day of the menstrual cycle. Rats maintained in animal rooms with light : dark cycles of 14 hr: 10 hr and with the lights going on at 5.00 AM and off at 7.00 PM have 4 day cycle. In such rats FSH is probably released simultaneously with LH on the day of proestrus between 2.00 and 4.00 PM. Hilliard has shown that in the rabbit a positive feedback system dependent on the LH stimulated secretion of 20α-hydroxypregn-4-en-3-one supports the secretion of LH until ovulation occurs.

Effects of follicle-stimulating hormone

The role of FSH in males in unclear since FSH given in small doses to hypophysectomized rats does not maintain either the germinal epithelium or testicular weight. However, FSH and LH administered together in ratios of 4 : 1 to 400 :1 do maintain testicular function. The combined action of these hormones probably involves stimulation of androgen secretion as final maturation of the sperm requires the presence of androgens.

In females, FSH acts on growing follicles to stimulate their maturation. The transition from an immature oocyte surrounded by a single layer of granulosa cells to a mature structure involves stimulation of mitotic activity and consequent growth of three different cell types: (1) connective tissue cells that form the outer layers of follicles; (2) theca internal cells in the middle layer; and (3) granulosa cells in the innermost layer.

Effects of luteinizing hormone

In males, ICSH causes the interstitial cells to secrete testosterone. There is little evidence to indicate cyclic release of ICSH from the adenohypophysis, except in seasonal breeders, for example, deer. In animals such as rabbit, dog, rat, and man, there is continuous stimulation of the interstitial cells.

In females, ovulation may either occur spontaneously as part of a cycle, or as a reflex in response to a stimulus (most commonly coitus). Rats, dogs, sheep, and most primates are spontaneous ovulators, whereas cats, ferrets, and rabbits are reflexive. The action of LH is more easily understood in the latter group. The initial response of a rabbit to LH occurs within seconds of the secretion of the hormone and constitutes a hyperemic reaction, perhaps due to a local release of histamine. The next response appears within minutes and is manifested by discharge of a progestin (20α-hydroxypregn-4-en-3-one) from the interstitial cells. The mature follicle is noticeably enlarged 6 hr after coitus and ruptures after 10 to 11 hr. Following the rupture, the granulosa cells hypertrophy and increase their secretory activities and the entire follicular space becomes filled with large lipid-containing cells, becoming a corpus luteum. Stimulation of steroid biosynthesis by ICSH (LH) in interstitial cells and the corpus luteum may involve the generation of c-AMP as a "second messenger" in a manner similar to the action of other hormones such as ACTH, TSH, vasopressin, and norepinephrine upon their targets.

Lipotrophins

In 1936 Best and Campbell described a fat-mobilizing effect of anterior pituitary extracts. Subsequent preparations having specific lipotrophic effects were obtained, and in 1965 and 1966 Li and co-workers isolated two lipolytic polypeptides and determined their structures: one, with 90 amino acids and a molecular weight of 9,500 was designated β-lipotrophin (β-LPH); the second, with 58 amino acids and a molecular weight of 5,810, was named γ-lipotrophin (γ-LPH). The γ-LPH molecule is represented in its entirety in the β-LPH molecule. These lipolytic polypeptides. ACTH, and MSH all share a common heptapeptide core, which exhibits MSH-type activity.

In vitro MSH activity of β-LPH is about twice as great as that of γ-LPH: the former has an activity of about 1.9×10^{11} units/mole and the latter about 0.9×10^{11} units/mole. The same two-to-one relationship of the two lipotrophins is true for *in vitro* lipolytic activity in rabbit adipose tissue. In the rat, sheep γ-LPH is inactive while β-LPH is active at concentrations 10 to 50 times higher than in the rabbit by the lipolytic assay performed on adipose tissues of the two species.

Little is known about the regulation of secretion of the lipotrophic hormones or the role they play in the vertebrate organism.

Melanocyte-Stimulating Hormone

Chemistry

Melanocyte-stimulating hormone (MSH) has been demonstrated to be in the anterior lobe of the pituitary of all vertebrates and in the intermediate lobe of those vertebrates possessing such a structure. It is present in mammalian pituitaries in two forms designated by Lerner as α- and β-MSH. The former, a tridecapeptide, has greater biological activity and has been found in all mammals; β-MSH, has a variable chain length (the human hormone has 22 amino acids, whereas β-MSH in pigs, cattle, and horses has 18) and amino acid sequence. All the MSH molecules have a common heptapeptide core, which is also contained in ACTH and β- and γ-lipotrophin. Furthermore, α-MSH has a sequence of 13 amino acids identical to that at the N-terminal end of ACTH. However, the N-terminal seryl residue of α-MSH is acetylated, that of ACTH is not. The C-terminal valyl residue of α-MSH is in the amide form, while the corresponding valyl of ACTH connects to a continuing amino acid sequence.

Human β -MSH and bovine β-MSH have been synthesized and relationships between structure and activity have been explored. The heptapeptide core, met-glu-his-phe-arg-try-gly, has melanophore-dispersing activity, but the smallest sequence still possessing such activity is the pentapeptide, his-phe-arg-try-gly. Maximum activity is attained with the tridecapeptide sequence found in N-terminal acetylated α-MSH.

Regulation of secretion

The dispersion of pigment in the frog skin is caused by MSH; concentration of pigment may be controlled by melatonin, produced by the pineal gland. The presently conceived regulation of MSH secretion as being dependent on the release of an MSH-release-inhibiting hormone (MIH), produced in the supraoptic region of the hypothalamus.

Melatonin may act on the hypothalamus to stimulate secretion of MIH and directly on the melanophore to cause pigment concentration around the nucleus and blanching of the skin.

Effects

Melanin-containing cells are generally of two types. In cold-blooded vertebrates, melanin is contained in *melanophores*. Skin color is determined by the distribution of pigment in the cytoplasm; the skin is pale when the pigment is concentrated near the nucleus and dark when widely dispersed through the cytoplasm of the cell In birds and mammals, melanin is contained in *melanocytes*. These cells synthesize and store their pigment, and can contribute it to adjacent cells and structures. The α-MSH causes increased synthesis

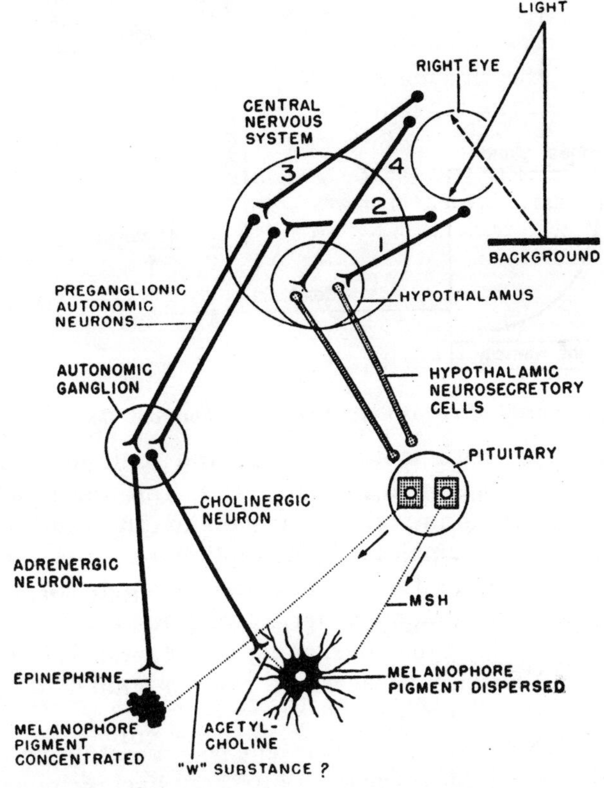

Fig. 2.12. Possible mechanisms involved in regulating the melanophores of the eel Anguilla.

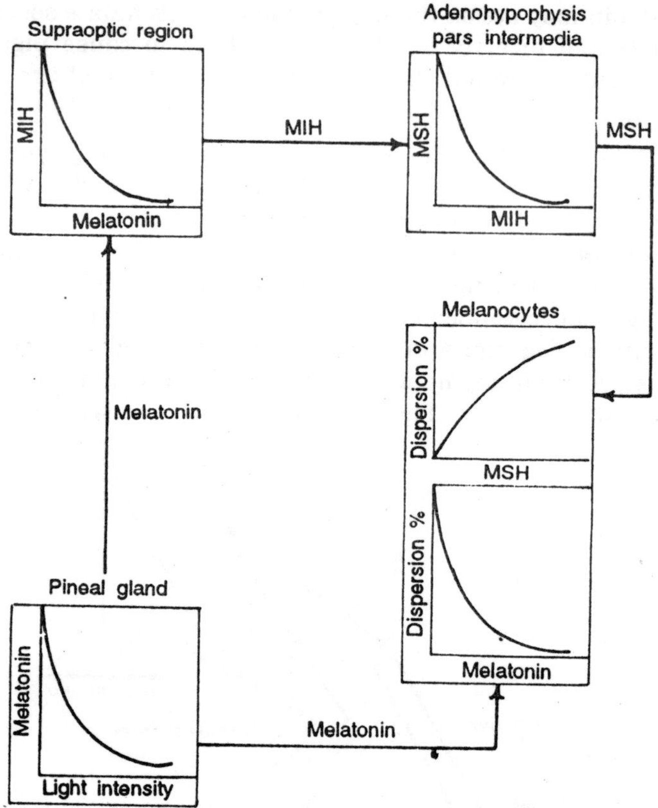

Fig. 2.13. Control of melanocyte-stimulating hormone (MSH) secretion.

of melanin in both melanophores and melanocytes and dispersion of melanin in melanophores. The role of α-MSH and β-MSH in mammals is not clear. In cold-blooded vertebrates—fish, amphibia, and reptiles—α-MSH serves a protective function; the animal becomes pale on a light background and dark on a dark background. Adrenocorticotrophin can also cause darkening of the skin (bronzing of the skin in Addison's disease) in a manner similar to that of α-MSH.

The α-MSH exerts a thyrotrophic action on thyroid tissue and both α- and β-MSH cause an increased lipolysis of adipose tissue in vitro. These effects may not be physiologically important.

Neurohypophyseal Hormones

The secretions associated with the neurohypophysis and produced in the hypothalamus. Seven natural cyclic nonapeptides have been identified in vertebrate neurohypophyseal extracts: oxytocin, vasotocin, mesotocin, ichthyotocin (isotocin), glumitocin, arginine vasopressin, and lysine vasopressin. In reptiles, birds, and mammals, these secretions originate in the supraoptic and paraventricular nuclei, while in fish and amphibia they originate in the preoptic nuclei. The hormones are transmitted through the axons of the neurons that compose these nuclei to the capillary beds in the neurohypophysis. There the hormones are stored in granules accumulated at the axon terminals; they must be released from these granules (*neurophysins*) before they can be secreted. The neurophysins may be involved in transport and storage of the hormones. Three hormone-binding proteins, neurophysin I, II and III, have been isolated. Neurophysin III appears in smaller amounts than either I or II. The mechanism that causes hormone release is not known; the axon terminals may abut on either the basement membranes of capillaries or cells surrounding the capillaries. In mammals, each hormone is secreted independently from its own hypothalamic nucleus; the supraoptic nucleus has been associated with vasopressin secretion and the paraventricular nucleus with oxytocin secretion.

Chemistry

Oxytocin and arginine vasopressin are present -in the posterior pituitaries of all mammals except the hog, which has lysine vasopressin. Oxytocin and arginine vasotocin are found in birds, reptiles, and some amphibians; vasotocin and ichthyotocin (isotocin) are found in some teleost fish. Mesotocin has been isolated from the posterior pituitary of some amphibians and lungfish and may be present in reptiles. Glumitocin has been identified in cartilaginous fish (ray fish and dogfish) and prepared from the neurohypophyses of rayfish. All the hormones contain a pentapeptide ring joined by a disulfide bond between two half-cystine molecules at the 1 and 6 positions. This ring structure containing 20 atoms is essential to biological activity. The hormones differ in amino acid composition at the 3, 4, and 8 positions. The amino acids at the 1, 2, 5, 6, 7, and 9 positions are constant in the hormones presently known.

These hormones have similar qualitative activities, but they differ remarkably in their quantitative effects. As an antidiuretic in the amphibian assay, arginine vasotocin (3-ile, 8-arg) is approximately 400 times as effective as arginine vasopressin (3-phe, 8arg). In rat antidiuresis assays, arginine vasopressin is about twice as active as lysine vasopressin. In the rabbit milk ejection assay, oxytocin (3-ile, 8-leu) is about eight times as effective as either arginine vasopressin or lysine vasopressin, whereas in the rat uterus assay oxytocin is four times as effective as the 8-lysyl analogs. Glumitocin shows a low level of oxytocin activity which is increased tenfold in the presence of magnesium ions.

Oxytocin is most active when the 3 and 8 positions are filled by isoleucine and leucine, respectively. Substitutions of other amino acids uniformly reduce oxytocin activity, more for uterine contraction than for milk ejection. Antidiuresis in rats is most effective when the side chain contains a basic amino acid such as arginine; the effect is decreased when lysine is substituted and even further decreased when citrulline is substituted for arginine.

Regulation of vasopressin secretion

Regulation of vasopressin (antidiuretic hormone) secretion may be initiated by a number of different pathways converging on the hypothalamus. Alteration of plasma osmotic pressure is probably the primary regulatory mechanism; however, alteration of blood volume or blood pressure, which activates volume receptors in the right atrium or baroceptors of the carotid arteries, also contributes to the control of vasopressin secretion. An increase in blood volume and consequent distention of the right atrium causes a diuresis, while hemorrhage, when accompanied by a reduced blood volume and pressure, causes water retention. Extrinsic factors such as pain, which affect the nervous system, induce an emotional stress inhibiting the secretion of vasopressin.

Regulation of oxytocin secretion

Although oxytocin has no demonstrated role in males, it has a well-established role in females. In pregnant animals

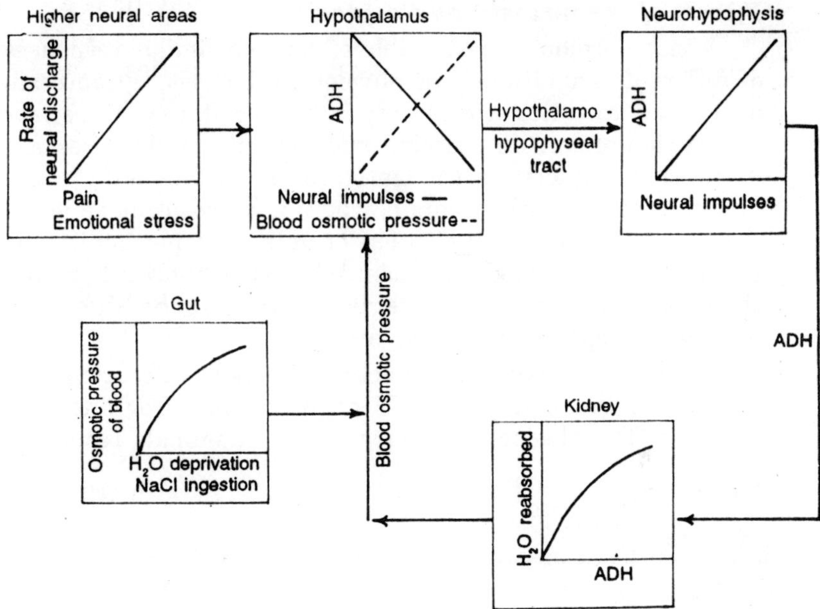

Fig. 2.14. Control of water reabsorption by vasopressin acting as the antidiuretic hormone (ADH).

parturition can be induced and brought to completion with oxytocin. During parturition the plasma oxytocin in cows increases from 55 μU/ml (microunits per milliliter) in the first stage of labor to between 10 and 20 times this amount during the second stage, and falls to control levels 2 hr after delivery. In second stage labor in women, there is a threefold increase in plasma oxytocin concentration (from 50 to 150 μU/ml). Oxytocin release is inhibited by progesterone in rats, rabbits, pigs and cows, but not in sheep or women.

Oxytocin also is important in the lactating female since in its absence no milk is released from the mammary gland. Ejection of milk occurs after administration of oxytocin or after the application of an electrical stimulus to the paraventricular nucleus, infundibulum, or any point on the hypothalamohypophyseal tract. Secretion of oxytocin can be prevented by anesthetizing the suckled teat, cutting the innervation of the gland, placing lesions in the paraventricular nucleus, or sectioning the pituitary stalk.

Effects of vasopressin and oxytocin

These hormones are capable of causing similar responses of different magnitudes. In physiological doses, vasopressin exerts an antidiuretic effect; in larger doses, it causes contraction of smooth muscle, especially in the blood vessels. At present, oxytocin is thought to be important only in the pregnant and lactating female, where it causes contraction of the uterine muscle and ejection of milk. The mechanism by which it acts on smooth muscle cells is not known; however, the speed of this response is thought to be due to a membrane-dependent phenomenon.

Vasopressin manifests its action largely by affecting water and urea permeability of renal nephrons, and frog and toad bladders. The absence of vasopressin in mammals results in

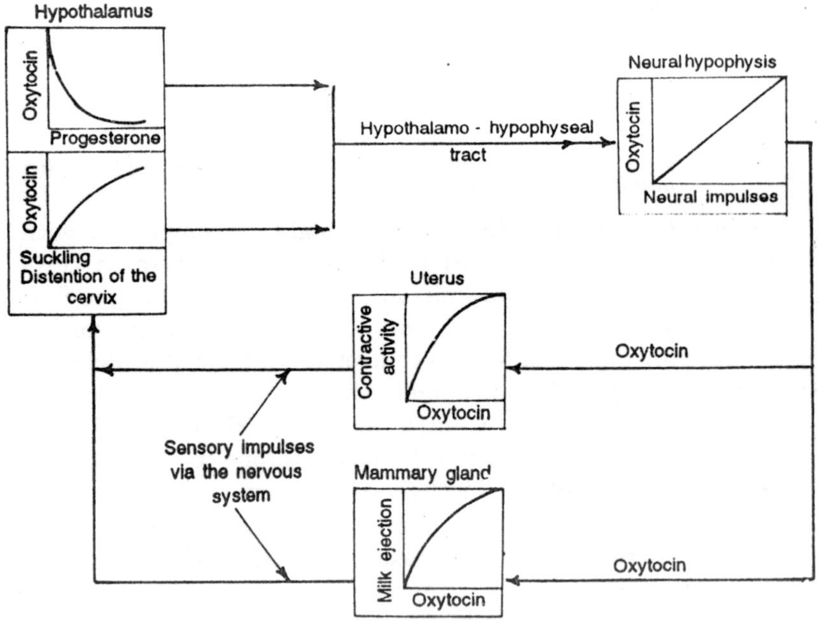

Fig. 2.15. Control of oxytocin secretion in the pregnant and lactating animal. In some mammals, the decline in the secretion of progesterone with aging of the placenta causes a secretion of oxytocin, which induces parturition. Once initiated, the process becomes a positive-feedback system, since distention of the cervix induces further secretion of oxytocin. The relationship of suckling to oxytocin is that of positive-feedback system.

diabetes insipidus, a condition in which large volumes of water are consumed and then lost in the urine. The most acceptable hypothesis to explain the action of vasopressin (and perhaps oxytocin) was proposed by Orloff and Handler, who suggested that the hormone triggers the activity of the adenyl cyclase system in cell membranes. This proposal is based on the observation that vasopressin increases concentrations of cAMP in susceptible tissues. (c-AMP induces effects similar to vasopressin as does theophylline, which protects c-AMP against destruction by phosphodiesterase).

3

Hypophyseal Abnormalities

Acromegaly

Acromegaly is a condition due to a hypersecretion of the growth hormone and is characterized by enlargement of the hands and feet, and of the bones and cutaneous tissues of the face, and by splanchnomegaly, etc.

History

In 1886 Pierre Marie, of the Salpetriere, described two cases, using the term 'acromegalie', and quoting five cases, with similar features, from the literature. In 1888 he collected further cases, bringing the number up to seventeen. However, Sternberg (1897) recognized acromegaly in Wier's account of a giantess as far back as 1567. In 1932 Atkinson tabulated and analysed 1,319 cases with 265 autopsies.

In his original paper Marie did not ascribe any cause to the disorder, but in 1887 Minkowski connected the disease with the pituitary gland. Benda (1900) was the first to detect an increase in the number of eosinophil cells in the anterior lobe of the pituitary; this was confirmed by Lewis (1905) in a case in which there was no enlargement of the pituitary. In 1895, Brissaud and Meige correctly postulated that giantism in childhood and adolescence corresponds to acromegaly in adults. In 1921 Evans and Long produced giantism in rats by injection of anterior pituitary, and in 1929 Putnam and others produced acromegalic changes in the head, skeleton, and viscera of English bulldogs.

Pathology

The pituitary gland is usually, although not necessarily, enlarged, and on section an eosinophil adenoma, which may be microscopic in size or very extensive, is found in the anterior lobe. When the pituitary appears to be normal, a differential cell count may nevertheless reveal a relative excess of eosinophil cells; or there may be a local hyperplasia of acidophilic elements (cushing and Davidoff, 1927). Occasionally a developmental pituitary nest of eosinophil cells is found at necropsy in the sphenoidal air sinuses (Erdheim) or, as in one of Cushing's cases, an enormous eosinophil adenoma may project outwards from the sella turcica, leaving within an apparently normal pituitary gland. Where there is a central eosinophil adenoma within the sella turcica, the rest of the anterior lobe may be greatly shrunken and degenerate. The pituitary gland may press on the optic chiasma, eroding or invading the floor or roof of the sella turcica, the cavernous sinus, and the brain. Cushing (1912) recorded a case of acromegaly in a man of 33, with two years history, in which a large cerebellar cyst was the primary lesion; this produced a secondary hydrocephalus, which apparently resulted in an enlarged pituitary anterior lobe, with hyperplasia of the eosinophil cells. The changes in other endocrine glands are dependent on variations in the secretion of their trophic hormones. Pressure of the acidophil tumour may induce a failure of gonadotrophin secretion, but hypersecretion of trophic hormones can occur, usually involving adrenocorticotrophin. However, growth hormone itself influences the size but not the development, or function of endocrine glands, and it is probably directly responsible for the frequent occurrence of adenomata in the endocrine glands of the acromegalic.

The cortex of both adrenals is usually hyperplastic and multiple adenomas are frequently present. The thyroid is enlarged in some 50 per cent. of cases, the usual change being an increase in vesicular colloid. A colloid goitre may be found in acromegaly even when complicated by thyrotoxicosis severe enough to call for thyroidectomy. The parathyroids may be enlarged, and adenomas have been recorded. Enlargement of the thymus and diffuse lymphoid hyperplasia, which also occurs both with Addison's disease and with adrenal cortical adenoma or hyperplasia, are common. The pancreas may be normal or atrophic, or may show hyperplasia, with, perhaps, adenomas of the islets of Langerhans. The ovaries and testes are usually atrophic. There is not much evidence on the histology of the ovaries, although amenorrhoea is common, but they may be cystic and degenerate, or fibrous, with only a few normal follicles. Cushing states that the mulberry ovaries (corpora lutea) of Evans' rats are not found in humans. The testes are often soft and flabby, the seminal vesicles disorganized, and the interstitial cells degenerate. The heart, lungs, liver, spleen, and kidneys are considerably enlarged, the stomach is often double its normal capacity and both the small and large intestine are considerably increased in length and circumference. Bony changes are considered in the clinical section.

Physiology

The essential feature is the continued over-secretion of growth hormone, producing skeletal and visceral changes comparable to those of experimental acromegaly. Excessive amounts of growth hormone have been detected in the blood of acromegalics.

Incidence

Acromegaly occurs in all parts of the world and in all races. Males and females are affected in equal proportions, the incidence among Jews and Swedes being relatively high. The maximum incidence is in the third decade of life. It may, however, begin at puberty, and Atkinson (1931) recorded one case at the age of 8.

Childhood acromegaly is a great rari and is inevitably associated with giantism. If one includes relatively mild manifestations of the disorder, acromegaly, as most other endocrine disorders, shows a family incidence: in one series in 30 per cent. of the cases. Acromegaly is more likely to develop in tall people. Thus Davidoff (1926) found in one series that the average height of men in whom the disease began before 20 was 6 ft. 2 in., and of women 5 ft. 6 in. This may well be due to an increased secretion growth hormone during the phase of growth preceding the excessive secretion which gives rise to acromegaly.

Aetiology

The cause of the development of an eosinophil hyperplasia, or adenoma, is rarely obvious. Occasionally pregnancy, or bilateral ovariectomy may precipitate the disorder. Although the highest incidence is in the third decade, many patients have had some manifestations at puberty, or in adolescence, and it is probable that the physiological endocrine changes of this period may fail to be autonomously controlled, or limited, in those patients developing acromegaly. The familial incidence in some 30 per cent of acromegalics suggests that this may be the case, and apart from the occasional onset of the major disorder at puberty and in pregnancy, transient fugitive acromegaly may occur at these periods. It is not generally recognized that mild acromegaly is not infrequently found at the climacteric, in certain types of women and men, but it is a slow insidious process. The onset in earlier life is also usually insidious and the disorder may progress for a decade or more before the patient realizes that he is suffering from a serious endocrine disturbance. Occasionally, general infection appears to be a precursor of the disorder, e.g. measles, typhoid.

Clinical features

As in many endocrine types, there is a considerable resemblance among all acromegalics. The large extremities, awkward movements, thickened features, and drooping shoulders with hands falling near the knees in advanced cases, give the picture of Simian man, and where giantism has preceded the acromegalic changes, of a primitive ape-like giant. Great strength, however, may give place to exhaustion and weakness in the later stages, and in order to understand the symptomatology, one must appreciate that Over-activity of the pituitary and related glands, e.g. adrenals, can be followed by exhaustion and underactivity; and further, that phases of activity may alternate with phases of inactivity before the terminal phase is reached.

Skeletal changes

As indicated by the name of the disease, the more obvious changes occur in the hands and feet, especially in the carpal and tarsal bones, which are enlarged and may fuse. The metacarpal and metatarsal bones are also thickened, and the heads of the phalanges may show outgrowths or 'tufting' on radiological examination. The considerable and progressive increase in the size of the hands and feet, which is due to the thickening of the soft tissues, as well as to skeletal changes, results in the characteristic need for an outsize in shoes and gloves. The fingers are thickened, and somewhat square in their termination, giving the appearance of podgy, spad-like hand, but where skeletal overgrowth has occurred before epiphyseal fusion the fingers may be very long. The long bones of the upper and lower limbs may show considerable periosteal thickening and deformity, or may appear normal.

The skull is considerably thickened, the ridges becoming very prominent, and the external occipital protuberance enlarged. The cranial sutures may be obliterated. Even more marked are the changes in the facial bones; thickening and enlargement of the zygomatic arches, of the malar bones, and especially of the lower jaw, which becomes prognathic through overgrowth and also through changes in the temporomandibular joint. The teeth become spaced wide apart as the jaw increases in width. The clavicles are thickened, and the antero-posterior diameter of the chest is greatly increased. The vertebrae undergo atrophy, hypertrophy, and partial fusion, with resulting kyphosis, lordosis, and scoliosis A thickening of the ridge, or bony prominence, occurs where muscles or tendons are attached to bones, and exostoses may appear near joints. Arthritis may follow changes at the articular surfaces of bones, and bony exostoses in the neighbourhood of the joints may severely limit movement. Occasionally a portion of the skeleton, such as one big toe, appears to be more susceptible to the growth hormone and enlarges quite out of proportion to the rest of the skeleton. This illustrates the principle that responsiveness of tissues, as well as the concentration of the hormone stimulus, determine the final result in endocrinopathies.

Muscular system

Hypertrophy of the muscular system associated with abnormal muscular strength, may occur in the initial stages. But though gigantic acromegalies may excel as wrestlers, boxers, or weight-lifters, their early prowess may be succeeded by muscular atrophy and atony. At any stage of the disease the increase of muscle bulk is seldom matched by a comparable increase in muscle power. Weight for weight, voluntary muscle in: acromegaly is less efficient than the normal.

Soft tissues

The tongue is greatly enlarged and the papillae prominent; and in spite of the increased buccal cavity, the tongue may be unable to find room within it, and may interfere with articulation and tend to obstruct the air passages in the recumbent position. The lips become thickened, protuberant, and negroid in appearance. These changes in the lips, and also in the nose, may occasionally precede the skeletal changes. The skin and the subcutaneous tissue are thick, the pores enlarged, and the sebaceous and sudoriferous glands hypertrophied. Fibromata mollusca may be an expression of local hypertrophy, of subcutaneous fibrous tissue. Excessive sweating (hyperidrosis) of the whole body may be troublesome and intractable and, together with the greasiness of the skin, results in persistent malodour. The hair on the trunk in both sexes may become abundant, and coarse and wiry in character, the thick, greasy, hairy skin contrasting with the fine, dry, hair-free (or covered with delicate hair) skin of the hypopituitary state. The enlargement of the hands and feet is partly that of soft tissue, and the presence of associated tissue oedema is suggested by the diminution in size within a few hours, following removal of a pituitary tumour.

Respiratory system

In both sexes the voice becomes deep and resonant, owing to the enlargement of the larynx and the increased width and resonance of the air sinuses, though the mucous membrane may be so thickened that respiratory obstruction may call for tracheotomy. Lungs are enlarged proportinately with the thorax. In the late asthenic stage of the disorder, death may follow phthisis or bronchopneumonia.

Cardiovascular system

The heart may be enormously enlarged and all coats of the peripheral blood vessels hypertrophied. Myocardial hypertrophy can occur with a normal blood pressure, but hypertension is common and poorly tolerated by the patient.

Indeed hypertensive heart failure or cerebrovascular accidents not infrequently terminate the disease. However, the occurrence of hypopituitarism will diminish the blood pressure and hypotension may be found.

The electrocardiogram may reveal evidence of left ventricular hypertrophy, hypertension or myocardial ischaemia (Hejtmancik *et al.,* 1951). The latter may be due to an abnormal myocardium rather than coronary artery disease.

Nervous system

Smell may be imparied owing to hypertrophy of the nasal turbinal bones. Pressure on the optic chiasma leads to optic atrophy, bitemporal hemianopia, and later to complete blindness of one or both eyes. Ocular palsies may result from pressure on the third, fourth, or sixth nerves; and involvement of the fifth nerve may produce pain and hyperaesthesia over one or more of its divisions. Deafness may be due to involvement of the auditory nerve or middle ear.

Headache may be very severe and bursting in character. It is occasionally migrainous in type, and associated with vomiting. Severe headaches are often quite intractable to medical therapy. Paraesthesias of the hands and legs may be early symptoms, though these disappear dramatically after operation or irradiation. But true neuritis follows when somatic nerves are caught in obliterated intervertebral foramina. Various types of chronic inflammation of the meninges of the skull and spinal cord, as well as of bony plates in the spinal dura, have been described. Areas of sclerosis may arise in the spinal cord with resulting ataxia and pseudotabes. True acromegaly may also be associated with syringomyelia, but this should not be confused with syringomyelitic bony deformities. Speech may be sluggish and slow, memory often being impaired, and the general behaviour characterized by apathy and lack of initiative. Depression, irritability, negativism, melancholia, mania, and delusional insanity may be additional symptoms. In the early stages, or in relatively mild cases, however, there may be great alertness, energy, and drive.

The cerebrospinal pressure may be considerable, headaches being temporarily relieved by withdrawing some 30 ml. of cerebrospinal fluid. In one case, Ellinger and Simpson (1937) demonstrated an antidiuretic hormone in the cerebrospinal fluid. This was associated with oliguria, positive water balance, and profuse sweating Radiation of the pituitary gland produced a gradual disappearance of the antidiuretic hormone from the cerebrospinal fluid, a concomitant disappearance of sweating, and a normal water balance. This case suggests that pressure, or a nervous mechanism, may produce over-activity of the pars nervosa; but, in other phases of the disorder, diabetes insipidus may be a complication through destruction of the nervosa by the encroaching eosinophil tumour.

Sexual system

Rarely an initial increase in libido may occur in both sexes, especially when acromegaly begins in adolescence. More commonly amenorrhoea and impotence are early features, and are almost invariably present in the later stages. Nevertheless interference with sex function may not be obvious for ten or more years, though skeletal and other manifestations are progressive. Normal menstruation, pregnancy, and parturition occasionally take place when the acromegalic process is well advanced. In adolescent acromegalics, the external genitals may be enlarged and puberty may be premature. In older acromegalics, however, impotence and amenorrhoea are often associated with atrophy of the genitals. The cause of initial sex stimulation in acromegalic adolescents may be ascribed to irritation of the basophil cells by the eosinophil tumour, whereas in the later stag the basophil cells are encroached upon and destroyed.

Carbohydrate metabolism

The demonstration of the diabetogenicity of growth hormone provides a clear explanation for the frequent occurrence of diabetes mellitus in the course of acromegaly. Indeed, it is more pertinent to comment on the fact that only about 30 per cent of acromegalies develop diabetes. During the active phase of acromegaly the only method by which the body could avoid diabetes in the presence of excessive growth hormone is the secretion of increased amounts of insulin. Such a mechanism has now been demonstrated by the constant finding of elevated plasma insulin levels in acromegaly.

The occurrence of diabetes mellitus during active acromegaly indicates that the effect of the growth hormone on carbohydrate metabolism has overcome even the increased level of insulin production. It is therefore not surprising that this type of diabetes is insensitive to injected insulin. Treatment of the acromegaly, with a fall in growth-hormone production, may result in a cure of the diabetes or, at least, its amelioration. This variability of carbohydrate tolerance with the phases of pituitary activity is of considerable practical importance as therapy for the diabetes requires constant supervision and adjustment. If hypopituitarism supervenes the patient will become unduly sensitive to the action of insulin and is unlikely to have diabetes.

Diabetes arising in the inactive phase of acromegaly or becoming permanent after an onset coincident with excess growth-hormone secretion is due to 'exhaustion' of the pancreas and a failure of insulin production. The course and treatment of this is essentially similar to that of idiopathic diabetes mellitus.

Manifestations of adrenal cortex hyperfunction

In both sexes there may be extensive growth of coarse oily hair over the trunk: In the female, abnormal hairiness sometimes develops on the face and extremities and the hair of the head falls on as in primary adrenal virilism: Although adrenal cortex over-activity itself produces amenorrhoea, probably by inhibition of pituitary activity through excessive androgen secretion, it is unlikely to be the initial cause of amenorrhoea in acromegaly, since amenorrhoea occurs early and often without any manifestations of virilism. Similar argument would also apply to impotence in the male.

A clear example of Cushing's syndrome has been observed as a concomitant of active acromegaly (Eliel and Pearson, 1951).

Thyroid disturbances

Excessive growth is accompanied by an increased metabolic rate; the basal metabolic rate of active acromegaly is above the range of normality in 50 per cent of cases.

A palpable enlargement of the thyroid, usually lobulated and with a tendency to the formation of adenomata, is present in about 20 per cent of acromegalics, but should not be confused with thyrotoxicosis merely because the metabolic rate is raised. Radioactive iodine studies reveal no alteration of function in the enlarged thyroid. However, true thyrotoxicosis with frank clinical signs, arises in about 5 per cent of acromegalics. Hypothyroidism may arise in the later phase of the disease, usually in association with other evidence of hypopituiturism. However, we have seen a colloid goitre become atrophic, leading to primary myxoedema. In this instance the thyroid was unresponsive to injected thyrotrophin and there was no evidence of hypopituitarism. Exophthalmos, in the absence of thyrotoxicosis, can occur, suggesting an increased secretion of thyrotrophic hormone.

Course and prognosis

Acromegaly is usually a chronic progressive disease, taking many years to develop. The changes may be sufficiently slow to last a lifetime without grave disability, though a patient can be completely incapacitated within a few years of the onset. Waves of remission and exacerbation occur in the more chronic types, and a stationary phase may last some years. The more active phase of the disease may ultimately be followed by an asthenic hypoactive phase in the same way as thyrotoxicosis may, after a variable course, be ultimately followed by myxoedema, even without surgical intervention.

Despite arrest of the disease by treatment some disability often remains. Personality changes commonly reduce the patient's capacity for work; headache may persist or osteoarthritis develop over the years to restrict movement.

Treatment

Surgical removal of the acidophil tumour is indicated if vision is severely threatened by pressure on the optic chiasm. It is wise to limit the operation to freeing the optic chiasm from surrounding tumour tissue rather than to make an all-out attempt to remove the whole adenoma.

Deep X-ray therapy is often effective in suppressing excessive growth-hormone production, without causing hypopituitarism (Hurxthal *et al.,* 1949). This is the treatment of choice for active acromegaly but there is no indication for its use in the inactive phase of the disease. The assessment of activity is mainly a matter of clinical observation over a period of time. Unfortunately no simple method of assay is at present suitable for the determination of the amount of circulating growth hormone Hoever, Reifenstein, Kinsell, and Albright (1946) have pointed out the importance of a raised serum phosphorus in active acromegaly; and a raised concentration of plasma insulin confirms a clinical diagnosis of acromegaly.

Oestrogens and androgens have been recommended for their inhibiting effect on the pituitary. Ethinyloestradiol, 0.1-0.3 mg daily, may be used in the female but its use should be restricted to the mild case as deep X-ray therapy is definitely indicated for all cases in which the disease shows considerable activity. Testosterone propiorite, 25 mg. daily, has been used in males, but we would prefer to restrict the use of androgens to the treatment of hypogonadism associated with acromegaly; for this, methyltestosterone, 1025 mg. daily, by mouth would suffice.

Giantism

Giantism is a condition of excessive height

Aetiology and pathology

Giantism may be due to: (1) an excessive secretion of the growth hormone before the epiphyses unit; (2) a delayed union of the epiphyses (eunuchoidism), a normal amount of growth hormone thus being permitted to act over an abnormally long period; and (3) a combination of (1) and (2). An additional factor, consisting of an inherent capacity of the bones to respond to the stimulus of the growth hormone, is probable. Giantism tends to run in families and to be common among certain races, e.g. the Swedish. Macroscopically the pituitary gland may be normal or enlarged. Microscopically it may not be possible to detect any abnormality, or there may be a relative preponderance of, or adenoma of, eosinophil cells.

Clinical picture

Rapidity of growth is generally noted in childhood, but may be most conspicuous during adolescence. Acromegalic manifestations occur in some 40 per cent of giants, and may be observed at puberty, in adolescence, or later in life.

Sexual development and libido sexualis may be normal or even supernormal at first, but after some years impotence may develop. In the primary eunchoid type, hypogonadism is an initial and persistent feature, the external genitals are small and the secondary sexual characteristics absent or deficient, and there is delayed union of the epiphyses. For this latter reason growth continues for some years longer than normal. Further, since castration results in over-activity of the pituitary there may be, with hypogonadism, an excessive secretion of the growth hormone. Excessive height is found among those Skopecs (a religious sect) who have been castrated before puberty. The fundi are usually normal, but in the presence of a pituitary eosinophil tumour, optic atrophy may be observed, with limitation of temporal fields of vision, as in acromegaly. In one patient a girl aged 14, brought to Out-Patients because of rapid growth (5 ft. $11^1/_2$ in.) and lack of energy (following on previous robust health), bilateral papilloedema was a surprising finding and a subsequent ventriculogram (T.G. I. James) showed symmetrical dilatation of the third ventricle. Later, autopsy revealed a huge hydrocephalus resulting from complete stenosis of the aqueduct of Sylvius, the lumen of which was completely obliterated by a subependymal gliosis. Although the pituitary gland showed no obvious abnormality on histological section, one must postulate excessive secretion of pituitary growth hormone. Other clinical features were plethoric countenance; big hands and feet (men's size 9 shoes); slight adiposity, but red lineae distensae of the abdomen and axillae; no hirsutism; and menstruation had not yet commenced. The blood count and carbohydrate tolerance were normal.

Occasionally, in addition to the general giantism, one part of the body—for example, a leg or a toe—may grow to a greater extent than the rest. Since the concentration of growth hormone is probably the same at all sites of the body, one must postulate a localized tissue hypersensitivity, the reason for which is as yet obscure. Giants may show supernormal muscular power, but after some years this may be followed by asthenia.

Abnormal growth before puberty may occur, with tumours of the adrenal cortex, the testis, and the ovary, and with other forms of sexual precocity, but premature union of the epiphyses results in a final height below normal. In the physiological section it has been noted that testosterone and oestradiol may produce excessive skeletal growth as long as the epiphyses remain ununited.

Diagnosis

Pituitary giantism is recognized by the acceleration of a previously normal growth-rate and the appearance of acromegalic stigmata; radiological demonstration of an enlarged sella turcica confirms the diagnosis. The growth spurt of the child suffering from sexual precocity is of course inevitably associated with the premature development of secondary sexual characters. On the other hand, eunuchoidal giantism does not appear until adult life as its development depends on a continued stimulus from growth hormone in the absence of epiphyseal closure by the action of sex hormones. Apart from the diagnostic signs of hypogonadism, the skeletal growth is disproportionate; the limbs being of an excessive length compared to the trunk.

Treatment

The disability of pituitary giantism arises from the distortion of the body with the attendant mental stress of becoming a freak. In children, scoliosis and kyphosis add to ,the troubles of a large frame. The aim of treatment is to prevent deformity without suppressing other aspects of pituitary function; indeed a nicely balanced programme which is difficult to carry out and requires great patience from physician and relatives. Deep X-ray therapy to the pituitary is the method of choice, using the lowest dose compatible with slowing the rate of growth. In consequence, before treatment is instituted, the patient must be observed until the rate of growth is established, and a further course, or courses, of X-ray treatment may be necessary if the initial therapy does not alter the growth curve within six to nine months.

In severe cases, where pituitary radiation has failed, surgical exploration of the pituitary gland should not be unduly delayed, even in the complete absence of optic atrophy (Simpson, 1958). In relatively mild cases, oestrogens—and to a less extent androgens—have been used to slow or arrest, skeletal growth as they depress pituitary function and accelerate epiphyseal union.

Dwarfish

There can be no absolute definition of dwarfism, but if the height is obviously below the arbitrarily accepted lower level of normal limits, dwarfism may be diagnosed. If the condition is to be diagnosed before adult life, the above definition must be qualified by phrase 'as compared with individuals of the same age'.

Causes

Some of the causes of dwarfism are as follows:

Genetic (uncomplicated)

In this group the only abnormality is a failure of skeletal growth, presumably due to a congenital deficiency in the number of eosinophil cells compare congenitally dwarf mice). This recessive gene may alternate with the contrasting one determining excessive height, since it is not infrequent to meet exceptionally tall embers of the same family.

Chronic illness in childhood

Syphilis, tuberculosis, malaria, pancreatic disease including diabetes mellitus, diabetes insipidus, coeliac disease, chronic diarrhoea (e.g. Crohn's disease), renal rickets, and von Gierke's disease, are all causes of deficient growth, with without some failure of sexual development. Malnutrition, qualitative or quantitative, also a factor which operates in civilized communities, and supplementary dietetic experiments in poor schools have confirmed this (Corry Mann, 1926). Even in the absence of poverty and ignorance the maladjusted child who refuses food or indulges in fads comes undernourished and fails to grow (Talbot *et al.,* 1947). In our opinion this type anorexia is an important cause of short stature. We would also stress that the clinical presentation of steatorrhoea may be dwarfism, in the absence of any disturbance of bowel function.

Achondroplasia

This is a congenital abnormality of cartilage bone formation arising in foetal life, and characterized by short legs and arms, a relatively long body, good intelligence, a big head and face, a square nose with a depressed bridge, and spade-like hands. Underlying endocrine defects have not been discovered. These people breed families of achondroplasiacs. They are active, strong, and acrobatic, with normal intelligence and rather vain, and are to be found on the stage and in circuses.

Prococious sexual maturity

This leads to a preliminary rapid skeletal growth but ultimate dwarfism, because of premature union of the epiphyses. This is well known in connexion with obvious pathological sexual precocity, but it is less well recognized that any girls and some boys have premature union of the epiphyses of the long bone: at e age of 14 or 15 instead of 17 to 18.

Diagnosis

In uncomplicated cases this is merely dependent upon the height of the individuals. It is obvious that many short people are otherwise perfectly formed and some occupy very prominent positions. In many cases short stature is merely an accompaniment of some generalized disease and is inessential to diagnosis or treatment. It is very likely that the failure of growth in such cases is due to secondary depression of pituitary function. However, primary failure of growth hormone secretion is extremely rare; a diagnosis of primary pituitary dwarfism in the absence of other endocrine dysfunction can only be made by excluding other disease processes that affect growth. Unfortunately there is no simple assay method available for the detection of normal or subnormal concentrations of growth hormone in plasma.

Treatment

Any underlying disease or causative lesion calls for its appropriate treatment. No treatment is of avail if the epiphyses are united, or if the patient's chronological age is much above that of adolescence. Pituitary growth hormone is theoretically specific, provided the nutritional status is adequate; but available preparations, although potent in laboratory animals, have proved most disappointing in clinical practice. However, the recent discovery that growth hormone prepared from rhesus monkeys, or from human pituitaries is potent in the human gives great hope for the future (Beck *et al.,* 1957). It appears that there is a species difference in growth hormones (Li and Papkoff, 1956) and that only material prepared from primate pituitaries will elicit a response in the human being. Thyroid by mouth is sometimes helpful, quite apart from cretinism. The treatment of the various endocrinopathies of which dwarfism is a feature, is discussed under separate headings. In boys, androgens will produce a rapid increase in height but at the same time induce sex development and accelerate epiphyseal closure. Such treatment is of value over a short period, provided the dose is kept at such a level that gross development of secondary sex characters does not occur; methyltestosterone, 10-15 mg daily by mouth, will suffice. However, long-term treatment merely closes the epiphyses and prevents further growth. The same problem is inherent in the use of protein anabolic steroids with minimal androgenic potency. Such substances as norethandrolone, 15-30 mg daily, can be used intermittently in girls without causing virilization.

Infantilism

Infantilism is a condition of somatic growth and sexual development corresponding to a normal individual several years younger than the patient, and never attaining adult physique or sexual maturity. The term Levi-Loraine syndrome is used synonymously with infantilism. Actually, Loraine wrote a preface to a paper by one of his pupils, Faneau de la Cour, who submitted his thesis In Paris in 1871 under the title 'Du Feminisme et de l'infantilisme chez les tuberculeux'. Failure of somatic and sexual development was thus observed as a result of chronic tuberculosis of various types commencing in childhood. Ettirore Levi (1908) described several cases of infantilism, and in particular its occurrence in two sisters, aged 15 and 20, in the elder of whom there was enlargement of the pituitary fossa and optic atrophy. He thus drew attention to a pituitary type of infantilism, although his account appears to envisage a polyglandular disturbance. He also described infantilism as secondary to rheumatic carditis in childhood. One of the important diagnostic features stressed by Levi was the ununited epiphyses, which we now know as a manifestation of any type of hypogonadism.

Brissaud described a thyroid type of 'infantilism' which is more correctly spoken of as 'cretinism' or, if commencing in childhood, as 'juvenile myxoedema'. However, one occasionally meets with a type of pituitary infantilism complicated by clinical signs of hypothyroidism, and then the term Brissaud's infantilism might be applicable.

Frohlich's syndrome, is by our definition, a form of infantilism, but since infantilism is usually applied only to those patients who are not fat, and who are usually thin, and since Frohlich's syndrome is said to have connotation of fatness as an essential feature, this disorder is described separately. Another atypical type of infantilism, associated with a primary gonadal defect, will be considered under a separate heading.

Dwarfism is a failure of one function only, namely, skeletal growth, and may be present with normal sexual development. The term is therefore not synonymous with infantilism.

Pathology

Unfortunately there is little or no evidence of the pathology or morbid anatomy of uncomplicated infantilism. Craniopharyngioma may produce infantilism, but complicated by other secondary endocrine disturbances (Rowlands and Simpson, 1942).

Clinical description

Apart from retarded skeletal growth and sex development, patients with infantilism usually-present a delicate and gracious appearance. They are pleasing to the eye, and in no sense whatsoever grotesque. Their childlike and often attractive appearance compels friendship and sympathy, and a desire to help and protect them. Their physique is on slender' and graceful lines, with narrow shoulders and narrow hips, slender tapering fingers, and—somewhat surprisingly—often relatively long, slender, well-shaped legs.

The skin is of smooth and delicate texture and the complexion good. In males the penis and testes remain infantile, and there is an absence of sexual hair on the face and body. Rarely there is a small growth of pubic and axillary hair, but the pubic hair never extends onto the abdomen in male triangular fashion. The voice does not break. In females there is amenorrhoea, the uterus is infantile, the breasts do not develop, and the pelvis does not become wide at puberty.

Intellect is in no sense impaired, although the emotional and behaviour pattern often retains childish characteristics. Some people with infantilism, especially incomplete forms, attain intellectual brilliance and high scientific and cultural distinction.

As in nearly all endocrine disorders an incomplete form of infantilism must be recognized, especially as it is probably much more common than the classical complete type- With partial infantilism the general and somatic descriptions given above are usually applicable, but hypogonadism is not complete, and, contrary to our definition, may be inconspicuous. In the male the penis and testes tend to be subnormal in size and the upper margin pubic hair horizontal; the voice is rather high-pitched, the skin of delicate, feminine texture, and facial hair slight in amount, so that shaving every other day may be sufficient. However, such patients may be fertile- In the female the pelvis remains narrow and the breasts small; menstruation is often scanty, but conception may occur. These varieties of incomplete infantilism have their parallel in strains of dwarf mice, in which failure of somatic growth and incomplete sexual development are genetic disturbances associated with pituitary defects

Clinical types

1. *Pituitary infantilism.* Clinical or radiological evidence of a pituitary defect is a rarity; however, some cases of the syndrome are due to suprasellar cysts. The patients bear a striking resemblance to each other with their childlike features, delicate textured skin, and graceful fragile appearance. If Peter Pan is a fanciful name for them, there is certainly something Barriesque about their sweet temper and delicate air.
2. *Idiopathic Infantilism.* This is the largest group. Comparison with known cases of pituitary origin, together with our physiological knowledge, justifies—by inference—a pituitary aetiology. It is confirmed by the response of the gonads, more especially in males, to pituitary, or chorionic, gonadotrophic hormone.
3. *Chronic Infection.* This includes the tuberculous group of de la Cour and Loraine. The tuberculosis can be of bone, glandular, abdominal, or pulmonary, but must have continued in a chronic form during several years of childhood. The French writers also postulate congenital syphilis as a cause. Chronic malaria in childhood and, in fact, any chronic infection may be aetiological. The mechanism is obscure. By some ill-understood alteration of the hypothalamic-pituitary relationship secretion of both growth hormone and gonadotrophins is impaired. From

the teleological point of view, growth and sex development are sacrificed to concentrate all pituitary function on survival of the organism in terms of maintaining adrenal and thyroid function.

4. *Chronic diarrhoea.* Any cause of chronic diarrhoea in childhood, e.g. steatorrhoea, coeliac disease, pancreatic disease, ulcerative colitis and Crohn's disease (Snapper, 1937), may result in infantilism. Poor absorption of food appears to be the most important factor, and chronic inanition may also result in failure of development. It must be stressed that malabsorption from disorders of the small intestine may be severe in the absence of any disturbance of bowel function.
5. *Hypothalamic infantilism.* We have met with several cases of arrested development following severe shock, or concussion, or a severe virus disease such as measles complicated by encephalitic symptoms. It would appear that in all these cases a hypothalamic-pituitary mechanism is inhibited, or brought into play.
6. *Metabolic disorders.* Severe diabetes mellitus, diabetes insipidus, von Gierke's disease, and renal rickets in childhood may also produce infantilism. Diabetes mellitus, however, may develop some years after infantilism has been diagnosed.
7. *Cardiac disease.* Congenital cyanotic heart disease and severe rheumatic carditis in childhood may result in infantilism.

Diagnosis

This depends upon the essential features of subnormal growth, infantile genitals, absence of secondary sex characters, and delayed union of epiphyses. The diagnosis cannot be made with certainty before the age of normal puberty, when maturation of sex characters is not a physiological event. Primary gonadal agenesis, which is associated genetically with short stature, is differentiated by the presence of various deformities, such as a webbed neck, cubitus valgus and congenital cardiac lesions, and the growth of sparse pubic hair. The excessive gonadotrophin excretion of these cases indicates active pituitary sex function. Acquired atrophy of the gonads prior to puberty is difficult to differentiate in that the normal puberty growth spurt does not occur, but growth continues slowly until the patient shows the typical tall thin eunuchoid habitus, with abnormally long limbs. Failure to respond to injected gonadotrophins makes it clear that the primary lesion is of the gonads.

Infantilism secondary to some generalized disease process provides no diagnostic difficulty; nor does the rare case of craniopharyngioma, with attendant localizing signs of a space-filling lesion in the region of the pituitary. However, delay in the onset of normal puberty mimics infantilism, until sexual development occurs spontaneously. For this reason infantilism cannot be diagnosed with confidence under the age of 18 years.

Treatment

The endocrine treatment of infantilism is often disappointing. The result so far obtained with the available preparations of growth hormone are not satisfactory but should not prejudice the future use of new potent preparations obtained from monkey or human pituitaries.

The use of sex hormones will produce satisfactory development of secondary sex characteristics, and increase in somatic growth and a deepening of the emotions. Such treatment, although limited in its results because neither fertility nor normal menstruation will be induced, is of the greatest importance in transforming the awkward, shy, childlike person into a normal citizen, adjusted to society and capable of normal sexual relationships.

Frohlich's Syndrome

Babinski-Frohlich syndrome-adiposogenital dystrophy.

History

In June 1900 Joseph Babinski, the famous French neurologist and pupil of Charcot, presented to the *Societe de Neurologie de Paris* a paper with illustrations of a girl aged 17 years, with failure of sexual development and obesity, apparently due to a craniopharyngioma. The title of this paper was 'Tumour of the Pituitary body without Acromegaly, and with arrest of Development of the Genital Organs'. In the following year, Alfred Frohlich, of Vienna, described a similar case due to a craniopharyngioma in a boy of 14 years, under the heading, "A Case of Tumour of the Hypophysis" without Acromegaly, and owing to historical researches into probable cases described in the preceding fifty years, and fuller description, the disorder is usually called after him. His account was translated into English by Bruch (1939). The French give precedence to Babinski's name, and the Germans call it Frohlich-Babinski's syndrome. The descriptive label adiposogenital dystrophy was introduced by Bartels in 1906.

Both he and Erdheim (1904) pointed out that the obesity should not be considered as directly related to the pituitary but rather to a lesion of the hypothalamus.

Definition

Owing to a number of conditions allied to and confused with this syndrome, it is important to lay down a definition which is quite clear, based upon the original descriptions and on clinical studies. The following definition is suggested: A syndrome characterized by a failure of normal maturation of the gonads, subnormal height, and adiposity, the usual cause being a pituitary or parapituitary destructive lesion (craniopharyngioma) involving also the hypothalamus. Primary lesions of the hypothalamus as described below may also cause the syndrome. Diabetes insipidus may or may not be present, and cannot be considered as essential for the diagnosis.

Pathology

According to Kraus (1945) the following lesions have been known to cause the adiposogenital syndrome; craniopharygioma (intrasellar or suprasellar); chromophobe adenoma; third ventricle tumours; chronic hydrocephalus; chronic tuberculous, or syphilitic, meningoencephalitis; epidemic encephalitis, and meningoencephalitis; bullet wounds; fracture of the skull. He further states that 'in cases of severe and long-standing hydrocephalus, the hypophysis shows depletion of the chromophilic cells and finally atrophy of the whole organ'. However, the hypothalamic-pituitary cause of the syndrome is more often a matter of inference based on experimental knowledge rather than one of histological demonstration. Unfortunately the known pathology of the allied Laurence-Moon-Biedl syndrome does not add appreciable to our knowledge of the pathology of Frohlich's syndrome.

Incidence

The adiposogenital syndrome, as defined here, must be regarded as one of the rarest syndromes in endocrinology;

the label is usually applied erroneously. Conditions which may be confused with it are common, and will be discussed.

Clinical features

Failure of sexual maturation is an essential feature. As proper maturity in normal subjects does not occur before puberty, the sexual factor can only have a relative significance before puberty in the male and even less in the female. In the body it may be possible to state from appearance that the penis and gonads are retarded in their development and subnormal in size. At the time of normal puberty, and with greater certainty in adolescence, it will become clear that the penis and testes retain infantile or childish proportions, and no secondary sexual characteristics develop. Pubic and axillary hair does not develop, and the face is tree from any sign of male hair. In the female menstruation does not occur.

Frohlich's description included the following paragraph: "The penis, which is otherwise [ubrigens] normally developed, appears to be embedded in accumulations of fat to such an extent that the genitals approach the female type. The testes are palpable in the depth of the *fatty* tissue, and show infantile conditions [Verhaltnisse]. In the neighbourhood of the breasts, there are also considerable accumulations of fat. In the mammary glands several nodules are palpable, but fluid cannot be expressed. Hairs in the axillae are lacking, and only occasional small hairs are present in the genital region, There is at least a suspicion of a myxoedematous condition."

The cause of hypogonadism is failure of the pituitary gonadotrophic secretion, even when the initial lesion is hypothalamic, because apart from morbid anatomy, (1) patients (at least males) respond to a gonadotrophic stimulus, and (2) biological assays show an absence of gonadotrophic hormone in the urine. The cause of the adiposity in Frohlich's syndrome is probably hypothalamic in origin, as is indicated by the work of Bailey and Brewer (1921). Especially is this likely to be the case when diabetes insipidus is a complication. Contrary to general opinion, the adiposity in the original cases of Babinski and Frohlich was only moderate in amount, and it is doubtful if adiposity of marked degree is a feature of the syndrome. However, the distribution is characteristic, namely, face and neck, breasts, abdomen, pubis, and thighs. The face is not plethoric, and may be pale as in destructive pituitary lesions. The general appearance of the male patients is feminine, and the pelvis is gynaecoid. Drowsiness, polyuria, and polydipsia, when present are regarded as hypothalamic symptoms. Very little is known of the metabolism and biochemistry of the adiposogenital syndrome, and the significance of such studies as have been made (Werner, 1941) depends upon the criteria of diagnosis. A craving for sweet things may be present and associated with a tendency to hypoglycaemia.

Dwarfism, or subnormal height, constitutes the triad of the adiposogenital syndrome, and always permits a diagnosis with greater confidence. In our opinion a height above normal rules out this condition. Ununited epiphyses are a manifestation of hypogonadism. The muscles tend to be hypotonic and hyperextensibility of joints may be noted.

In one of our cases, now 32 years old, the obesity has remained static for the past ten years but some pubic hair has grown. This is probably due to adrenal androgens as the testes remain atrophic. The causative lesion in this case was a craniopharyngioma.

Diagnosis

This depends essentially upon definition, and should not be made with any dogmatism before puberty. Hypogonadism, adiposity, and dwarfism are the triad of syndromes upon which a firm diagnosis can be made. The following conditions should be differentiated:

Constitutional familial adiposity

In this condition the genital development is normal or slightly retarded. In boys the penis is often embedded in the pubic fat and therefore appears smaller than it is. This condition is frequently diagnosed as Frohlich's syndrome. In the absence of pathological knowledge one can only theorize as to conditions intermediate between this and Frohlich's syndrome. Frohlich himself referred to the masking effect of the suprapubic fat on the size of the penis is his own patient, so that this in itself obviously does not exclude the diagnosis.

Cushing's syndrome

If this occurs at or before puberty, sexual development may be delayed. The facies is congested, or plethoric, and the lineae distensae are red or violet rather than white. The obesity spares the limbs, whose muscles are often wasted, and involves face and trunk. Stunting of growth and hypertension also occur.

Adipose gynandrism

This is a condition with which Frohlich's syndrome has been most frequently confused. It is described in the adrenal section and is essentially different in the following respects: (a) the height is usually above normal; (b) plethora, cyanosis and red lineae distensae are usually present; (c) there are no gross intracranial lesions; (d) sexual maturation is delayed but occurs spontaneously.

Infantilism

This should not be confused with Frohlich's syndrome if it is appreciated that by definition infantilism cannot be associated with adiposity.

Laurence-Moon-Biedl syndrome

This disorder is characterized by adiposity, hypogonadism, polydactylism, mental deficiency, and retinitis pigmentosa. It tends to occur in families.

Prognosis

Babinski's and Frohlich's patients with cranipharyngiomas both died in the second decade. The prognosis in regard to life depends upon the nature of the cranial lesion. The genitals will not mature spontaneously, but respond to gonadotrophic hormone.

Treatment

Any local intracerebral condition calls for appropriate surgical treatment. Adiposity receives the general treatment of the condition, namely a low calorie diet and appetite depressors such as dexamphet amine. Chorionic gonadotrophin, 500 units injected intramuscularly, thrice weekly, will promote sexual maturation in the male but is not effective in the female. Methyltestosterone, 15-30 mg daily, will develop male secondary sex characters but leave the testes atrophic.

Laurence-Moon-Biedl Syndrome

A syndrome occurring sporadically and, in families, as an autosomal recessive, characterized by polydactyly, mental retardation, retinitis pigmentosa, with hypogonadism and adiposity; and relative dwarfism in some, if not all, members of the family affected.

History

Laurence and Moon first described the condition in the *London Ophthalmic Review* of 1866. Four of ten children of non-consanguineous parents were affected. It is recorded that one of the adults affected measured only 54 inches. Sievert and von Jaksch noted optic atrophy associated with the condition, and realized the possible relationship with Frohlich's syndrome, but it was Bardet (1920) and Biedi (1922) who postulated this connexion more clearly. Professor Sorsby (1935), to whom we owe a great deal for modern knowledge of the disorder, although his original approach was as an ophthalmologist, suggests that Bardet's name should be linked with Laurence and Moon's. Cockayne, Krestin, and Sorsby (1935) record that during the period 1925-35 thirty isolated cases had been reported in addition to fifteen familial groups.

Pathology

No autopsy had been recorded until 1936, and since that time four fairly complete and two incomplete autopsies have been described, and are summarized by Anderson (1941). In the first autopsy the pituitary was apparently normal; in the second, the sella turcica was enlarged and occupied by a large cyst, only a small portion of the glandular tissue remaining; in the third there was a strikingly high proportion of basophil cells, and a relative paucity of eosinophil cells in the pituitary; in the fourth and fifth cases (Riggs), two males aged 19 and 24 years, infantile testes were present, and certain changes were noted in the brain, but not in the hypothalamus or pituitary. Anderson's case was a 15 year-old body with hypertension secondary to polycystic kidneys and fatal uraemia. The adrenals, thyroid, parathyroids, and pancreas were grossly normal and the thymus was atrophic. The pituitary and pineal bodies and the brain were, macroscopically, normal In the testes there was no evidence of maturation of spermatozoa, and the interstitial cells were decreased in number. Microscopically the adrenals, parathyroids, and pancreas were apparently normal The thyroid contained many dilated acini filled with colloid and lined with cuboidal epithelium, and was considered to be a typical colloid goitre'. The pituitary was histologically remarkable. The anterior lobe was composed chiefly of basophil and eosinophilic cells. The marked predominance of the basophil cells coloured the sections blue to the naked eye. In serial sections no adenomata were seen. Differential cell counts showed 42 per cent basophil, 36 per cent eosinophil, and 22 per cent chromophobe as compared with the normal 11, 37 and 52 per cent respectively. There was no evidence of any hyaline change in the basophil cells. In the pars intermedia no colloid was seen; however, there was moderate invasion of the posterior lobe by basophilic cells. No definite changes were found in the hypothalamus or brain microscopically, except for diminution of Purkinje cells and loss of nerve cells in the granular layer of the cerebellum. Anderson (1941) believes the significance of a preponderance of basophil cells in his and Griffiths' cases may be significant. In both, however, there was chronic kidney disease, and in one, severe hypertension. Autopsies have not really given decisive evidence about the exact pathology or whether the primary lesion is hypothalamic or pituitary.

Clinical features

A characteristic case suffers from mental defectiveness, polydactyly, pigmentary degeneration of the retina (retinitis pigmentosa), adiposity, and hypogenitalism. Adiposity is nearly always constant, but of two brothers with the disorder one was fat and the other thin; and one adult became very thin after a fat childhood. Adiposity is often of moderate amount but may be extreme. Hypogonadism is usual, but not constant. Mental retardation is present in the great majority of cases, ranging from mental deficiency to idiocy. Polydactyly is nearly always present, but may be absent in otherwise typical cases. Retinal degeneration (retinitis pigmentosa) is reported by all observers. Generally visual defect is noted in early childhood, but in quite a number of cases it does not become apparent till later. Optic atrophy is occasionally found. Dwarfism is usual, but normal height may be found. Other rare complications are congenital heart disease, microcephaly, head-nodding, choreiform movements, and muscular weakness. One member of the family may have the complete syndrome, and another perhaps one or two stigmata of it. Symptoms are usually present from birth.

Diagnosis

This is easy, as defined above. The condition is generally thought of as a special variation of the Frohlich syndrome but in so far as patients may be plethoric rather than pallid it should be considered also in relation to adipose gynism and gynandrism.

Prognosis

This is poor, and intercurrent infection not infrequent. Congenital renal abnormalities may cause an early death in uraemia.

Treatment

No fundamental treatment is possible.

Simmonds' Disease

Panhypopituitarism-Anterior Pituitary Deficiency

Definition

Simmonds' disease is a condition of anterior pituitary deficiency and secondary involution and hypofunction of the thyroid, adrenal, and sex glands, manifested in typical examples by asthenia, apathy, amenorrhoea or impotence, hypersensitivity to cold and to insulin, a low basal metabolism, and a very low level of 17-ketosteroids in the urine. Morris Simmonds, a Hamburg pathologist, constructed a clinical picture in retrospect, after observing at autopsy, examples of inchaemic necrosis of large areas of the anterior pituitary gland in patients who died after parturition. However, the subsequent clinical features and obstetric background of severe parturition haemorrhage were illuminatingly discussed by Simpson as long ago as 1883.

Pathology and aetiology

The essential lesion is destruction of the anterior lobe of the pituitary gland, and the commonest cause is thrombosis of the pituitary vessels following parturition associated with severe haemorrhage and, not infrequently, retained placenta. Infection may be present, but is not regarded by Sheehan (1937-9) as a fundamental factor in producing the thrombosis.

Simmonds (1914) pointed out that the arteries to the anterior pituitary are end arteries, and that emboli would therefore lead to infection and necrosis. Reye (1926) thought the vascular process was thrombosis, which he contended, was especially liable to happen in an organ like the pituitary, which underwent involution at parturition after being hypertrophied during pregnancy. The penetrating and painstaking researches of H.L. Sheehan (1937 onwards) have resulted in a valuable pathological and clinical elucidation. He pointed out that 'the illustrations of Simmonds show what appear to be thrombi *in situ* in the capillary sinuses' and suggests that 'it seems probable that in these reports the word "embolus" is used in the sense of a few organisms being carried in the circulation to the pituitary and there forming the nucleus of thrombus formation. No satisfactory description has been given to suggest that the actual thrombus found obstructing the anterior pituitary vessel had been carried thereby the arterial stream.' In fact, in the absence of infective endocarditis, or a patent foramen ovale, it is difficult to see how an embolus could be carried to the pituitary blood vessels.

In the majority of cases Sheehan states that the thrombi are not infective, and predisposing factors are a peculiar distribution of the blood supply to the anterior lobe during pregnancy, the rapid involution of the anterior lobe after delivery, the increased coagulability of the blood during the puerperium, and especially a sudden large haemorrhage during delivery. Puerperal sepsis may be a complication. The thrombosed sinuses, as found by Sheehan, were small, and it was not possible to say whether they were functionally arterial or venous. 'In the absence of any other cause for the necrosis, it would appear that the thrombosis is the primary lesion.' In all cases, even in those dying some time afterwards, the necroses all appear to date from the time of delivery. Characteristic necrotic findings are rarely discernible however, if death takes place earlier than fourteen hours after parturition. The necrosis may be small, large or almost complete, but it usually spares the pars tuberalis, the part just in front of the attachment of the stalk, the region of the pars intermedia, and a thin layer, or small scattered islets just beneath the capsule. Sheehan writes: 'adopting arbitrary figures, there will probably be no symptoms with a loss of less than 50 per cent of the gland, the symptoms will be slight with a 60 per cent loss, moderate with a 75 per cent loss, and severe with a 95 per cent loss'. Judging from the histology of the pituitary in patients who died several years later from other causes it seems remarkable how small a portion of gland need remain undestroyed in order to carry on normal, adequate function, but this is quite comparable with our knowledge of other endocrine glands, e.g. thyroid and pancreas.

Although ischaemic necrosis accounts for the majority of female cases of Simmonds' disease, many other pathological processes have been described. Sheehan and Summers (1949), reviewing the literature, accepted the following lesions as causing anterior pituitary failure:

1. Ischaemic necrosis
2. Chronic fibroid lesions-aetiology unknown.
3. Tuberculosis, syphilis, and giant-cell granulomata.
4. Trauma to base of skull.
5. Trauma of pituitary or of the suprasellar region.

It is obvious from the varied pathology that recognition of the disorder of function is not enough in itself, for the causal lesion may dominate the ultimate prognosis. Tumours arising from suprasellar lesions may be associated with diabetes insipidus or narcolepsy. As the lesion progresses to destroy the anterior pituitary, polyuria will diminish.

Other endocrine glands

Thyroid. Partial or complete atrophy of the thyroid is found, and its weight is less than half of the normal, 25 g (Sheehan, 1939). The alveoli are scanty and atrophic, with little or no colloid therein, and there is gross fibrosis, round-celled infiltration, and numerous lymphoid follicles.

Suprarenals. The cortex is atrophied, and the zona glomerulosa and reticularis are very thin and sometimes absent. There is rarely any fibrosis, although the capsule may be thickened. Lipoid is usually present in the cortex, and sometimes in normal amounts, but there may be none. The medulla is normal.

Gonads. The ovaries and uterus may be completely atrophic and fibrotic, or incomplete maturation of Graafian follicles may be found. The testes may be minute and atrophic, with complete disorganization of the normal structure, and no evidence of the presence of spermatozoa or spermatogonia.

Parathyroids. Sheehan (1939) found the parathyroids 'rather small', or very fatty, or fibrous.

Pancreas. The islets of Langerhans might be very small, or apparently normal

Viscera

In contrast to acromegaly the stomach, intestines, lungs, heart, liver, and spleen are smaller than average without necessarily showing any pathological changes. The heart may show brown atrophy. There is a striking absence of fat.

Incidence

In view of the importance of parturition haemorrhage as an aetiological factor, the disease is much more frequent in women than in men. The most common aetiology in the male is a neoplasm in the pituitary region. The disease has always been regarded as rare, but more accurate diagnosis shows that the disturbance is one of the more common major disorders of the pituitary. Not all cases exhibit total pituitary failure; partial failure usually affects growth hormone and gonadotrophins rather than thyrotrophin or adrenocorticotrophin.

Clinical features

The condition is usually one of adult life but rarely—when, for example, it is due to slow-growing craniopharyngioma, as in a case under observation—the initial symptoms are seen in childhood as a failure of normal, somatic, and sexual development (Rowlands and Simpson, 1942) but even so, the characteristic picture of Simmonds' disease is not fully developed until adult life.

In the majority of patients the disease starts from a parturition which is associated with severe haemorrhage and collapse, frequently with a retained placenta and sometimes with puerperal sepsis. These serious concomitants overshadow any specific manifestations of a sudden destruction of pituitary function, but it is soon observed that the recovery of the patient does not take a normal course.

More specifically, there is a complete absence of mammary activity and of lactation in many patients, and Levy-Solal (1932) has described hypoglycaemic shock after delivery, when the blood sugar may be as low as 50 mg per 100 ml. Failure of the pubic hair to regrow after its removal prior to labour is an other indication of pituitary insufficiency. However, in the majority of patients there may be no specific diagnostic features, and the condition mayor may not be suspect until the more chronic picture evolves.

General symptoms and signs

Behaviour is often characterized by apathy and inertia, and sometimes by phases of irritability or of somnolence associated with hypoglycaemia. The patient appears indifferent to her surroundings and normal social activities. She lacks initiative and spontaneity, and may become careless in her dress, sluggish in thought, and slow in speech. Questions may not be answered until after a long latent period. She sometimes appears stupid or absent-minded, and may stare into space for long periods or lie in bed. Sheehan states that 'more marked mental symptoms are rather frequent in the last few months, ranging from oddness, or alternating attacks of excitement or depression, up to definite insanity'. In moderate cases the patient complains of excessive weariness on slight exertion. Muscular atrophy and atony are concomitant features. Sensitivity to cold may be very marked even when warmly dressed. The patient tends to get completely under the bedclothes. The face takes on a pallid appearance, even with absence of anaemia—a pallor which may be regarded as in striking contrast to the plethora of Cushing's syndrome. The skin may be delicate and fine in texture, with an involution of sweat and sebaceous glands. Delicate wrinkling around the eyes and mouth contrasts with the soft hairless skin to give a curious mixture of youth and age. In others there is a tendency to mucoid infiltration of the skin as in myxoedema. The hair of the head loses its lustre and becomes dry, thin, and brittle. Its colour may also change, e.g. from a dark brown to a light brown. The eyebrows become thin, especially in the outer half. The pubic and axillary hair is usually completely lost. It may be partially retained but thinned in the earlier stages of the disease.

Contrary to the initial concepts of the syndrome, cachexia is not essential to the diagnosis. The early stages of the disease are very rarely associated with rapid loss of weight; indeed obesity may be present if hypothyroidism is the predominant disorder of function. In the later stages, when apathy and anorexia are marked, the patient may become cachectic. This is in contrast to the initial loss of appetite with rapid loss of weight in anorexia nervosa, which is followed by a secondary depression of pituitary function.

Constipation is common and achlorhydria occurs in 50 per cent of patients. A raised sedimentation rate is not uncommon in the absence of infection. Some degree of anaemia is the rule, usually normochromic in nature, but the pallor of the skin is commonly in contrast to the pink mucous membranes.

Summers (1952) found that the anaemia was either of a normocytic hypochromic or a macrocytic hypochromic type; in both the marrow was normoblastic. Glossitis and perleche are rare accompaniments. The normoblastic marrow indicates that there is neither folic acid nor vitamin B_{12} deficiency; the administration of iron, orally or parenterally, is ineffective. Thyroid extract has surprisingly little effect on the blood picture, indicating that the anaemia of hypopituitarism is not the same as the anaemia of myxoedema. However both cortisone and ACTH improve the blood picture considerably; the best results require thyroid and probably testosterone in addition to cortisone.

Gonadal function

In the female, amenorrhoea is the commonest manifestation of gonadal failure and menstruation may never return after parturition. In other patients menstruation is scanty, and at long, irregular intervals. Very rarely it may be normal, and another spontaneous pregnancy occur. In the male, libido and potency disappear; the testes become atrophic. Atrophy of the uterus and lower genital tract is usually marked in the female; dyspareunia is the rule and libido is absent.

Cardiovascular

Bradycardia is almost invariable, and the blood pressure is characteristically low but occasionally normal. The electrocardiogram shows a low voltage.

Pulmonary

There are no abnormal pulmonary changes in uncomplicated cases, but tuberculosis sometimes supervenes. It may then be insidious and silent, and is often unsuspected, especially as the pulse rate tends to remain slow and the temperature subnormal. Bronchopneumonia may also supervene in the later stages and prove fatal.

Skeletal

Children, when affected with a craniopharyngioma, fail to grow, and a type of dwarfism results. X-rays show a delay, or complete absence, of union of the epiphyses of the long bones. In contradistinction to acromegaly, the hands and feet are diminutive, and the fingers delicate and tapering. Bony absorption and retraction of the lower jaw may occur in adults, so that the opposite condition to prognathism results, the teeth decaying and falling out.

Metabolism

The basal metabolism is almost invariably low. Sheehan states that it lies between minus 25 and minus 33 per cent. in the first 15 years after onset, and later may be minus 40 per cent. In our experience the degree of depression of metabolism does not depend upon such duration. Further, it is important to realize that in some patients showing a definite but incomplete syndrome after parturition, the basal metabolism may be only slightly lowered, e.g. minus 12 per cent. The temperature is usually subnormal, and may be very low, 96°F. Blood cholesterol values are sometimes raised but, not, in our experience, raised to the same extent as in myxoedema, even when the basal metabolism is comparably lowered.

Thyroid function

Thyroid hypofunction is partly responsible for the lowered metabolism. Radioactive iodine studies show a slow accumulation of iodine by the thyroid to a maximum of about 10-15 per cent of the dose. Often even less iodine is retained in the thyroid, and the urinary excretion of the iodine is high, particularly for 24-72 hours after the dose, in contrast to the minimal secretion at that time by the normal subject. Administration of thyrotrophic hormone over 2-5 days results in an increase in the rate of thyroidal iodine uptake and retention of a larger percentage of the dose. By this means hypothyroidism of pituitary origin may be differentiated from true myxoedema of primary thyroid atrophy. The exact calibration of the degree of stimulus and response is still lacking; moreover Simmonds' disease of many years standing may show no thyroidal response to the test because of the atrophic state of the thyroid

Carbohydrate metabolism

Carbohydrate tolerance curves are not constant in Simmonds' disease. The fasting blood sugar is usually low, and the most characteristic curve is a flat one, showing a prolonged plateau at values only slightly above normal, and not returning to subnormal values until after two hours.

The characteristic abnormality of carbohydrate metabolism is elicited best by the intravenous injection of insulin. As Fraser and Smith (1942) pointed out, the speed and degree of fall in blood sugar after the injection of insulin is not the important point. The true abnormality, due to deficient adrenocortical function and lack of the diabetogenic effect of growth hormone, is the failure of the blood sugar to regain normal levels within 2 hours. This tardiness in the recovery from hypoglycaemia is termed 'hypoglycaemic unresponsiveness'.

The patient should be well fed (as far as is possible) for three days before the test. Then after taking a fasting blood sugar specimen, insulin is injected intravenously. For normal people 0.1 units of insulin per kilogram body-weight is the dosage, but for suspected Simmonds' disease a characteristic curve will be given with half this dosage (or even a third). The reduction of the standard insulin dosage is advisable to avoid too severe a hypoglycaemic response, endangering life. It may be necessary to interrupt the test to give glucose or inject adrenaline. Capillary blood is taken at intervals of 20, 30, 45, 60,90 and 120 minutes after the insulin injection. Typical values (mg per 100 ml) are:

Normal control	100 (fasting):	45,	52,	60,	82,	95,	105
Simmonds' disease	70 (fasting):	36,	38,	42,	52,	54,	58

The initial fall of blood sugar is normal, allowing for the lower fasting basis by calculating as a percentage of the initial blood sugar value; but there is a marked delay in returning to the initial blood sugar values, and in severe cases, especially if the standard test dose of insulin is not drastically reduced, the patient's own resources may be quite unable to maintain, let along increase, the low blood sugar glycaemic crisis may develop. Since, however, these patients usually respond to adrenaline, hepatic glycogen stores are apparently present and mobilizable by an exogenous stimulus.

With anorexia nervosa the insulin sensitivity test is usually normal, but in severe cases it may closely resemble that of Simmonds' disease.

In myxoedema there may be some retardation of the subsequent rise of blood sugar (relative hypoglycaemic responsiveness), but there is also a delayed initial fall of blood sugar concentration.

The above observations on the disturbance of carbohydrate metabolism in Simmonds' disease are of great clinical importance, since spontaneous attacks of hypoglycaemic crisis may occur at any time and may prove fatal. Some peculiarities of behaviour and attacks of somnolence are also ascribable to hypoglycaemia, and in the course of conducting insulin sensitivity tests we have observed their production. In one case the patient had amnesia for the whole period of the test although she had answered questions during it and had not lost consciousness.

Adrenal cortex insufficiency

The assay of urinary 17-ketosteroids measures in blunderbuss fashion a variety of androgens derived from gonads and adrenal cortex. Normal values vary to some extent with the technique employed, but may be taken as being 5 to 18 mg per day.

In Simmonds' disease Fraser and Smith (1942) found extremely low values, under 0.5 mg per day; but in mild cases values are below 2 mg. In their four cases of anorexia nervosa assays ranged between 2.7 and 14.7 mg per twenty-four hours.

Very low levels of 17-ketosteroid excretion have been observed by us in patients with anorexia nervosa who are grossly malnourished. However, values of 2 mg or less per twenty-four hours in a case of suspected Simmonds disease are a strong supporting factor in the diagnosis. Low values are also recorded for urinary reducing steroids and for blood corticoids. Low values for all these steroid estimations are found in Addison's disease. In primary myxoedema, Fraser and Smith (1942) found values varying from zero to 1.7 mg per twenty-four hours, an observation which we would confirm in some but by no means all cases. Administration of thyroid to these cases is followed by a return to a normal excretion of 17-ketosteroids.

The serum values for sodium, potassium, and chloride rarely approximate to those found in Addison's disease, but the Kepler test of adrenal function not infrequently gives values as low as those found in Addison's disease. The reason for this is the marked inability of the patient with Simmonds' disease to excrete a water-load, a function dependent on the glucocorticoid and not on the mineralocorticoid activity. The serum values for sodium and chloride may be normal, but may be low, sometimes even as low as, or even lower than, those found in Addison's disease. This is not due, however, to an abnormal loss of sodium but to hydration and haemodilution following cortisone deficiency. In contrast with the severe phases of Addison's disease, the serum potassium values are not raised. The Kepler test is not without danger in these circumstances as drowsiness, headache and even coma from water poisoning may follow ingestion of large volume of water (Wynn and Garrod, 1955). The overnight fast in preparation for the test may induce hypoglycaemic symptoms as well.

Renal salt loss and potassium retention, characteristic of Addison's disease, is extremely rare in. Simmonds' disease, despite the secondary adrenocortical failure. This is now explicable by the discovery that aldosterone, the most potent mineralocorticoid, differs from the other adrenal steroids in that its rate of secretion is not directly related to the degree of ACTH stimulation. Therefore failure of anterior pituitary function is associated for many months with a normal output of aldosterone (Luetcher and Axelrad, 1954) but later, as actual atrophy of the adrenal cortex develops, the rate of secretion falls to subnormal levels.

Natural history

Severe cases may live a twilight existence for many years before a fatal crisis supervenes. The length of history, despite the degree of disability, is in sharp contrast to the often catastrophically short history of untreated Addison's disease.

However, the patient with well-developed anterior pituitary failure is always likely to succumb to intercurrent infection or to become grossly psychotic. Chronic dementia may mask the endocrine nature of the disease and lead to certification.

Variations in natural history depend on the degree of pituitary damage, a point well illustrated by ischaemic necrosis of the gland. Following such a severe post-partum haemorrhage that complete anuria ensued, one patient failed to lactate and her pubic hair failed to regrow for two months. Tests of pituitary function indicated both thyroid and adrenal insufficiency, yet convalescence was uneventful and she made a complete recovery with resumption of normal menstruation within four months of labour. She represents the mildest type of temporary pituitary insufficiency, which would not be recognized clinically without an intensive follow-up. In two other patients the only permanent endocrine dysfunction was amenorrhoea with extreme atrophy of the uterus and lower genital tract. However, the pituitary is not always involved in terms of gonadotrophin secretion. One of our patients resumed menstruation when her hypothyroidism and adrenocortical insufficiency were treated. She conceived successfully and, during pregnancy, remained well in the absence of hormonal therapy. Within three weeks of a normal delivery, clinical and biochemical evidence of pituitary insufficiency was apparent once more.

The relapse is noteworthy because Sheehan and Murdoch (1938) have suggested that pregnancy in the course of the disease is followed by spontaneous cure. Obviously this is not an invariable rule, but spontaneous recovery can occur in the absence of pregnancy as illustrated by a patient who made a complete recovery after treatment for a period of five years. However, such an event is very unusual.

Partial but permanent lesions do not involve all pituitary hormones to the same extent. Occasionally, failure of ACTH or thyrotrophin secretion is the only severe deficiency, but, as a general rule, gonadotrophin and growth-hormone secretion are affected with preservation of the more vital control of adrenal and thyroid. Thus young patients display dwarfism and sexual infantilism in the absence of thyroid and adrenal failure.

Recently more attention has been focused on the symptomatology and physiopathology of hypopituitary crises (Caughey and Garrod, 1954). Not only are these episodes likely to end in death unless immediately recognized and energetically treated, but the underlying disturbances of function are an acute exaggeration of those responsible for the chronic symptomatology of the condition and therefore illustrate the basic disorder of hypopituitarism. In a protected stable environment the body's primitive metabolic arrangements continue to function slowly in the absence of the pituitary but they are deprived of the necessary range of activity and adaptability which is normally provided by the stimulus of the adenohypophysis as an essential to good health. Consequently a breakdown in this limited homeostasis will follow even minor changes in environment and lead to the severe symptoms of a crisis. Trauma, infection, starvation, and excessive cold can all be the precipitating factors. Operations on the pituitary or adjacent areas are now not uncommon causes of crisis in patients already suffering from panhypopituitarism due to neoplasm in some cases the preoperative evidence of pituitary insufficiency is slight and the operation, by acutely inhibiting the remainder of pituitary function, will be followed by the most severe disturbance we have also seen operations on other sites such as a colporrhaphy, cause crisis by virtue of the stress imposed on an inefficient endocrine system. All types of infection, varying, in our experience from an infected tooth socket to gastroenteritis, induce a crisis in the same manner. It must also be remembered that panhypopituitarism renders a patient extremely sensitive to morphine and barbiturates; even routine dosage of these drugs can be followed by prolonged stupor. Lastly a crisis may signify the terminal breakdown of metabolism after a slowly progressive downhill course.

The symptoms of crisis are less variable than the causative disturbances of physiology. Refusal of food, resentment at any interference, querulous complaints, commonly of arthralgia, give way to increasing lethargy and prolonged sleep. Stupor follows, sometimes interrupted by short maniaceal outbursts often associated with hallucinations. The last stage is one of deep progressive coma. The patient may lie flaccid with absent reflexes or curled up, responding violently to examination. Bradycardia is usual and may be extreme. The blood pressure is often unrecordable and the radial pulse absent. However, hypotension is not necessarily an early sign. An important and often neglected sign is the extremely low body temperature—often unrecognized because the thermometer has not been shaken down to its lowest level. Crisis following cerebral surgery can be differentiated from loss of consciousness due to cerebral injury by the absence of localizing signs in the central nervous system. But differentiation is not always easy. It is not enough to confine one's thoughts to traumatic causes of postoperative stupor and the possibility of hypopituitarism must always be borne in mind.

The disorders of physiology underlying the coma of Simmonds' disease can be identified as:

1. Hypothermia.
2. Hypothermia.
3. Water intoxication
4. Hypoglycaemia.
5. Electrolyte disturbance.

However it is seldom that one noxious factor can be isolated as the cause of a particular coma. In some, hypothermia is dominant and consciousness returns on gently heating the patient in others hypoglycaemia, apart from its obvious role in producing, acute states of unconsciousness, is associated with slowly progressive stupor. Yet administration of glucose is not always followed by a return to consciousness when the blood sugar is again normal. Water poisoning with coma may follow directly on a water-load, but it is seldom realized that patients with hypopituitarism are very easily overloaded with fluids and this is a common feature of many crises. The lowered serum sodium due to haemodilution must be distinguished from that of true sodium depiction. However, such Addisonian electrolyte changes are rare unless severe vomiting has preceded the coma. Cerebral anoxia may be induced by hypothermia or hypoglycaemia and hypotension will exaggerate it. Postural hypotension is a common cause of more transient attacks of faintness rather than stupor.

All these changes are due to a combination of thyroid and adrenal failure. Hypothyroidism, apart from its role in hypothermia, is seldom vital; indeed it may protect the body from the effects of adrenal failure because the administration of thyroid or even, as in our experience, the injection of thyrotrophin may raise the metabolic rate and precipitate a crisis. This does not occur if cortisone has been given concurrently. Glucocorticoid deficiency is undoubtedly the most important factor of all; the administration of cortisone or hydrocortisone in hypopituitary crisis produces results which are far superior to any other form of treatment. It is interesting that the electroencephalographic abnormalities, which are characteristic but not specific of a crisis, are corrected by cortisone but disappear some time after return of the clinical state to normal (Salmon, 1956). We have also observed this sequence of events in Addison's disease treated with cortisone but there is no correction following dexoycortone.

Diagnosis

Amenorrhoea, weakness, apathy, poor appetite with some loss of weight, bradycardia, hypotension, low basal metabolism and temperature, and hypersensitivity to cold, are present in a characteristic case; and frequently these symptoms follow a parturition associated with severe haemorrhage. It is now accepted that gross loss of weight is not a cardinal feature of the disease and the maintenance of normal nutrition is the rule rather than the exception. However, a loss of 4 to 8 lb. in weight is common and we have observed a gain in weight only in exceptional cases when hypothyroidism has dominated the picture.

It is important to remember that many cases are examples of an incomplete syndrome, and may well be missed unless the condition is kept in mind as a sequel of complicated parturition. Amenorrhoea, for example, was formerly considered to be an invariable feature, but this is not so, and undoubtedly cases have even become pregnant. However, the presence of normal menstruation makes the diagnosis much less certain. We attach importance to bradycardia, but it is claimed by other observers that it is not invariable. The same applies to low blood pressure, and the basal metabolism is not necessarily markedly lowered.

Simmonds' disease may be due to neoplasms or granuloma of the pituitary gland, and it will be remembered, therefore, that parturition is not the only cause, and that the disorder may also occur in men. If the condition starts in childhood, e.g. a slow-growing destructive craniopharyngioma, infantilism will be the background on which the adult syndrome will be superimposed.

The disease is often insidious in onset, even when due to a post-parturition thrombosis, but it may be acute and fatal, e.g. haemorrhage into a chromophobe adenoma. On the other hand, it may be present and pass unrecognized for thirty or more years.

A number of conditions will be separately considered in the following differential diagnosis, and in some it will be indicated that there we are dealing with a question of nomenclature rather than differential diagnosis:

Differentials diagnosis

This includes:

1. *Addison's disease*. From the fact that both in Addison's and Simmonds' disease the adrenal cortex is functioning poorly, much of the symptomatology is common to both diseases'. Pigmentation is usual in Addisons' disease and is widespread, usually including the mucous membranes; but in acute cases, or in fair people, pigmentation may be minimal. In Simmonds disease pigmentation is absent and pallor of the skin is often a striking feature. Bradycardia is rare in Addison's disease and the basal metabolism is not appreciably lowered except in crisis. Pubic and axillary hair is seldom absent in Addison's disease, and menstruation is not infrequently normal.
2. *Myxoedema*. This diagnosis is often erroneously made. Thyroid insufficiency is one feature of the disease but not the whole clinical picture (see above). The 17-ketosteroids are below 2.0 mg per day in both conditions. The blood sugar is usually not low in myxoedema, there is a slower fall in blood sugar after intravenous insulin, but a normal or slightly delayed subsequent rate of recovery of blood sugar values. Pubic and axillary hair is not lost. Examples of Simmonds' disease have been described under the title of 'pituitary myxoedema' —a confusing term which should not be used.

Myxoedema will respond dramatically to thyroid, whereas in Simmonds' disease the response is only partial and sometimes the patient is made worse. A more experimental approach will show that the iodine uptake of the thyroid will respond to thyrotrophic hormone in Simmonds' disease, whereas in myxoedema it is completely uninfluenced, the thyroid being unable to respond.

It is important to realize that severe untreated myxoedema of many years' duration will result in a depression of pituitary function. The rise in thyrotrophin secretion which follows destruction of the thyroid is succeeded by a gross fall in the output of this hormone, together with a diminution of gonadotrophin secretion. The subsequent administration of thyroid extract is followed by a return to normal secretion of both these hormones.

3. *Anaemia.* In Simmonds' disease the blood picture may be normal, but anaemia of moderate degree is the rule. Some 50 per cent of cases have achlorhydria. These findings, together with the pollor of the skin, which is disproportionate to the pinkness of the mucosae may lead to a diagnosis of anaemia, without recognition of the endocrine condition.

4. *Anorexia nervosa.* This condition is a neurosis and not an endocrine disorder. Autospy findings are quite different from Simmonds' disease in that the endocrine glands are normal apart from involution of the ovaries and uterus (Brosin and Apfelbach, 1941). The importance of anorexia nervosa as a differential diagnosis to Simmonds' disease has receded, now that it is clear that cachexia is not a cardinal part of the clinical picture of pituitary destruction.

Amenorrhoea occurs early, due to the mental disturbances mediated by the hypothalamus. At the stage gonadotrophin secretion is still present. In the later stages of inanition gonadotrophin secretion diminishes or ceases. The fall in metabolic rate is related to starvation and not initially to thyroid hypofunction, as radioactive iodine uptake by the thyroid is normal; later a raised blood cholesterol and a low radioactive iodine uptake indicate secondary hypothyroidism. In marked contrast to Simmonds' disease the patient remains intensely active, often with obsessional traits, despite the extreme cachexia. Anorexia is of course an early and predominant feature. Body hair tends to increase over a rough dry skin, although in the later stages pubic hair may become scanty. It is apparent from Ley ton's (1946) studies in starvation in prisoners of war that amenorrhoea, impotence, bradycardia, hypotension, low basal metabolic rate, and some lowering of the blood sugar result from inanition. Certainly a depression of pituitary function accompanies these changes, but it is best to avoid the term "functional Simmonds' disease" (Wilson 1954).

Treatment

Surgical conditions, such as intracranial neoplasms, require surgery, but the majority of cases need hormone substitution therapy, and the latter is also called for after surgical hypophysectomy.

Since nearly all the endocrine glands are affected, it is difficult to decide in anyone case which form of therapy use, and it is technically and practically impossible to employ complete replacement therapy. Our own experience with several cases is as follows:

Cortisone

The availability of cortisone acetate, for oral use, or intramuscular injection, has marked a signal advance in the treatment of Simmonds' disease. Previous experience had shown that testosterone, for either sex, was of great efficacy in the restoration of strength and well-being. But those patients who have changed from testosterone to cortisone have been amazed at their new state of well-being. The oral administration of 12.5-37.5 mg per day ($^1/_2$ tablet once to three times daily) is dramatic in its effect and renders the patient less prone to crisis after infection or hypoglycaemia on fasting. The use of adrenocorticotrophin is more logical for hypopituitarism, but the disadvantage of repeated injections (even if the long-acting ACTH gel is injected once or twice weekly) makes it less suitable for maintenance therapy. However, adrenocorticotrophin is of use in initial therapy, and we are impressed by the responsiveness of the adrenals even if they have been inactive for years. However, we administer both adrenocorticotrophin and cortisone with considerable caution to the long-standing case of Simmonds' disease as we have observed that these hormones may induce a temporary but severe psychosis, usually of the manic type. Careful adjustment of the dose allows long-term therapy to continue. In states of crisis, intravenous injection of 50 mg of hydrocortisone, prepared specially for this route of administration, is preferable to other forms of therapy. Cortisone intramuscularly provides the necessary supportive therapy.

Testosterone

We have found testosterone propionate by injection (e.g. 25 mg on alternate clays) or implantation (0.4 G. of testosterone) a very useful therapeutic measure. Oral administration of methyltestosterone, 5-10 mg three times a day, is also effective, not only for men but for women, in whom the secretion of adrenal androgens, which is common to both sexes, is very low, as judged by urinary excretion. This makes testosterone a logical substitution therapy. Further, androgens are protein anabolic hormones, producing nitrogen retention, increasing strength and weight, and preventing osteoporosis.

In the male, already on cortisone therapy, testosterone is necessary to restore potency, penile erections, and a normal volume of ejaculate. In the female, large doses have the disadvantage of producing facial hirsutes, acne, and a deepening voice, but small doses in conjunction with cortisone stimulate protein anabolism and offset the catabolic action of cortisone.

Thyroid

Thyroid extract or L-thyroxine by mouth helps partially in some patients, but in others proves harmful and may even precipitate a crisis if given in large doses. It should therefore be used with caution, commencing with small doses, and it is best deferred in severe cases until general improvement is made by other means. An initial dose of thyroid extract, $^1/_2$ gr. daily, or L-thyroxine, 0.05 mg daily, should be doubled after 2 weeks and, after a similar interval, the maintenance dose of 2-3 gr. thyroid extract can be reached. We have noted that patients already taking thyroid require some increase in dosage if they are treated with cortisone.

Oestrogens

Oestrogens by mouth may not be well tolerated, but by injection or subcutaneous implantation they proved helpful generally, apart from the possibility of restoring the uterus to normal size and producing a bleeding. They should be given in large doses, e.g. 5 mg oestradiol benzoate injected daily. Restoration of fluid to the subcutaneous tissues appears to be a good effect.

The problem of dyspareunia from an atrophic vagina is solved by oestrogens. This important point is seldom considered before treatment, because of the patient's complete loss of libido. When normal sex desire returns with treatment some patients are too reticent to complain of pain or intercourse.

Deoxycortone acetate (DOCA) is never required. Salt loss is seldom a feature of Simmonds' disease, and can usually be met by a high salt diet. Cortisone itself causes little salt retention but continued aldosterone secretion makes mineralocorticoid therapy unnecessary. Indeed prednisone and prednisolone can be used for their potent glucocorticoid properties despite the absence of salt-retaining power. Thyrotrophin is of diagnostic use but it is not recommended in therapy. Gonadotrophins are likewise of no practical importance.

Treatment of crisis

Early diagnosis and the increase of glucocorticoid dosage in anticipation of factors such as operation or infection, will do much to diminish the incidence of crisis. When such as event has occurred, treatment must be guided by the dominant physiological abnormality. Hypothermia is always important, and should be countered by gentle heating of the patient; drastic methods with heat cradles are dangerous. A warm bath is advocated by Sheehan and Summers (1952) despite its obvious practical difficulties. Hypoglycaemia can be corrected rapidly with intravenous dextrose (50 per cent) and frequent glucose drinks should be given later to prevent a subsequent fall in the blood sugar. In rare instances sodium depletion will require the intravenous administration of normal saline, but at all times the fear of water intoxication should limit the volume of intravenous fluid.

Undoubtedly glucocorticoid therapy is of paramount importance in all cases. Intravenous administration of hydrocortisone (as the 'free alcohol' or as the hemisuccinate) should aim at an initial dose of 50 mg to be followed by a continuous intravenous infusion delivery of 10 mg hourly. Intramuscular or oral cortisone can be given from the outset in the milder case or started when the condition is improving: about 100 mg daily are required. The slow response of the adrenals to ACTH makes this hormone of little use in crisis. The immediate administration of thyroid extract or its equivalents is dangerous. Glucocorticoid therapy should always be started first. Hours later, triiodothyronine, 20 μg, three times a day, can be given if hypothyroidism appears to be severe.

Diabetes Insipidus

Diabetes insipidus is a disturbance of water balance, characterized by primary polyuria and secondary polydipsia, due to a destructive lesion of the posterior lobe of the pituitary gland or the adjacent part of the hypothalamus. An experimental puncture lesion of the tuber cinereum, or nucleus supraopticus, is followed by atrophy of the posterior pituitary lobe.

Cushing (1933) stated that: 'Its secretory product (pars nervosa) is unmistakably derived from the investing pars intermedia whose cells become basophilic when ripened and under nervous impulses from hypothalamic nuclei, the basophil cells invade the pars nervosa and become transformed into "hyaline bodies", which pass into the infundibular cavity'. However, experimentally it is clear that diabetes insipidus can occur in the presence of a normal pars intermedia, and that the antidiuretic hormone can be extracted from pars nervosa (posterior pituitary) but not from pars intermedia (Pickford, 1945). Fisher, Ingram, and Ranson (1938) showed that the essential lesion was anywhere in the supraoptic tract, and Richer (1935) reported that the diuresis preceded excessive intake of fluid. The diuresis will continue for a while even if fluid is completely withheld. The recent application of histochemical techniques has shown that the hormones of the supraoptico-hypophyseal tract are actually produced in the upper part of the tract, passing downwards to be released into the circulation from the posterior lobe of the pituitary; the function of this lobe is one of storage and secretion but not synthesis of hormones. Excessive quantities of antidiuretic hormone are found in the urine in normal animals subjected to water deprivation or to emotional stimuli. The amount of antidiuretic hormone is greatly reduced by a water-load so that 'water diuresis may be fitly and accurately described as a condition of physiological diabetes insipidus; and there can be little doubt that the antidiuretic secretion of the neurohypophysis is a hormone in the physiological sense, its liberation being continuously governed by the contemporary concentration of chloride, and possibly of other osmotically active substances, in the arterial plasma'.

Verney's conception of the supraoptico-hypophyseal tract as a regulator of blood osmolarity, responding to an increasing osmotic pressure by the release of antidiuretic hormone, must be expanded to include the overall control of antidiuretic hormone released by nervous impulses from the hypothalamus. Harris has shown that electrical stimulation of the hypothalamus results in secretion of antidiuretic hormone. Furthermore nicotine applied directly to the supraoptic nuclei produces the same effect. Even injection or inhalation of nicotine (from a cigarette) will cause hormone secretion; a physiological observation which has led to the diagnostic use of nicotine when a water diuresis has been induced in a patient with possible diabetes insipidus.

The chemistry of posterior pituitary hormones has now been revealed by the brilliance of du Vigneau *et. al.* (1953), who not only identified vasopressin and oxytocin as two distinct but closely related cyclic polypeptides, but actually synthesized the two hormones. Vasopressin is the antidiuretic hormone. Its site of action is the renal tubule, thereby regulating the renal excretion of water. Despite rather conflicting experimental work there is no evidence that vasopressin exerts any direct action on electrolyte excretion in man.

The renal effect of vasopressin is modified by the action of the adrenal cortex and thyroid; the hormones of these glands are necessary for the full development of polyuria when the secretion of antidiuretic hormone is stopped (i.e. by section of the supraopticohypophyseal tract). Coincident damage of the anterior and posterior lobes of the pituitary does not cause full diabetes insipidus because the secretion of adrenocorticotrophin and thyrotrophin is diminished as well as the secretion of vasopressin. Under these circumstances the administration of hydrocortisone and thyroxine is followed by the appearance of gross diabetes insipidus. Similarly thyroidectomy or removal of the anterior pituitary lobe ameliorates severe diabetes insipidus. If the supraoptico-hypophyseal tract is left intact and the anterior lobe of the pituitary destroyed, renal elimination of a water-load is impaired and water poisoning may occur. The same situation arises in Addison's disease. It is of interest that the maintenance of body water balance is related to glucocorticoid and not mineralocorticoid hormones.

Incidence

Diabetes insipidus may be familial, and then occurs in young people, males more commonly than females. It may, however occur at any age.

Aetiology

The causes of diabetes insipidus are as follows:

1. Familial.
2. Idiopathic.
3. Tumours adjacent to hypothalamus or pituitary.
4. Inflammatory lesions in the parapituitary area, e.g. encephalitis lethargica, basal meningitis, syphilis.
5. Fractured skull or surgical trauma to supraoptico-hypophyseal tract.
6. Xanthomatosis (Schuller-Christian disease).

Clinical picture

The essential feature is the passage of large quantities (10 litres a day) of pale urine of low specific gravity (usually less than 1.004); and good experimental evidence indicates that this is the primary disorder, the resulting unquenchable thirst being a secondary phenomenon. The tissues become dehydrated; there is no obvious perspiration; and the skin and mucous membranes are very dry. The faeces are hard and desiccated, and the patient constipated. Nocturia prevents sleep.

Course and progress

The natural history is dependent on the causative lesion. The physiological disturbance itself is benign and, in the absence of a lethal cause, the prognosis is very good. However, it is important to realize that not a few cases of so-called idiopathic diabetes insipidus later show evidence of an intracranial tumour as the origin of the disease. Hence the initial prognosis should always be guarded.

Some psychological disturbances may derange the hypothalamic control of body water; episodes of extreme polyuria and thirst alternating with a suppression of urinary output and water retention. It may be inferred that temporary inhibition of secretion gives way to an excessive secretion of antidiuretic hormone, an event which is in accord with the experimental evidence of increased antidiuresis in dogs who are frightened.

Diagnosis

The differential diagnosis is as follows:

1. Diabetes mellitus—glycosuria.
2. Chronic nephritis—albuminuria, renal casts, and hypertension.
3. Neurosis or hysteria.
4. Caffeine idiosyncrasy—this runs in families and polyuria ceases if water only is drunk.
5. Nephrogenic diabetes insipidus—renal tubular failure to resorb water. Seen in babies. Uninfluenced by vasopressin.
6. Hypercalcuria—gross polyuria associated with high urinary calcium excretion. Seen in rare cases of hyperparathyroidism.

In practice the only difficulty is diagnosis lies in the distinction of diabetes insipidus from hysterical polydipsia. The most reliable diagnostic procedure at the moment is the administration of nicotine when a water diuresis has been induced. Prompt cessation of the diuresis indicates an intact supraoptico-hypophyseal tract while a continued diuresis indicates diabetes insipidus.

Treatment

Vasopressin (*Pitressin*), 10-20 Units, in 1 ml aqueous solution, injected subcutaneously, is effective. Its duration of action is only a few hours and an injection of 1 ml (5 Units) vasopressin tannate in oily solution has proved of greater convenience, the effect of a single injection lasting for twenty-four hours owing to slower absorption at the site of injection. A dried powder of posterior pituitary, e.g. *Piton*, or *Di-Sipidin*, is also effective when used as a nasal snuff. The dose is 25-50 mg. three times a day. The patient learns well enough to estimate the required dose without weighing. Excessive dosage may cause a feeling of faintness, or intestinal spasm, and in one patient it caused ureteric spasm and dysuria. Vasopressin sclution by the nasal route, as a spray, is also used. Voluntary or compulsory restriction of fluid is illogical, since the primary trouble is excessive excretion, not intake. In severe cases the patient will drink any fluid obtainable, even his own urine, and will stop at nothing to get fluid.

4

THYROID GLAND

The thyroid gland is a bilobed structure located in the neck. A narrow central portion, the *isthmus*, crosses anterior to the trachea and connects two lateral lobes. An additional *pyramidal lobe* commonly extends superiorly from the isthmus. The thyroid weighs about 40 g.

Embryologically, the thyroid is of endodermal origin, arising from a median downward growth in the floor of the pharyngeal gut near the base of the tongue. The parafollicular cells are derived from neural crest cells that migrate into the ultimobranchial bodies of the fourth pharyngeal pouches. The thyroid gland, which begins functioning at about the 10th week of fetal life, plays an essential role in regulating tissue metabolism and general growth and development. Deficiencies in the thyroid hormones during fetal development result in irreversible damage including the development of fewer and smaller neurons, defective myelination, stunted body growth, and mental retardation.

Body Supply and Nerves

The thyroid gland receives its blood supply primarily through the superior and inferior thyroid arteries. These unusually large vessels, which have numerous anastomoses, provide the gland with a rich blood supply. The veins of the gland form a venous plexus, associated with the capsule of the gland. Unusually extensive blood and lymphatic capillary plexuses surround the parenchymal tissue within the gland. As is typical of most endocrine tissue, the endothelium of the blood capilalries is fenestrated. In the thyroid, the lymphatic capillaries provide an important route for conveying the hormones from the gland.

Most of the nerve fibers entering the thyroid with the larger branches of the thyroid arteries are postganglionic sympathetic fibers from the superior, middle, and inferior cervical ganglia. These nerves are believed to function primarily in regulating blood flow in the gland, but some evidence suggests that the sympathetic fibers may stimulate secretion of thyroid hormones.

STRUCTURE

The thyroid is surrounded by a thin, connective tissue capsule from which septa extend into the gland, dividing it into irregular lobes and lobules. The parenchymal cells are organized in spherical or short, blind ending cylindrical masses called *follicles*. The follicles are cyst-like structures consisting of a single layer of epithelial cells surrounding a central mass of gelatinous colloidal material. In humans, the follicles vary in size from 0.02 to 0.9 mm in diameter. Rich blood and lymphatic capillary plexuses are present in the connective tissue that separates the follicles.

The follicular epithelial cells, which rest on a very thin basal lamina, vary in height from cuboidal to squamous. The height of the epithelium is indicative of the functional activity; the squamous follicular cells are referred to as "resting." Two

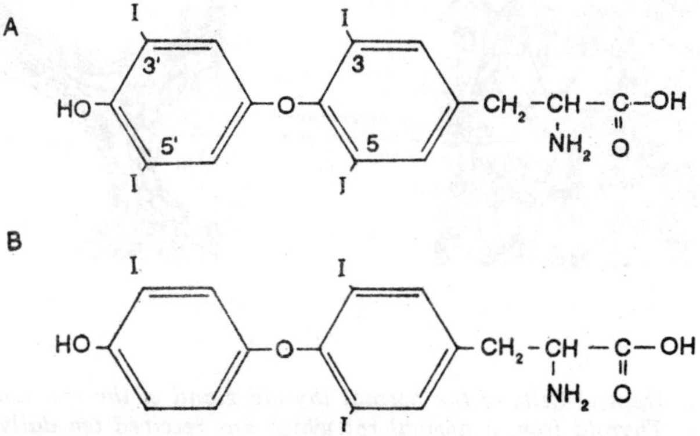

Fig. 4.1. Structural formulas of the thyroid hormones. A—Thyroxine (T_4) or 3, 5, 3', 5'-tetraiodothyronine; B—3, 5, 3'-triiodothyronine (T_3).

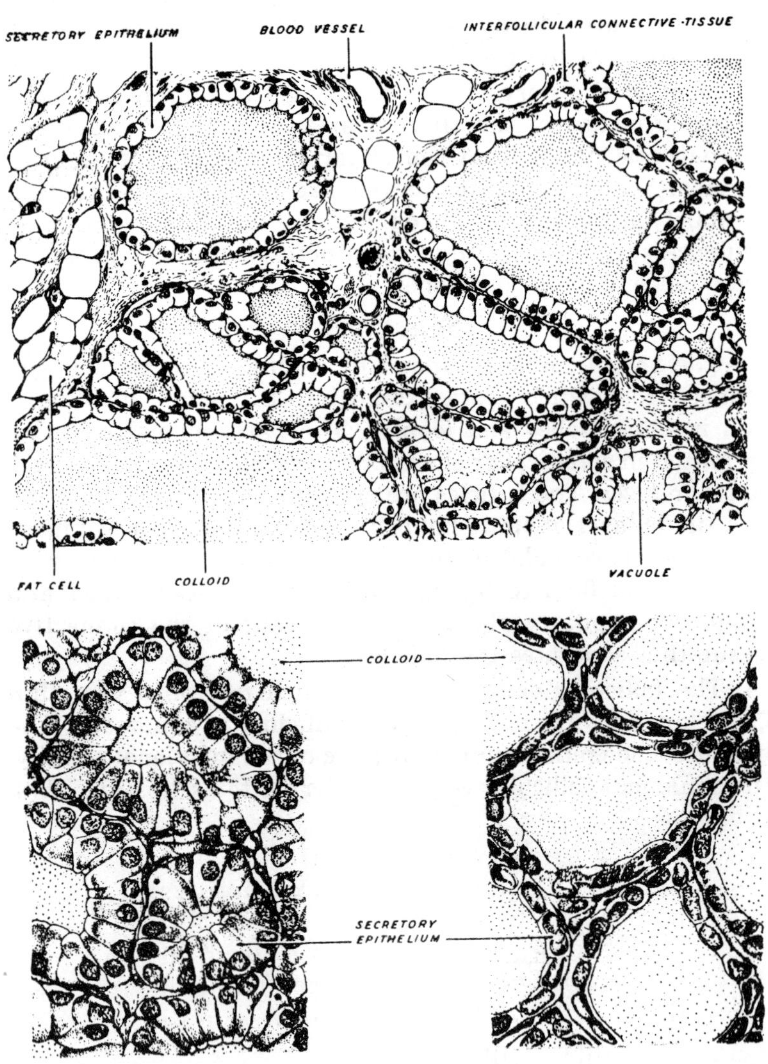

Fig. 4.1. Upper—Cells of the normal thyroid gland of the rat. Lower left—Thyroid from a normal rat which has received ten daily injection of thyrotropin. Lower right—Thyroid from a rat six months after complete removal of the pituitary gland.

cell types identified in the follicles are *principal* or *follicular cells* and *parafollicular* or *C cells*. The luminal colloid stains with both basic and acidic dyes and is PAS-positive. The colloid consists of proteolytic enzymes, mucoproteins, and a glycoprotein called *thyroglobulin*, which is the storage from the thyroid hormone. It should be noted that the thyroid is unique among the endocrine glands in its ability to store its secretory product.

Principal Cells

The activities of the *principal cells* include the synthesis of thyroglobulin, iodination, and storage of thyroglobulin within the follicular lumen, resorption and hydrolysis of thyroglobulin, and release of the thyroid hormones into the blood and lymphatic capillaries.

Most of the follicular epithelial cells are *principal cell* resting on the basal lamina with their apices directed inward toward the follicular cavity. The principal cells vary in height from columnar to squamous. Their nuclei are spherical and contain one or more prominent nucleoli. With the light microscope a supranuclear Golgi complex, lipid droplets, and PAS-positive droplets can be identified in the basophilic cytoplasm. Ultrastructural studies of the active cuboidal or columnar cells reveal the presence of organelles commonly associated with both secretory and absorptive cells schematically (1) typical junctional complexes at the apical end of the lateral plasma membrane; (2) short microvilli on the apical surface of the cells; (3) numerous profiles of rough endoplasmic reticulum in the basal region; (4) a well-developed, supranuclear Golgi complex; (5) numerous small vesicles in the apical cytoplasm, which are morphologically similar to vesicles associated with the Golgi; (6) abundant lysosomes and multivesicular bodies; and (7) membrane-limited vesicles, identified as *colloidal resorption droplets*, in the apical region.

Parafollicular Cells

The *parafollicular cells* may be present in the follicular epithelium or scattered in the connective tissue between the follicles. When present in the follicle, they are located near the basal lamina and do not extend to the follicular lumen. Ultrastructurally, the most characteristic feature of these cells is the presence of numerous, dense, membrane-bound secretory granules measuring 0.1 to 0.5 mm in diameter.

Formation and Secretion of the Thyroid Hormone

Physiologic Anatomy of the Thyroid Gland

The thyroid gland is composed of large numbers of encysted *follicles* filled with a substance called *colloid* and lined with *cuboidal epithelioid cells* that secrete into the interior of the follicles. Once the secretion has entered the follicles, it must be absorbed back through the follicular epithelium into the blood before it can function in the body.

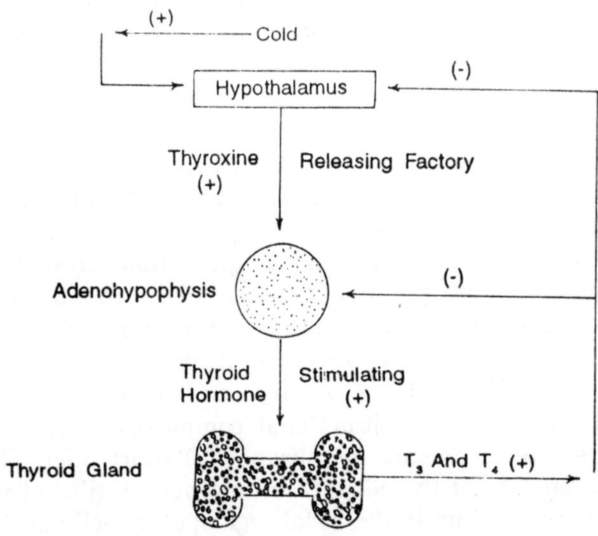

Fig. 4.3. Feedback mechanism whereby physiologic alterations in the environmental temperature alter thyroid secretion.

Requirement of Iodine for Formation of Thyroxine

To form normal quantities of thyroxine, approximately 35 to 50 mg. of ingested iodine are required each year, or approximately 1 mg. per week. To prevent iodine deficiency, common table salt is iodized with one part sodium iodide to every 100,000 parts sodium chloride.

The Iodide Pump

The thyroid cell membranes have a specific ability to transport iodides actively to the interior of the cell and then into the follicle; this is called the *iodide pump*, or *iodide trapping*. In a normal gland, the iodide pump can concentrate iodide to 25 to 50 times its concentration in the blood. However, when the thyroid gland becomes maximally active, the concentration can rise to as high as 350 times that in the blood.

Chemistry of Thyroxine and Triiodothyronine Formation

Oxidation of the Iodides

The first stage in the formation of the thyroid hormones is believed to be oxidative conversion of iodides either to *elemental iodine* or to some other oxidized form of iodine that is then capable fo combining with the amino acid *tyrosine* to form the thyroid hormone. The reason for believing this is that iodination of tyrosine can be effected readily even in vitro by elemental iodine but not by iodides. Furthermore, large amounts of the enzyme *peroxidase* are present in the thyroid glandular cells. This enzyme is capable of oxidizing iodides, and, when it is absent from the cells, the rate of formation of thyroid hormones is greatly decreased.

Iodination of Tyroxine and Formation of the Thyroid Hormone

The successive stages of iodination of tyrosine and final formation of the two most important thyroid hormones, *thyroxine* and *triiodothyronine*. Tyrosine is first iodized to *monoiodotyrosine* and then to *diiodotyrosine*. Two molecules of diiodotyrosine then conjugate, with the loss of the amino acid alanine, to form one molecule of *thyroxine*. Or one molecule of monoiodotyrosine combines with one molecule of diiodotyrosine to form *triiodothyronine*.

Storage of Thyroid Hormones in Thyroglobulin

In addition to synthesizing the thyroid hormones, the thyroid cells also secrete a large globulin called *thyroglobulin* directly into the thyroid follicles and not into the circulating blood. Furthermore, it is the amino acid tyrosine in this molecule that is converted into monoiodotyrosine, diiodotyrosine, triiodothyronine, and thyroxine. This conversion of tyrosine probably can occur both before and after the thyroglobulin has been secreted into the follicles. It is in the form of this complex molecule thyroglobulin that thyroid hormones are stored in the thyroid gland for a period of several weeks before being released into circulating blood.

Release of Thyroxine from the Follicles

Thyroglobulin itself is not released into the circulating blood but instead is digested by *proteinases* that are also secreted into the follicles by the thyroid cells. The proteinases split thyroxine and minute quantities of triiodothyronine from the thyroglobulin complex and allow these to diffuse through the glandular cells into the blood outside the follicles. The rate of activity of the proteinases is controlled by thyrotropin from the adenohypophysis.

Transport of Thyroxine and Triiodothyronine to the Tissues

Quantitatively, at least 95 per cent, and probably closer to 100 per cent, of the active thyroid hormones entering the blood are thyroxine, and the remainder is almost entirely triiodothyronine.

Binding of Thyroxine and triiodothyronine with the Plasma Proteins. On entering the blood, both thyroxine and triiodothyronine immediately combine with several of the plasma proteins, mainly with a globulin called *thyroxine-binding globulin.*

The affinity of thyroxine-binding globulin for thyroxine is approximately three times as great as its affinity for triiodothyronine. For this reason, whenever these two hormones are injected intravenously or secreted by the thyroid gland in equal quantities, the amount of *free* thyroxine that remains in the plasma is only one-third the amount of free

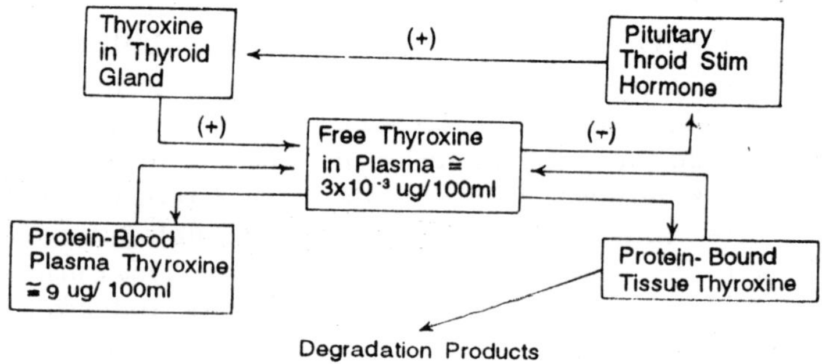

Fig. 4.4. Thyroxine fractions in the human body.

triiodothyronine. This allows far more rapid entry of triiodothyronine into the cells than of thyroxine, which causes triiodothyronine to have a potency about three to five times that of thyroxine.

About half the thyroxine is the blood is released to the tissue cells every eight days. Since thyroxine constitutes the major bulk of the thyroid hormones, this slow release of thyroxine causes the very slow action of thyroxine.

Functions of Thyroxine in the Tissues

General Increase in Metabolic Rate

The principal effect of thyroxine is to increase the metabolic activities of essentially all tissues of the body. The basal metabolic rate increases to as much as 60 to 100 per cent above normal when large quantities of thyroxine are secreted. The rate of utilization of foods for energy is greatly

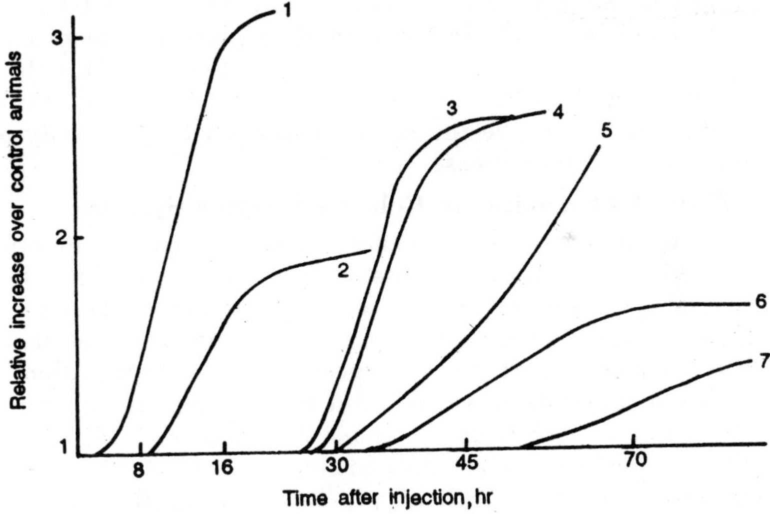

Fig. 4.5. Proposed temporal relationship among the responses of some hepatic cellular activities, the basal metabolic rate, and liver growth in thyroidectomized rats following a single injection of 15-25 μg of triiodothyronine: (1) synthesis of rapidly labelled nuclear RNA; (2) synthesis of DNA-dependent RNA polymerase of the nucleus; (3) mitochondrial and microsomal incorporation of amino acid into protein; (4) synthesis of mitochondrial cytochrome oxidase; (5) synthesis of microsomal NADPH-cytochrome c reductase; (6) basal metabolic rate; (7) liver weight.

accelerated. The rate of protein synthesis is at times increased, while at the same time the rate of protein catabolism is also increased. The growth rate of young persons is greatly accelerated. The mental processes are excited, and activity of many other endocrine glands is often increased. Yet despite the fact that these many changes in metabolism are under the influence of thyroxine, the basic mechanism by which thyroxine acts is almost completely unknown. Some of the theories are the following :

Effect of Thyroxine on Mitochondira

When thyroxine is given to an animal, the mitochondria in most cells of the body increase in size and also in number. Furthermore, the total membrane surface of all the mitochondria increases almost directly in proportion to the increased metabolic rate of the whole animal. Therefore, it is an obvious deduction that the principal function of thyroxine might be simply to increase the number and activity of mitochondria, and that these in turn increase the rate of formation of ATP to energize cellular function. Unfortunately, though, the increase in number and activity of mitochondria could just as well be the *result* of increased cell activity as the cause of the increase.

Effect of Thyroxine on Cellular Enzyme Systems

Vast numbers of intracellular enzymes increase in quantity a week or so following administration of thyroxine. For instance, one enzyme, α-glycero-phosphate dehydrogenase, is often increased to an activity six times its normal level. Since this enzyme is particularly important in the degradation of carbohydrates, its increase could help to explain the rapid utilization of carbohydrates under the influence of thyroxine. Also, the oxidative enzymes and the elements of the electron transport system, both of which are normally found in mitochondria, are greatly increased.

Since the enzymes that are increased are mainly those found also in mitochondria, it is possible that this effect is indirectly caused by the increase in mitochondria.

Latency and Duration of Thyroid Hormone Activity

After injection of a relatively large quantity of thyroxine into a human being, essentially no effect on the metabolic rate can be discerned during the first 24 hours, which illustrates that there is a long *latent period* before thyroxine activity begins. Once activity does begin, it increases progressively and reaches a maximum in about 12 days in the human being. Thereafter, it decreases with a half-time of about 15 days. Some of the activity still persists as long as six weeks to two months later.

The long time required for maximum effect of thyroxine to occur is believed to be caused by the time required for the cells to accumulate increased numbers of mitochondria and to synthesize the enzymes whose development is promoted by the thyroid hormones. Once these elements have been formed, they presumably continue to act for many weeks, which would explain the long duration of thyroid hormone effect even after the thyroid hormone itself is gone.

PHYSIOLOGIC EFFECTS OF THYROXINE ON DIFFERENT BODILY MECHANISMS

Effect on Basal Metabolic Rate

Excessive quantities of thyroxine can occasionally increase the basal metabolic rate to as much as 100 per cent above normal. However, in most patients with severe hyperthyroidism the basal metabolic rate ranges between 40 to 60 per cent above normal. On the other hand, when no thyroid hormone is produced, the basal metabolic rate falls to —40 to —50.

Effect on Body Weight

Greatly increased thyroxine production almost always decreases the body weight, and greatly decreased thyroxine production almost always increases the body weight; but these effects do not always occur, because thyroxine increases the appetite, which may overbalance the change in metabolic rate.

Effect on Growth

Because anabolism of proteins cannot occur normally in the absence of thyroxine, the growth effect of growth hormone form the pituitary gland is insignificant without concurrent presence of thyroxine in the body fluids. There are two conditions in which his effect on growth is especially evident: First, in growing children who are hypothyroid, the rate of growth is greatly retarded. Second, in growing children who are hyperthyroid, excessive skeletal growth often occurs, causing the child to become considerably taller than otherwise.

Effect on the Cardiovascular System

Because increased metabolism induced by thyroxine increases the demand of the tissues for nutrient substances, activity of the cardiovascular system is increased in all ways. Especially, there is increase in local blood flow in each tissue, and concomitant incrase in cardiac output sometimes to two times normal. Other effects are an increase in heart rate, systolic arterial pressure, pulse pressure, and blood volume.

Effect on Respiration

The increased rate of metabolism caused by thyroid hormone increases the utilization of oxygen and the formation of carbon dioxide; these effects activate all the mechanisms that increase the rate and depth of respiration.

Effect on the Gastrointestinal Tract

In addition to increased rate of absorption of foodstuffs, which has been discussed, thyroxine increases both the rate of secretion of the digestive juices and the motility of the gastrointestinal tract. Often, diarrhea results. Also, associated with this increased secretion and motility is increased appetite, so that the food intake usually increases. Lack of thyroxine causes constipation.

Effect on the Central Nervous System

In general, thyroxine increases the rapidity of cerebration, while, on the other hand, lack of thyroid hormone decreases this function. The hyperthyroid individual is likely to develop extreme nervousness and is likely to have many psychoneurotic tendencies such as anxiety complexes, extreme worry, or paranoias.

Effect on the Function of the Muscles

Moderate increase in thyroxine usually makes the muscles react with considerable vigor, but when the quantity of thyroxine become extreme, the muscles become weakened because of excess protein catabolism. On the other hand, lack of thyroxine causes the muscles to become extremely sluggish; they contract readily but relax slowly after a contraction.

Muscle Tremor

One of the most characteristic signs of hyperthyroidism is a fine muscle tremor. This is not the coarse tremor that occurs in Parkinson's disease or in shivering, for it occurs at the rapid frequency of 8 to 15 times per second. The tremor can be observed easily by placing a sheet of paper on the extended fingers and noting the degree of vibration of the paper. The cause of this tremor is not definitely known, but it is possibly due to increased activity in the areas of the cord that control muscle tone. The tremor is an excellent means for assessing the degree of thyroxine effect on the central nervous system.

Effect on Sleep

Because of the exhausting effect of thyroxine on the musculature and on the central nervous system, the hyperthyroid subject often has a feeling of constant tiredness; but, also, because of the excitable effects of thyroxine on the synapses, it is difficult for him to sleep. On the other hand, extreme somnolence is characteristic of hypothyroidism.

5

PARATHYROID GLANDS

The parathyroid glands, named because of their association with the thyroid gland, are usually found embedded in the connective tissue capsule on the posterior surface of the thyroid. In the human usually two pairs of small, yellowish-brown, oval structures are present. They typically measure about 6 mm in length, 3 to 4 mm in width, and 1 to 2 mm in thickness. They are identified by their location as *superior* and *inferior parathyroid glands*. In some individuals, additional glands are associated with the thymus.

Embryologically, the inferior parathyroid glands and the thymus are derived from the third branchial pouch and the superior glands from the fourth branchial pouch. Both the inferior parathyroids and the thymus migrate caudally during development. Normally, the inferior parathyroids separate from the thymus and come to lie below the superior parathyroids. Failure of these structures to separate results in the atypical association of the parathyroids with the thymus in the adult. The parathyroids function in the regulation of blood ion levels, especially calcium and phosphate. These glands are essential for life. Therefore, care must be taken during thyroidectomy to avoid removing the parathyroids. If the glands are removed, death will ensue as muscles, go into tetanic contraction as the blood calcium level falls.

Blood Supply and Innervation

The parathyroids receive their bood supply from the inferior thyroid arteries or anastomoses between the superior

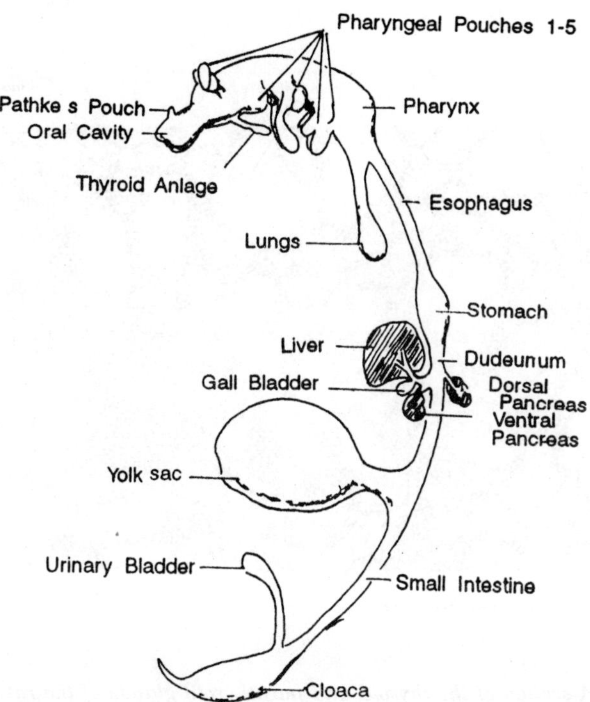

Fig. 5.1. A diagram illustrating the parts and outgrowth of the embryonic digestive tube.

and inferior thyroid arteries. The parenchyma of the gland is surrounded by rich blood and lymphatic capillary plexuses. Fenestrations are present in the endothelial cells of the blood capillary plexus.

The autonomic nerve fibers present in the gland appear to be involved in the regulation of blood flow and are not secretomotor in function.

Structure

Each glandular mass is surrounded by a thin, connective tissue capsule from which septa extend into the gland, dividing it into poorly defined lobules and separating the densely packed parenchymal cells into anastomosing cords. The extent of the connective tissue becomes more obvious in the adult as the number of fat cells increases with age. Their number increases dramatically at puberty and continues to increase throughout life. In the elderly adult, 60 to 70 percent of the

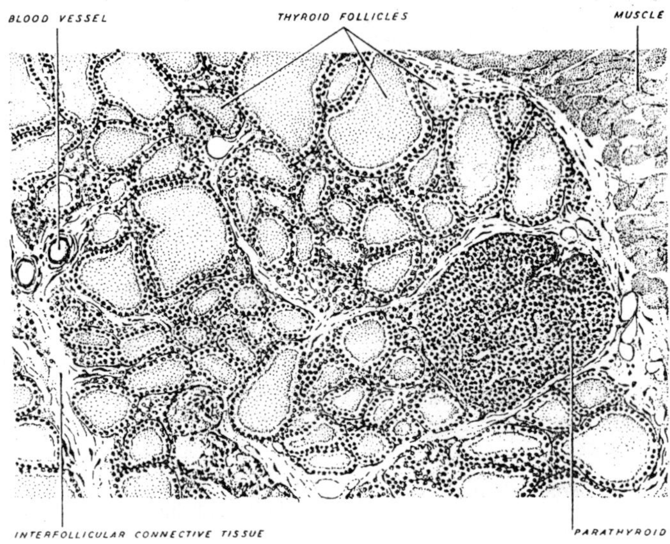

Fig. 5.2. A section of the thyroid and parathyroid glands of the rat as seen under low power of the microscope.

glandular mass may be occupied by fat cells. The septa provide a route of access for the blood and lymphatic vessels and nerves.

Two basic cell types are identified in the glandular epithelium : *principal* or *chief cells* and *oxyphil cells*. Some investigators identify a third, transitional cell type that is intermediate in structure between the two, however, the features associated with these cells may result from poor fixation of the tissue.

The *principal cells* are the most common parenchymal cell in the parathyroid. They are small, polygonal cells, measuring about 7 to 10 µm in diameter. They have a centrally located, vesicular nucleus and a pale-staining, slightly acidophilic cytoplasm. Lipofuscin granules, large accumulations of glycogen, lipid droplets, and small, dense membrane-limited granules are seen in addition to the typical cytoplasmic organelles.

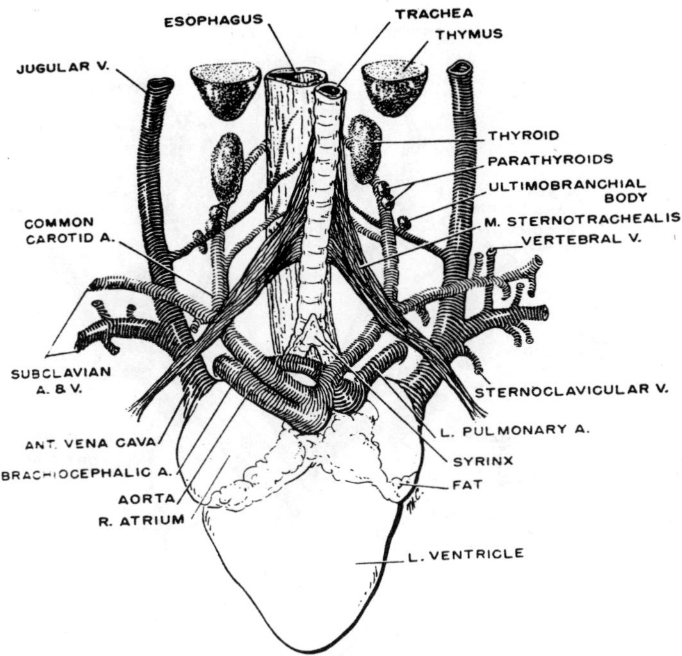

Fig. 5.3. Dissection of the thorax of the domestic fowl, showing the thyroids, parathyroids, and ultimobranchial bodies in relation to the heart and major blood vessels.

The *oxyphil cells* constitute a minor portion of the parenchymal cells. They are distinctly larger than the principal cells and have intensely eosinophilic cytoplasm. Their nuclei are often smaller and more heterochromatic than those of the principal cells. Ultrastructural studies have revealed that the eosinophilic staining of the cytoplasm is due to the presence of numerous large mitochondria. Morphologically, the cells are further characterized by limited amounts of endoplasmic reticulum, a small Golgi complex, and the absence of secretory granules.

Effect of Parathyroid Hormone on Calcium and Phosphate Concentrations in the Extracellular Fluid

The effect of injecting parathyroid hormone subcutaneously into a human being, showing marked elevation in calcium ion concentration of the extracellular fluids and depression of phosphate concentration. The rise in calcium ion concentration

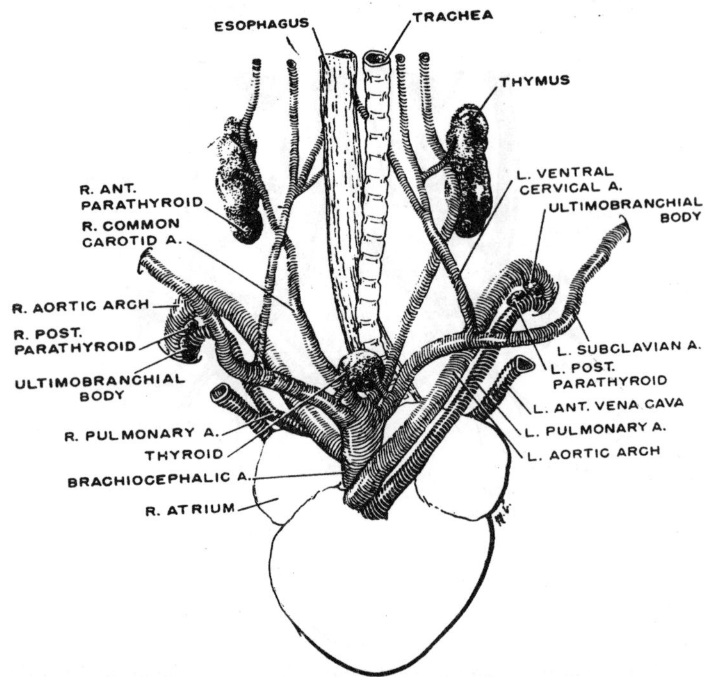

Fig. 5.4. Heart and major blood vessels of the turtle (Pseudemys scipta), showing the thyroid gland and the usual locations of parathyroid glands and ultimo-branchial bodies.

is caused principally by a direct effect of parathyroid hormone to increase bone absorption. The decline in phosphate concentration, on the other hand, is caused principally by an effect of parathyroid hormone on the kidneys to increase the rate of phosphate excretion.

Bone Absorption Caused by Parathyroid Hormone

The absorptive effect of parathyroid hormone in bones is believed to result from the ability of this hormone to convert osteoblasts and osteocytes in bone into osteoclasts and, also, to increase their osteoclastic activity. In turn, the osteoclasts are believed to secrete either enzymes or acids that absorb the bone.

Bone contains such great amounts of calcium in comparison with the total amount in all the extracellular fluids that even when parathyroid hormone causes a great rise in calcium concentration in the fluids, it is impossible to discern

Fig. 5.5. Homogeneous osteoporosis produced in the dog by the surgical removal of the stomach.

any immediate effect at all on the bones. Yet prolonged administration of parathyroid hormone finally results in evident absorption in all the bones with development of large cavities filled with very large, multinucleated osteoclasts.

Effect of Parathyroid Hormone on Phosphate and Calcium Excretion by the Kidneys

Administration of parathyroid hormone causes immediate and rapid loss of phosphate in the urine. This effect is caused by diminished renal tubule reabsorption of phosphate ions. It was long believed that this was the most important function of parathyroid hormone. However, it is now known that this function is of secondary importance to the effect of parathyroid hormone on bone absorption.

Parathyroid hormone causes increased renal tubular reabsorption of calcium at the same time that it diminishes the rate of phosphate reabsorption. Were it not for this effect of parathyroid hormone on the kidneys to increase calcium reabsorption, the continual loss of calcium into the urine would eventually deplete the bones of this mineral.

Regulation of Extracellular Calcium ion Concentration by the Parathyroid Glands

The calcium ion concentration in the extracellular fluids under normal conditions rarely varies more than 5 per cent above or below the normal concentration. Therefore, calcium ion regulation is one of the most highly developed homeostatic mechanisms of the entire body. This regulation is effected almost entirely by the parathyroid glands in the following way :

Control of Parathyroid Secretion by Calcium Ion Concentration

Even the slightest decrease in calcium ion concentration in the extracellular fluid causes the parathyroid glands to increase their rate of secretion and to hypertrophy. For instance, the parathyroid glands become greatly enlarged in *rickets*, in which the level of calcium is usually depressed only a few per cent; also they become greatly enlarged in pregnancy, even though the decrease in calcium ion concentration in the mother's extracellular fluid is hardly measurable; and, they are greatly enlarged during lactation because calcium is used for milk formation.

Feedback Control of Calcium ion Concentration

It can now be understood how the parathyroid glands control the calcium ion concentration in the extracellular fluids, for reduced calcium concentration increases the rate of parathyroid secretion; and this in turn increases bone absorption, thereby raising the calcium ion concentration to normal. Conversely, increased calcium ion concentration inhibits the parathyroid glands, which automatically decreases the calcium ion concentration to normal.

Note particularly that the regulatory properties of parathyroid hormone do not derive only form its effect on bone, for in addition it *increases the absorption of calcium from the gut* and *from the renal tubules,* which also helps to raise the calcium ion concentration. These are particularly valuable effects when the bone salts have been so depleted that parathyroid hormone can no longer cause additional absorption of calcium from the bone.

Calcitonin

About ten years ago, a new hormone that has effect on blood calcium opposite to those of parathyroid hormone was discovered in several lower animals. This hormone was named *calcitonin* because it reduces the blood calcium ion concentration. In the human being calcitonin is secreted by the thyroid gland by the so-called *parafollicular cells* in the interstitial tissue between the follicles.

Physiologic Effects of Calcitonin

Calcitonin inhibits the absorption of bone. It presumably acts at the same site in bone as does parathyroid hormone, because it even inhibits the absorption of bone after administration of large quantities of parathyroid hormone. Unfortunately, the mechanism of this inhibition is unknown, though it occurs extremely rapidly. It causes almost total cessation of bone salt reabsorption within minutes; as a result, the calcium ion concentration in the body fluids begins to fall also within minutes. Thus, the action of calcitonin on calcium ion concentration occurs many times as rapidly as the effect of parathyroid hormone.

Role of Calcitonin in Feedback Homeostasis of Calcium Ion Concentration

Perfusion of the thyroid gland with plasma containing even slightly elevated calcium ion concentration causes an immediate, several-fold increase in rate of secretion of calcitonin. Conversely, decreased plasma calcium ion concentration can decrease calcitonin secretion to as little as one-fifth normal. Therefore, the calcitonin mechanism, like the parathyroid mechanism, can also act in a feedback system to control calcium ion concentration.

However, there is one major difference between the calcitonin and the parathyroid feedback systems: the calcitonin mechanism operates extremely rapidly, reaching peak activity in less than an hour; this is in contrast to the many hours required for peak activity to be attained following parathyroid secretion.

Physiology of Parathyroid and Bone Diseases

Hypoparathyroidism

When the parathyroid glands do not secrete sufficient parathyroid hormone, the osteoclasts of the bone become almost totally inactive, and the calcium level in the blood falls from the normal value of 10 mg. per cent to 7 mg. per cent within two to three days. When this level is reached, the usual signs of hypocalcemic tetany deelòp. Among the muscles of the body most sensitive to tetanic spasm are the laryngeal muscles. Spasm of these obstructs respiration, which is the usual cause of death in tetany unless appropriate treatment is applied.

Treatment of Hypoparathyroidism

Parathyroid hormone. Parathyroid hormone is occasionally used for treating hypoparathyroidism. However, because of the expense of this hormone, because its effect lasts only about 24 to 36 hours, and because the tendency of the body to develop immune bodies against it makes it progressively less active in the body, treatment of hypoparathyroidism with parathyroid hormone is rare in present-day therapy.

Dihydrotachysterol and Vitamin D. In addition to its ability to cause increased absorption of calcium from the gastrointestinal tract, vitamin D also causes a weak effect similar to that of parathyroid hormone in promoting calcium and phosphate absorption from bones. Therefore, a person with hypoparathyroidism can be treated satisfactorily by administration of *large quantities* of vitamin D. One of the vitamin D compounds, dihydrotachysterol, has a more marked ability to cause bone absorption than do most of the other vitamin D compounds. Administration of calcium plus vitamin D three or more times a week can almost completely control the calcium level in the extracellular fluid of a hypoparathyroid person.

Hyperparathyroidism

The cause of hyperparathyroidism ordinarily is a tumor of one of the parathyroid glands, and such tumors occur much more frequently in women than in men or children. Consequently, it is believed that pregnancy, lactation, and perhaps other causes of prolonged low calcium levels, all of which stimulate the parathyroid gland, may predispose to the development of such a tumor. Treatment is always simply to remove the tumor, though this may be difficult to do because these tumors are sometimes only pea-sized and cannot be found at operation.

In hyperparathyroidism extreme osteoclastic activity occurs in the bones, and this elevates the calcium ion concentration in the extracellular fluid while usually depressing slightly the concentration of phosphate ions.

Bone disease in hyperparathyroidism

In severe hyper-parathyroidism, osteoclastic absorption of bone soon far outstrips osteoblastic deposition, and the bone may be eaten away almost entirely. Indeed, the reason a hyperparathyroid person comes to the doctor is often a bone that has broken without cause.

Effects of hypercalcemia in hyperparathyroidism

Hyperparathyroidism can at times cause the plasma calcium level to rise to as high as 15 to 17 mg. percent. The effects of such elevated calcium levels are mild to moderate depression of the central and peripheral nervous systems, muscular weakness, constipation, abdominal pain, peptic ulcer, lack of appetite, and depressed relaxation of the heart during diastole.

Rickets

Rickets occurs mainly in children as a result of calcium or phosphate deficiency in the extracellular fluids. Ordinarily, rickets is caused by lack of vitamin D rather than lack of calcium or phosphate in the diet. If the child is properly exposed to sunlight, at 7-dehydrocholesterol in his skin becomes activated by the ultraviiolet rays and forms vitamin D_3, which prevents rickets by promoting calcium and phosphate absorption from the intestines.

Children who remain indoors through the winter generally do not receive adequate quantities of vitamin D without some supplementary therapy in the diet. Rickets tends to occur especially in the spring months because vitamin D formed during the preceding summer can be stored for several months in the liver, and, also, calcium and phosphorus absorption from the bones must take place for several months before clinical signs of rickets become apparent.

Effect of Rickets on the Bone

During prolonged deficiency of calcium and phosphate in the body fluids, increased parathyroid hormone secretion protects the body against hypocalcemia by causing osteoclastic absorption of the bone; however, this causes the bone to become progressively weaker and also imposes marked physical strain on the bone, resulting in rapid osteoblastic activity as well as osteoclastic activity. The osteoblasts lay down large quantities of organic bone matrix, which does not become calcified because the calcium and phosphate ions are insufficient to cause calcification. Consequently, the newly formed, uncalcified organic matrix gradually takes the place of other bone that is being reabsorbed.

Obviously, hyperplasia of the parathyroid glands is marked in rickets because of the decreased blood calcium level.

Tetany in Rickets

In the early stages of rickets, tetany almost never occurs because the parathyroid glands continually stimulate osteoclastic absorption of bone and therefore maintain an almost normal level of calcium in the body fluids. However, when the bones become exhausted of calcium, the level of calcium may fall rapidly. As the blood level of calcium falls below 7 mg. per cent, the usual signs of tetany develop, and the child may die of tetanic respiratory spasm unless intravenous calcium is administered, which relieves the tetany within seconds.

6

Parathyroid Anomalies

Hyperparathyroidism

Hyperparathyroidism is *primary* when there is an adenoma or, much less frequently, idiopathic hyperplasia of the parathyroid glands; and *secondary* when increased parathyroid secretion exists as a compensatory mechanism maintaining the blood calcium level in various diseases which tend to lower that level.

Primary Hyperparathyroidism

In primary hyperparathyroidism a secreting parathyroid adenoma causes generalized osteitis fibrosa (von Recklinghausen's osteitis fibrosa), metastatic calcification, renal calculi, and disturbances of calcium and phosphorus metabolism. Largely as a result of Albright's work, it is now recognized that renal calculi may be the only outward manifestation parathyroid adenoma.

History

In 1891 von Recklinghausen included a probable case of this disease among an undifferentiated motley of bone disorders, but the first recorded case is said to have been described by Courtial in 1700. The first report of a parathyroid tumour, in a patient suffering from generalized osteitis fibrosa, was by Askanazy in 1904; and in 1906 Erdheim observed enlargement of the parathyroid glands in patients with osteomalacia and supposed the hyperplasia to be a secondary compensatory reaction. However, it is probable that some of these cases were actually osteitis fibrosa and Schlagenhaufer (1915), having found parathyroid tumours in two fatal cases of this disease, regarded them as being the aetiological factor and recommended parathyroidectomy as the correct treatment Nevertheless, it was not until 1926 that Mandl removed a parathyroid tumour from a patient with generalized osteitis fibrosa and recorded the improvement which followed.

Although MacCallum had reported as far back as 1905 the presence of a tumour of a parathyroid gland in association with chronic renal disease, it was not until 1934 that Albright and Bloomberg drew attention to the frequent occurrence of renal calculi in hyperparathyroidism and pointed out that such calculi might be the only symptoms of a parathyroid tumour.

Aeitology and pathogenesis

It is not known what, in general, causes tumour formation or primary hyperplasia of the parathyroid glands. Though it has been suggested that the condition can be compared with toxic goitre and is due to excessive secretion of pituitary parathyrotrophic hormone, there is grave doubt that such a hormone has any significance—or even exists. Cases of hyperparathyroidism following pregnancy have been reported. Davies *et al.* (1956) have reported two patients in whom hyperparathyroidism appears to have resulted from chronic steatorrhoea; is one of these, two parathyroid adenomata were found so that in this patient at least, 'primary' hyperparathyroidism was caused by the steatorrhoea. A familial aspect may also be of significance; Frohner and Wolgamot (1954) found five cases in a single family and suggest that it would not be inappropriate to screen the families of all the cases of hyperparathyroidism.

The excessive secretion of parathyroid hormone leads to a rise in the level of plasma calcium, a fall in that of plasma phosphate, and an increased excretion of calcium and phosphate urine. This constitutes the biochemical syndrome of hyperparathyroidism.

The increased calcium excretion by the kidney requires increased water excretion, so leading to polyuria and polydipsia—the so-called 'calcium diabetes'. This condition may in fact be mistaken for diabetes insipidus. The excessive excretion of calcium by the kidney leads sooner or later to renal damage; calcium may be deposited in the collecting tubules and give rise to nephrocalcinosis, or renal calculi may be formed, producing nephrolithiasis. The progression of this renal syndrome may lead to death through renal failure without any of the bone lesions of hyperparathyroidism having become manifest. This aspect of the disease, which probably results when the intake of calcium in the diet is adequate to maintain the raised serum calcium, has been called 'hyperparathyroidism-without-bone-disease' by Albright. In the Massachusetts General Hospital, 5 per cent of all cases of renal calculi were found to be associated with hyperparathyroidism and it may therefore be emphasized that all cases of renal stone should have at least a plasma calcium estimation performed in order to exclude this possible cause of the Stone. Since the prime cause of death in cases of hyperparathyroidism is renal failure and the prognosis depends mainly on how soon the renal condition is recognized and treated, the importance of detecting the disorder in its earliest phases cannot be overemphasized.

It seems probable that the bone disease associated with hyperparathyroidism depends largely upon the availability of calcium, which in turn is determined by the amount of calcium in the diet and the efficiency of calcium absorption from the gut. If adequate supplies of calcium from the gut are not available then the calcium loss from the bone consequent upon the action of the parathyroid hormone is not made good and the manifestations of osteitis fibrosa cystica generalisata appear. Excessive osteoclast activity is probably a direct result of parathyroid hormone action. It leads to decalcification and proliferation of fibrous tissue in the Haversian canals. This is followed by cyst formation and this in turn may be followed by the formation of osteoclasts among fibrocytes that are less differentiated than in the remainder of the fibrous marrow Osteoclastomata are not malignant and they disappear following removal of parathyroid adenoma. Because there is an accompanying marked increase in osteoblast activity, a raised level of serum alkaline phosphatase is found.

The parathyroid tumour is single in the great majority of cases but cases have been reported in which there were two tumours. The remaining parathyroid glands may show involution comparable to that found in the opposite adrenal gland when an adrenal cortex tumour is present. Tumours weighing up to 34 g have been reported but the severity of the disease is not related to the size of adenoma and a small tumour weighing only 1 g has been known to produce crippling disease. In a series of twenty-five cases seen at the Lahey Clinic hyperplasia of the parathyroids was found in only one, a single tumour being present in all the others.

Incidence

The disease is usually found in the third or fourth decades but may be present at any age. It is said to be more common in women than in men but, owing to its rarity, such preponderance is not necessarily encountered in any group of cases.

Clinical features

The onset is usually gradual and insidious, although the first major manifestation may be sudden and dramatic, e.g. a pathological spontaneous fracture, or renal colic. A not uncommon presentation is a bone tumour which, on section, is seen to be an osteoclastoma; or the long bones, the chest and spine, may become thickened, bent, or deformed Increasing asthenia and wasting, with muscular and bony pains may be the predominating features and give no obvious indication of the underlying disorder. A more complete clinical picture can be synthesized from consideration of the various systems affected.

Skeleton

The skeleton changes are characteristically well marked and generalized but not necessarily uniform. In some patients, however, the bony changes, as seen by X-ray, may be minimal The more usual finding of extensive decalcification results in rarefaction and softening of the bones, which become bent and deformed. The lower limbs tend to become bowed, and walking becomes increasingly difficult, the gait being awkward and waddling. In the later stages the patient is bedridden. Osteoclastomata form swellings at the ends and in the shaft of long bones, on the jaw, and anywhere on the skull. The vertebrae tend to be absorbed, and collapse with resulting kyphosis and considerable loss of height. Pathological fractures occur and delayed union is frequent. The jaw may become massive and its deformity leads to displacement and falling out of teeth, but the latter are not decalcified. The bones are osteoclastomatous swellings are often tender, especially upon pressure. Pain in the lower back may be severe. Clubbing of fingers may occur from resorption of the terminal phalanges. It is most important, however, to remember that the bone changes may be quite inconspicuous.

Radiography

All the bones may show great rarefaction, Osteoclastomatous swelling and cyst formation may be seen in various parts of the skeleton. The skull may show a granular appearance and lack of differentiation between the tables. Recent and old fractures of the long bones are revealed. Subperiosteal erosions of the phalanges, and especially of the terminal phalanges, are quite typical radiographic findings in hyperparathyroidism. Loss of the lamina dura of the teeth is also pathognomonic.

Muscles

Hypotonicity of the muscles is present, in contrast to the hypertonicity and irritability of the muscles in the opposite condition of tetany. The patient feels weak and is incapable of muscular effort. Considerable wasting of the muscles is not infrequent. Spontaneous pains occur, simulating rheumatism, pseudo-arthritis, and neuritis, and the muscles may be tender to touch. The joints may be excessively mobile.

Gastro-intestinal system

Anorexia and intermittent nausea, with progressive loss of weight, are frequent features. Vomiting may be intractable and is ascribed by Snapper (1940) to the hypercalcaemia. Abdominal pains and cramps are met with, and gastric ulcer or appendicitis may be simulated. Constipation is usual. Dry mouth and polydipsia are the result of the polyuria, due to the high calcium excretion.

Heart and Lungs

Dyspnoea and cynosis may result from calcium deposition in the heart and lungs, as well as from bony chest deformities and muscular weakness. Tachycardia is not uncommon.

Blood

The viscosity of the blood is increased. Destruction of the bone marrow may result in a secondary anaemia and in leucopenia.

Nervous system

Though patients may be nervous and irritable, being sometimes confused mentally, organic lesions of the central nervous system are not present. Apathy is not infrequent. Symptoms of tetany may be encountered after removal of a parathyroid tumour; or may rarely be present terminally in hyperparathyroidism complicated by renal insufficiency.

Metabolism

This has already been discussed. The characteristic chemical findings are a raised plasma calcium (e.g. above 11 mg per 100 ml), a lowered plasma phosphorus (usually below 2.5 mg. per 100 ml), and a raised plasma alkaline phosphatase (normal 4 to 8 King-Armstrong units). There is a negative calcium balance. If renal function is impaired the plasma phosphorus may be normal or raised. In the latter case the plasma calcium values may not be appreciably elevated. Rose, working with C.E. Dent, has described a method for determining ionized calcium and he has fond that the plasma-ionized calcium is always high in primary hyperparathyroidism, even when the total plasma calcium is within normal limits. This is not so in secondary hyperparathyroidism.

Kidneys and renal function

Polyuria and polydipsia with nocturia are frequent symptoms and do not necessarily indicate renal disease, being manifestations of the attempt to excrete increased calcium is soluble form. However, since the kidneys are frequently involved in hyperparathyroidism, the first manifestation often is renal, e.g. haematuria, renal colic, renal calculi, or chronic nephritis. The excess of calcium in the blood may be deposited in all the soft tissues of the body or selectively in the kidneys or in the arterioles. When the kidneys are involved the calcium deposition may be in diffusely scattered particles (calcinosis), or multiple calculi, often very large, may form either in the kidney substance or the pelvis, or both. The changes are sometimes found in the absence of gross skeletal changes.

Where diffuse calcinosis is present, the kidneys undergo fibrosis and hyalinization and the ultimate clinical and pathological picture is that of chronic interstitial nephritis. The deposition of calcium in the arterioles may lead to their degeneration and hyaline occlusion. Hypertension develops and may be severe. The patient dies in uraemia or from intercurrent infection. Although no impairment of renal function may be observed at the time of removal of a parathyroid tumour, and although such an operation is usually followed by a complete control of the primary hyperparathyroidism, yet chronic nephritis may develop some years later, and one must assume that initial irreversible changes were sufficiently severe to progress to fibrosis without further calcium deposition. In a case reported by Simpson and Wilson the patient died from renal insufficiency four years after the removal of a parathyroid adenoma, having developed hyperplasia of the remaining parathyroid glands secondary to the renal insufficiency. After remove of a parathyroid tumour, renal calculi may disintegrate rapidly, though this is not always the case.

A further complication of the relationship between the parathyroids and the kidneys is the fact that, in chronic renal disease of any aetiology, parathyroid hypertrophy and hyperplasia may be a secondary effect in an endeavour to excrete the retained phosphorus. Such secondary hyperparathyroidism will be discussed later.

Diagnosis

The disease may present itself in many forms, e.g. rheumatism, spontaneous fracture, a bony swelling (osteoclastoma) of the long bones, jaw, or skull, weakness and fatigability, gastro-intestinal disorder, renal colic, haematuria, or nephritis. Most cases show generalized osteoporosis on radiographic examination but such bone changes may be minimal. The skull frequently gives a characteristic granular radiographic picture and the phalanges show subperiosteal bone resorption. The lamina dura of the teeth is absent. Calcium deposits may be seen in the kidneys and in other viscera. The plasma calcium is raised and the plasma phosphorus is low, while calcium-balance experiments show a negative balance with excessive urinary excretion of calcium.

The plasma phosphatase is almost invariably raised. However, an atypical chemical picture as a result of renal disease may cause some difficulty in diagnosis. It was further pointed out by Lui (1940) that where hyperparathyroidism is associated with hypovitaminosis D, the plasma calcium may be normal and no calcium may be excreted in the urine. However, the giving of calciferol in adequate dosage to correct the hypovitaminosis will be followed by a return to the characteristic chemical picture of hyperparathyroidism.

Unlike a thyroid adenoma, a parathyroid adenoma is usually not palpable and must be sought for at operation. Rarely, however, such a parathyroid adenoma may even be palpable clinically before operation.

A number of conditions which bear a possible resemblance in one or more clinical, or radiographical, or biochemical features to generalized osteitis fibrosa are considered below:

Albright's Syndrome

This was described by Albright and colleagues in 1937 under the title 'Syndrome characterized by osteitis fibrosa disseninata, areas of pigmentation and endocrine dysfunction with precocious puberty in females'. Falconer *et al.* (1942) showed that such precocious puberty also occurred in boys. The main characteristics are: (a) multiple bone cysts which have a distribution suggesting a relation to nerve roots, or to an embryological defect in the myotomes; (b) areas of pigmentation which have a distribution suggesting some connexion with the bone cysts, the pigment being melanin; (c) precocious puberty with premature union of the epiphyses and ultimate dwarfism. In Albright's four cases, the first menstruation was observed at 1, $2^1/_2$, $3^1/_2$, and 7 years of age respectively. Normal pregnancy may occur in adult patients. The blood calcium and phosphorus and the calcium balance are normal, but the phosphatase may be raised. Falconer deserved other features, e.g., massive bony deformity of the head, with resulting optic atrophy; and of the jaw, resembling leontiasis ossea; acromegalic features; enlargement of the thyroid, unilateral exophthalmos, and wide separation of the eyes.

Focal osteitis fibrosa

This condition may be found in one or more bones, and its incidence is chiefly in adolescence, although it may appear at any age. The commonest clinical manifestation is spontaneous fracture. General symptoms and constitutional disturbance are absent The cause is unknown and there is certainly no obvious cannexion with the parathyroids, since the calcium and phosphorus levels in the plasma, and the calcium balance are normal. Where several bones are affected the plasma phosphatase may be raised.

Paget's osteitis Deformans

Paget, who described the disease in 1877, believed the condition to be inflammatory in origin and, although this is not generally held to be the case today, no aetiological cause, metabolic or otherwise, has been substantiated. Kay, Simpson, and Riddoch (1934) observed and investigated thirty-four cases of this disease. Its onset is usually in middle life, although it may be first observed in the third or eighth decade of life. The average age of onset in this series was 46, the youngest 30 and the oldest 60 years of age. There were 18 females and 16 males. A brother and sister and their mother constituted the only example in this series of a familial incidence, but others have observed such familial tendencies in the minority of cases.

The onset of osteitis deformans is usually insidious and the progress ,gradual. The disease may be limited to one tibia for some years, but more usually is generalized. The vault of the skull is considerably thickened, enlarged, and deformed, and the face seems in comparison to the top-heavy head. Owing to phases of softness of bones, the spine and shoulders bend, and in advanced cases the patient's stance is simian in character. The legs become bowed and may cross over in scissor fashion, so that the patient may become bedridden, contrary to Paget's original observation that 'the limbs, however mis-shapen, remain strong and fit to support the trunk'. Generalized pains of the bones are common and may be very severe, especially in the legs. Fractures following slight trauma occurred in five of the patients mentioned above. Sarcomatous change in a long bone rarely supervenes. Plasma calcium and phosphorus levels are normal, but a negative calcium balance is not infrequent, and the plasma phosphatase is always raised. No pathological changes in the parathyroids have been recorded. Radiologically, area of rarefaction and increased density are seen side by side, and a characteristic fluffy, cotton, wool appearance may be seen in the pelvis and skull. Complications are arterial degeneration and, rarely, spastic paraplegia or optic atrophy, due to bony compression of nerve tissue.

Carcinomatosis of bone

If metastases from an undetected primary carcinoma, e.g. prostate, are numerous, a radiological resemblance to generalized osteitis fibrosa may be observed. However, the chemical picture is not characteristic of hyperparathyroidism, the plasma calcium and phosphorus being normal. The phosphatase is raised, as in most generalized bone disturbances.

Multiple myelomatosis

The presence of multiple marrow tumour may simulate generalized osteitis fibrosa. Progressive anaemia and cachexia, associated with a high incidence of fractures, are major features. If there is much bone destruction, the plasma calcium may be high and a negative calcium balance exists. The plasma phosphorus is never low. It is normal or above normal when the kidneys are involved. Bence Jones protein in the urine is characteristic.

Osteomalacia

This vitamin D deficiency disorder is a rarity in England but may be found as a result of gastro-intestinal dysfunction not permitting adequate absorption, e.g. chronic steatorrhoea. The plasma calcium and phosphorus are both low, or one may be normal, but never are they above normal. The urinary calcium is low and the faecal calcium high. The plasma phosphatase is raised. As previously mentioned, two cases of chronic steatorrhoea, leading to the development of hyperparathyroidism, have been described by Davies *et al.* (1956).

Generalized osteoporosis resulting from a low renal threshold

Hunter (1935) recorded two such cases, due to the kidneys permitting an excess of calcium to escape into the urine. The serum calcium, however, was not raised and the renal aetiology was put forward as a suggestion.

Thyrotoxic osteoporosis

The clinical picture is that of thyrotoxicosis, but some 40 percent of patients show some rarefaction of bones and a negative calcium balance with normal plasma calcium and phosphorus values. Thyroidectomy, or antithyroid drugs, correct the osteoporosis and negative calcium balance, which are related not to the parathyroids but to the increased metabolism of hyperthyroidism.

Senile osteoporosis

Generalized osteoporosis may be found in senile people, or in younger people confined to immobility for long periods. The cause is undetermined but the diminution of gonadal secretion, especially oestrogens, has been postulated.

Eunuchoid osteoporosis

Osteoporosis may accompany the other skeletal deformities found in some cases of gonadal dysgenesis.

Fragilitas ossium

Multiple fractures occur *in utero*, in infancy, or in childhood. There is no disturbance of calcium or phosphorus metabolism and no known endocrine cause. The sclerotics of the eye are blue.

Course and prognosis

In the absence of correct treatment hyperparathyroidism is inevitably progressive. Weakness and deformity of the bones eventually render the patient bedridden and intercurrent infection, such as pneumonia, leads to death. Sometimes the kidney lesion is the most severe feature and terminal uraemia results.

Treatment

The condition cannot be controlled medically. If the clinical and biochemical pictures indicate hyperparathyroidism, the neck should be explored and the adenoma removed. The difficulty of the operative procedure is finding the tumour, since sometimes this is deeply embedded in the thyroid gland, or may lie in the mediastinum or behind the oesophagus. Re-exploration, if the first operation fails to find the tumour, is always a far more difficult procedure because of the scar tissue resulting from the previous operation; it is therefore necessary that the surgeon should make every effort to succeed at the first attempt.

It is wise to administer a high calcium diet for two or three weeks before the operation, in order to make good, as far as possible, the depletion of the calcium reserves. The diet, therefore, should be rich in milk, cheese, and eggs; and in addition, calciferol should be administered in a dose of 50,000 Units daily. Careful post-operative attention is necessary, since latent or manifest tetany, due to a pronounced fall in plasma calcium, may occur. This is due to disuse atrophy of the remaining glands, and is most likely to develop in patients with generalized osteitis fibrosa and a high serum phosphatase; it is rare in patients without bony changes. Usually it is only temporary, but in some cases it has led to death. As a precaution against this occurrence, 2-4 G. of calcium lactate can be given three times a day for several weeks and the administration of calciferol should be continued for several months. On the first suspicion of tetany, intravenous calcium gluconate solution, 10-20 ml of a 10 or 20 per cent solution, should be given and repeated, even several times a day, if necessary. Parathormone, should not be given, since it is obviously illogical to mobilize calcium from bones already decalcified. The manifestations of subnormal calcium concentration may be psychotic rather than tetanic, and similarly are abolished when the plasma calcium reaches normal values again.

The more immediate results of operation are a disappearance of pain in the limbs an increase in strength the loss of gastro-intestinal symptoms, and the cessation of polyuria and polydipsia. Recalcification of bone may take several months or more and, although further deformity or fractures are unlikely, the existing malformation may prevent proper ambulation. Osteoclastomata tend to disappear in a few weeks. Renal calculi may disintegrate and be passed as gravel. Chronic nephritis, if of recent onset, may improve—but it may also continue to deteriorate and eventually cause death through renal failure. Cataract or lenticular opacities may occur if calcium is not given in adequate doses after operation. Menstruation, if previously scanty or absent, may return to normal. The plasma calcium falls immediately and the subnormal levels reached may lead to tetany; the plasma phosphorus tends to return to normal and the increased calcium excretion ceases. The plasma phosphatase usually remains raised, only gradually falling after a period of months or more; it is therefore more a measure of the degree and extent of bony change rather than of the excess of parathyroid hormone.

Secondary Hyperparathyroidism

It has been pointed out that in primary hyperparathyroidism the cause of the parathyroid adenoma is usually undiscoverable. In secondary hyperparathyroidism, however, hyperplasia of the parathyroid glands occurs as a compensatory mechanism which helps to maintain a normal concentration of calcium in the body fluids in various conditions in which the level of blood calcium tends to be decreased. Since a fall in plasma calcium leads to increased parathyroid hormone secretion, it follows that hyperplasia of glands can be expected in those conditions which lead to a reduction in the calcium, level—such as rickets, osteomalacia, steatorrhoea, pregnancy, lactation, and long-standing renal insufficiency. The cause of the low plasma calcium in rickets, osteomalacia, and steatorrhoea is insufficient or defective absorption of calcium from the intestinal tract; during pregnancy and lactation it results from the continuous drain on maternal calcium. The increased parathyroid activity removes calcium and phosphorus from the bones and increases the excretion of phosphorus, so that in the blood the calcium level remains normal, while that of phosphorus falls. In the meantime, of course, the bones decalcify. Albright (1941) has argued that when, in vitamin D deficiency, the plasma calcium is low and the plasma phosphorus is normal, compensatory hyperparathyroidism has failed to take place; while if both calcium and phosphorus levels are low, increased parathyroid activity has occurred, though in a degree insufficient to raise the plasma calcium to normal level.

The enlargement of the parathyroid glands, which not uncommonly accompanies chronic nephritis, is considered by Albright to result from the following sequence of events: phosphorus is retained because of the renal failure; this leads to lowering of the blood calcium; parathyroid hyperfunction is the consequence. Were it not for this secondary hyperparathyroidism it seems probable that many patients with renal insufficiency would have severe tetany. Albright has defined a syndrome of 'renal osteitis fibrosa cystica' in which marked renal insufficiency has lasted a long time, with phosphate retention and a high plasma inorganic phosphorus level, a slight reduction in the plasma calcium level, marked acidosis, calcium deposits in the neighborhood of joints, extreme calcification of the media of all arteries, generalized osteitis fibrosa of all bones, and enormous enlargement of all parathyroid tissue. He further considers that 'renal rickets' is really a variety of this syndrome in childhood, the bone changes' being secondary to chronic nephritis and produced by secondary parathyroid hyperplasia.

Acute Hyperparathyrodism

Only seven cases of this extremely rare condition have been reported. It is caused by a chief-cell adenoma of the parathyroid gland but it differs from chronic hyperparathyroidism in so far as the clinical picture shows a very rapid onset of symptoms. These consist of anorexia, vomiting constipation, pains in the bones, loss of weight, and lassitude developing into increasing drowsiness and asthenia. The abdominal pain may be severe, resembling acute abdominal disease. There is slight fever, with a disproportionately high pulse rate, and there may be evidence of impaired renal function. In some cases there have been multiple thromboses. The metabolic disturbances typical of hyperparathyroidism are present, but bony change may be absent. There is widespread visceral and vascular calcification. Death usually occurs after several days.

As cause of death in acute hyperparathyroidism, Oliver has suggested dehydration, loss of electrolytes, and failure of the brain to utilize glucose in the absence of diffusible phosphorus. It is this latter which may be the explanation of the gradual onset of coma. In the case described by Hanes (1939), it was felt that death was due either to severe myocardial lesions or to lack of diffusible phosphorus in the presence of a very high plasma calcium. It is the opinion of Oliver that prompt recognition and treatment of the disease might lead to good results. The treatment recommended consists of a continuous withdrawal of blood from a vein and its replacement by normal saline solution, followed, as soon as possible, by removal of the tumour.

Hypoparathyroidism

Hypoparathyroidism results from destruction, or operative removal, of the parathyroid glands. The resulting fall in the level of the blood calcium leads to tetany, a condition characterized by prolonged muscular spasms, due to increased neuromuscular excitability.

History

Tetany was first described clinically in 1815 by John Clarke. However, it was not until the end of the nineteenth century that it was brought into prominence as a result of its occurrence after partial thyroidectomy, an operation which was then initiated for goitre. Post-thyroidectomy tetany was produced experimentally by Schiff in 1884 and Horsley in 1885. In the meantime, however, Trousseau had described his sign in 1862 and Chvostek recorded his in 1876. The chemical basis of the parathyroid deficiency was recognized in the early part of this century by a series of workers, especially, MacCallum and Voegtlin (1908-9). The guanidine theory of Paton and Findlay (1916) became popular for some time but was refuted in 1925 by Collip and Clark who showed that guanidine intoxication, which causes symptoms similar to those of tetany, could be produced even in the presence of excess of parathyroid hormone.

Aeitology of tetany

Tetany can be due to many causes, but the ultimate basis is a fall in the plasma-ionized calcium. Although in alkalosis normal plasma calcium values may be found, there is nevertheless a decreased proportion of ionized calcium.

Interference with, or removal of, parathyroid tissue

1. After removal of a parathyroid adenoma for hyperparathyroidism, most patients have some immediate post-operative symptoms, due to the fall of plasma calcium, but after some days or weeks these tend to disappear and it is assumed that the remaining parathyroid glands undergo compensatory hyperplasia, or return to normal if previously involuted.
2. After thyroidectomy the symptoms may occur at once, or after a latent period of some weeks—the so-called tetania parathyropriva. They may rarely persist in a chronic form. In a series of 277 patients operated on for thyrotoxicosis, Lachman (1941) found that 14 per cent showed transient evidence of post-operative parathyroid insufficiency. This figure seems a high one, even if only half required any

treatment and more recent writers consider that post-thyroidectomy hypoparathyroidism should not occur in 1 per cent of cases under optimum conditions. Tetany is more frequent after total thyroidectomy for cardiac disease, but should not be a deterrent to operation in suitable cases, since it is usually easily controlled. Post-thyroidectomy tetany occurs practically only in women; even allowing for the increased incidence of thyrotoxicosis among women, it is probable that this fact is not without significance.

Since the finding of parathyroid tissue in thyroid glands removed at operation for thyrotoxicosis is exceptionally rare, it has been postulated that post-thyroidectomy tetany is due to interference with the blood supply of the parathyroids. It is also possible that removal of large amounts of thyroids tissue may in itself cause a transient fall in plasma calcium, since the opposite condition of thyroid hyperplasia may be associated with an increased urinary excretion of calcium and osteoporosis. In a study of post-thyroidectomy metabolism. Robertson (1941) found that the urinary excretion of phosphorus falls in cases of tetany following subtotal thyroidectomy for thyrotoxicosis, but is diminished to the same extent even after uncomplicated subtotal thyroidectomy

Idiopathic hypoparathyroidism

This condition is considered to be rare, only some 60 or 70 cases having been described in the literature. However, de Mowbray (1953) considers that more cases might come to light if the diagnosis were entertained in patients presenting with paraesthesiae, pains, cramps, and spasms in the extremities; in cases of epilepsy; and of cataracts occurring in young people; while cases associated with trophic changes in skin, hair, and nails might be found in skin clinics. Moreover, three-quarters of the cases of idiopathic hypoparathyroidism reported in the literature have had their onset of symptoms in childhood, so that diagnosis should be considered in cases of infantile convulsions. In these cases of idiopathic hypoparathyroidism the gland cells of the parathyroids are replaced by fibrous or fatty tissue, or show marked round cell infiltration. In some cases the parathyroids are congenitally absent.

Rickets, osteomalacia (adult rickets), steatorrhoea

In these conditions there is deficient absorption and utilization of calcium. Infantile spasmophilia may also be included.

Pregnancy, lactation, and menstruation

The increased demands for calcium in these conditions probably render a latent tetany manifest. Deficient calcium in the diet and deficient sunlight and vitamin D may be the predisposing causes.

Alkalosis

This acts by reducing the ratio of ionized to non-ionized calcium. It may result from excessive vomiting or gastric lavage, e.g. pyloric or intestinal obstruction; hyperpnoea, which may be hysterical, voluntary, or due to encephalitis lethargic a; excessive alkaline treatment of peptic ulcer; and the use of alkalis in nephritis.

Miscellaneous

The following causes of tetany have occasionally been encountered: infection, in association with deficient diet; epidemic and occupational situations in which the causative factors have probably been deficient diet and sunlight; and dietary excess of phosphorus. This latter is a somewhat theoretical cause in man though in cattle it is well recognized and is presumed to act by depression of the blood calcium level.

Symptoms generally do not appear until the plasma calcium level is below 8 mg per 100 ml, and the plasma inorganic phosphorus level rises to above 5 mg. per 100 ml. In complete aparathyroidism, the plasma calcium may be as low as 4 mg and the plasma inorganic phosphorus as high as 12 mg per 100 ml. The fall in blood calcium level cannot be the whole explanation, for it is well known that patients with hypoparathyroidism may have periods of freedom from symptous, even though the calcium and phosphorus levels of the blood remain the same as those during periods of active tetany. What further factors may be involved in determining the abnormal neuromuscular activity are still largely speculative.

Clinical features

The presenting symptoms are: Tetany, in some 70 per cent of cases, epilepsy, or generalized convulsions, in over 40 per cent; laryngeal spasm, ectodermal lesions, and failing vision, due to cataracts, each in some 10 per cent of cases. The major manifestations may be classified under seven headings. (1) Paraesthesiae and muscular cramps; (2) convulsions; (3) gastro-intestinal disturbances; (4) respiratory disturbances; (5) .neurological disturbances; (6) psychoneurosis and psychosis; (7) trophic disturbances. One, or all, of these manifestations may be met with, and in latent tetany ectodermal trophic changes may be the only clinical manifestation.

Paraesthesiae and muscular spasms

The only disturbance in mild tetany may be numbness and tingling of the fingers. A burning sensation sometimes occurs, or the fingers may feel stiff. Cramps of the calf muscles tend to become worse at night. The muscles may be in a state of chronic tonic contraction, and if the facial muscles are so involved, a characteristic facies develops, e.g. a mask-like face with the comers of the mouth drawn downwards and the nasolabial folds accentuated; the forehead may be wrinkled and the eyes wide open. Tonic muscular contractions are occasionally associated with violent and unbearable pain.

Fibrillary muscular twitchings may be localized to groups of muscles, or widespread. Gross muscular spasms are often very painful and the patient then cries out with pain and perturbation. They last for minutes or hours, and may occur many times a day. Characteristically, the muscles of the hands and feet are involved, producing carpopedal spasm. The fingers are flexed at the metacarpophalangeal joints and extended at the interphalangeal joints, while the thumb is flexed across the palm, producing the obstetric hand. The wrists and elbows may also be flexed, bring the hands across the body. The toes are flexed, the ankle joint extended, and the sole of the foot inverted. Facial and neck muscles are also involved in some patients.

Convulsions

Epileptiform attacks, both 'petit mal' and 'grand mal' are known to occur, even in the absence of both signs, and patients have been treated in neurological clinics for years as idiopathic epileptics. Kowallis, describes an attack in a woman of 32 suffering from idiopathic hypoparathyroidism as follows:

"Seizures commenced in the sixth postpartum week and subsequently occurred every two weeks. They were characterized by sudden onset, without warning, at any time of the day or night. There were loss of consciousness, biting of the tongue, clonic movements of all extremities and vesical incontinence. The seizure was followed by a deep sleep. The whole episode involved about an hour. The patient was not conscious of what happened during the attacks, although she remembered the residual headaches and mental depression. After the onset of her convulsions, definite personality changes developed, primarily irritability, stubbornness and forgefulness."

In this case, the plasma calcium was 5.1 mg, and the phosphorus 4.8 mg per 100 ml. In a case described by Himsworth and Maizels (1940) a boy of 12, carpopedal spasms or 'grand mal' attacks occurred and were completely controlled by calciferol, providing the plasma calcium was not allowed to fall below 7 mg per 100 ml. Even 'grand mal' attacks were avoided so long as the calcium level remained above 5.5 mg per 100 ml. Characteristic changes may be shown in the electroencephalogram in patients with tetany; there may be abnormal slow waves or epileptic patterns. Apart from headaches associated with epilepsy, typical migranine may be a manifestation of tetany.

Gastro-intestinal discrders

Spasm of the gastro-intestinal musculature produces abdominal cramps, pain, and vomiting. Laparatomy has been performed for suspected perforated peptic ulcer, or appendicitis and in fact, ileal spasm has produced a fatal ileus. Spasm of bile passages produces a clinical picture of gallstone colic and, rarely even transitory jaundice.

Respiratory disorders

Spasm of the larynx is liable to occur, more especially in rickety children with spasmophilia. The attacks and sudden in onset and alarming in their dramatic presentation. Spasm of the bronchial musculature simulates bronchial asthma.

Neurological disturbances

In addition to the psychiatric disturbances mentioned below, papilloedema, cerebellar dysfunction, and cerebral calcification may be found. At times the changes may mimic brain tumour to the extent of precipitating neurosurgical diagnostic exploration.

Psychiatric disturbances

In addition to mental retardation, which may be found in juvenile patients with hypoparathyroidism, a variety of psychoneurotic and psychotic changes may be encountered. These include anxiety symptoms, irritability and depression, and impairment of memory and intellectual capacity. More rarely, hallucinations, confusion states, manic-depressive conditions, paranoia, schizoid personality, and dementia may be present in chronic tetany. In a series of eighteen examples of parathyroid insufficiency, no fewer than five were cases of psychosis. This is most commonly of a toxic delirious type, occurring during the first few months of the parathyroid deficiency, and may be its only manifestation. With adequate treatment of the parathyroid insufficiency, the prognosis is considered to be good, although the response may not be immediate.

Trophic changes

The skin may be dry, rough, and scaly, and there may be various skin disorders, such as dermatitis herpetiformis, pustules, acrodermatitis pustulosa continua, psoriasis, and chloasma. The hair may show diffuse thinning and occasionally even complete loss. The eyebrows and eyelashes may become thin, and there may be some loss of pubic and axillary hair. The nails show fraying, brittleness, grooving, necrosis, and detachment. There have been a number of reports of cases of idiopathic hypoparathyroidism in which the skin, nails, mouth and tongue have been infected with moniliasis. Whittaker *et. al.* (1956) have described a syndrome of familial juvenile hypo-adrenocorticism, hypoparathyroidism, and superficial moniliasis, based on a description of their own patient and twelve similar patients reported in the literature.

When the parathyroid deficiency arises in childhood, the teeth may be defective in enamel and dentine, and the enamel shows transverse ridging. Dental caries is frequent. The teeth may fall out.

Cataracts may occur in tetany from any cause. They may be an early feature in hypoparathyroidism, even in the absence of manifest tetany. They may also develop after tetany has become controlled by treatment. The history or finding of 'senile cataract' in any person below middle age, or after thyroidectomy, should immediately arouse suspicions of hypoparathyroidism.

Diagnosis

This is not difficult in a well-developed case and, once made, the detection of minor manifestations soon follows. Trousseau's sign consists of producing carpal spasm by maintaining a sphygmomanometer above the systolic pressure for 1 to 5 minutes. Chvostek's sign is the production of contraction of a facial muscles by tapping the facial nerve just below the zygoma and in front of the parotid gland Erb's sign is an exaggerated muscular contraction in response to minimal electrical stimuli.

The plasma calcium level is reduced, always below 8 mg per 100 ml and occasionally as low as 4 or 5 mg per 100 ml. The plasma inorganic phosphorus level is raised, occasionally as high as 10-12 mg per 100 ml. No direct relation exists between level of the plasma calcium and the severity of the tetany. Severe tetany probably occurs more often in post-operative cases, in which the level falls rapidly to between 7 and 8 mg per 100 ml; on the other hand, in chronic cases, tetany may be absent or only very mild, even when the level is as low as 5-6 mg per 100 ml. The alkaline phosphatase level is usually normal or reduced. In steatorrhoea or rickets, the plasma phosphorus level is normal. Bleeding and coagulation times are prolonged, owing to the low blood calcium. Tetanus is, of course, in no way related to tetany and is an infective condition, following a contaminated wound. Opisthotonos does not occur in tetany.

Course and prognosis

Hypoparathyroidism can be controlled, but in severe cases the borderline is easily crossed and intermittent attacks of tetany may occur. Tetany, due to disorders such as rickets or alkalosis, can be cured by appropriate treatment of the underlying disorder.

Treatment

Acute attacks

The immediate treatment is the slow intravenous injection of 10 or 20 ml of 10 per cent solution of calcium gluconate. Soon after the intravenous injection the same amount of calcium gluconate may be given intramuscularly in order to prolong the effect. Calcium chloride, 20 ml of a 5 per cent solution, may also be used intravenously but intramuscular injection may lead to necrosis and ulceration and so should be avoided. As an emergency measure in very severe cases, the intravenous injection of parathormone, in a dose of 20-60 units, has been advocated. It is doubtful, however, if parathormone has any advantage over calcium; it further demineralizes bones where osteitis fibrosa is already present; it is expensive; it is liable to give rise to reactions; and it leads to antihormone formation after repeated injections and thereby becomes ineffective. When parathormone is used the behaviour of the plasma calcium must be observed closely in order to avoid overdosage. The early symptoms of overdosage are anorexia, nausea, and vomiting, and later thirst and drowsiness which may proceed to coma. On the other hand, the hypercalcaemia may cause no symptoms at all. Overdosage has to be treated by discontinuing the injections, and lowering the increased viscosity of the blood, caused by the hypercalcaemia, by withdrawing a pint of blood from a vein and replacing it with twice the amount of normal saline solution.

When, by reason of the convulsions, the intravenous administration of the drugs mentioned above is difficult, it is necessary to sedate or even to anaesthetize the patient first.

Subacute or chronic tetany

The mainstay of treatment is the administration of an irradiated ergosterol derivative, either calciferol (vitamin D_2), or dihydrotachysterol (A.T. 10). These substances are of similar chemical structure, the difference being that calciferol has a double bond between the C 9 and 10 atoms. The principal effect of A.T. 10 is to stimulate phosphate excretion in the urine, and it also increases calcium absorption from the intestine. It thus appears to resemble more closely parathyroid hormone in its action than calciferol, which acts principally by increasing calcium absorption, though it does have a weak action on the excretion of phosphorus in the urine. There is also some evidence that both calciferol and A.T. 10 liberate calcium from bone.

In so far as it resembles parathyroid hormone more closely in its action, A.T. 10 is theoretically preferable; on the other hand, calciferol is much cheaper, so that whereas with the latter, an average maintenance dose costs about a shilling a month, with the former the cost is about £1 per month. As regards their effects in controlling symptoms and plasma calcium level, there seems little to choose between calciferol and A.T. 10, though it is possible that the former is more likely to lead to hypercalcaemia in view of the fact that it acts predominantly upon calcium absorption and has a cumulative action in this respect. The initial dose is usually from 200,000 to 500,000 Units of calciferol (or even more) or 3 ml of A.T. 10, daily. The daily maintenance doses are usually from 50,000 to 200,000 Units, or 0.25-1 ml respectively.

It has been observed that some patients with chronic tetany, when taking a dose of calciferol that has been sufficient to maintain their plasma calcium levels, may, for no apparent reason, develop an insensitivity to the calciferol and remain unaffected, even when the dosage is markedly increased. Dent *et al.* (1955) made metabolic studies on two patients taking large doses of calciferol while in such a calciferol-insensitive phase. None of the usual metabolic actions of calciferol were manifest. In four patients who had become calciferol-insensitive, a change was made from large doses of calciferol to normal doses of A.T. 10, or of pure vitamin D_3. This led to a rapid disappearance of tetany and a return to normal of their plasma calcium and phosphorus levels. In two patients, in whom the change from calciferol to A.T. 10 or to vitamin D_3 was undertaken while they were under full metabolic control, it was found that the change of treatment coincided with the onset of normal vitamin D action. These authors therefore consider that in many such patients it may be necessary, from time to time, to change the form of vitamin D preparation in order to maintain a continued response.

As a guide to the correct dosage, the Sulkowitch reaction is a useful rough indication of the amount of calcium being excreted in the urine. If the calcium is precipitated as a fine white powder of calcium oxalate, it is probable that the plasma calcium is about normal; if there is no precipitate the urine is free from calcium and the plasma calcium level may be between 5 and 75 mg per 100 ml; if there is a heavy precipitate, so that a milky appearance is produced, the plasma calcium may be above 12 mg per 100 ml. Nevertheless, repeated estimations of the plasma calcium level are advisable, since instances are not uncommon where the plasma calcium" low when the Sulkowitch reaction is positive and it is known that symptoms of acute hypercalcaemia may appear even when there is no excess of calcium in the urine. The ideal use of the Sulkowitch reagent is by the patient herself, since it is easy to carry out, and a heavy positive reaction should warn her to reduce the dose, or to stop taking the drug, and to see her physician immediately; while an absence of any precipitate would indicate the need an increase of the dose.

In many cases repeated readjustment of the dosage, in response to infection according to mental and physical stresses and strains, may be necessary, the situation in fact being comparable to that of the treatment of diabetes mellitus. Although it is true that a high-calcium, low-phosphorus diet may restore the plasma calcium and phosphorus levels to normal, it is unlikely that many patients could a to such a diet with sufficient accuracy. On the whole, therefore, it is better not to place patients on a strict diet but to rely on the effectiveness of the vitamin D preparation. It is however, advisable to avoid an excess of milk in view of the fact that it has a high content of phosphorus as well as being rich in calcium. It is likewise generally better to do without calcium salts, since adjustment is much easier if only one variable, namely that of the vitamin D preparation, has to be taken into account.

Pseudo-hypoparathyroidism

This rare congenital disorder was first described by Albright *et al.* (1942) and the total number of cases in the literature is still below twenty. It consists of three independent, most probably genetically determined, disturbances:

1. There is an abnormal peripheral resistance to the action of parathyroid hormone (target-organ failure) which results in all the manifestation hypoparathyroidism.
2. There is a dyschondroplasia which results in dwarfism and in shortening of some of the metacarpals and metatarsals.
3. There is a tendency to metaplastic formation of bone in the soft tissues.

The manifestations of hypoparathyroidism can be controlled by vitamin D preparations as in hypoparathyroidism itself. There is, of course, no treatment for the other aspects of the disorder.

7

Suprarenal Gland

The adrenal glands were first described in man by Eustachius in 1563. Other anatomists, dissecting poorly preserved cadavers, identified these organs along the anterior borders of the kidneys and were impressed by the fact that they were generally filled with fluid. Not appreciating the effects of post mortem decay, the term suprarenal capsules was used to describe them. Cuvier recognized that each gland consists of an inner and outer region, now referred to as the medulla and cortex, respectively. The first hint that these glands might be of functional significance came from Thomas Addison's description of a clinical condition resulting from their deterioration, a syndrome now bearing his name.

Anatomy of Adrenals

This gland is a compound structure consisting of an outer cortex and an inner medulla. The hormones of the cortex are steroids, whereas those of the medulla are amines. The two components of the organ originate from different embryonic primordia. The cortex is derived from mesoderm in close association with the developing gonads. The medulla is ectodermal since it differentiates from neural crest cells along with the sympathetic ganglia. The medullary cells are modified ganglion cells and remain in intimate contact with the preganglionic fibres of the sympathetic system. Secretion of the medulla is regulated very largely by means of these nerves. The cortex, on the other hand, resembles the anterior hypophysis in being practically devoid of secretory nerve terminals. Like other endocrine glands, the adrenals receive a rich blood supply. The human organs are flattened bodies situated in the retroperitoneal tissue along the cranial ends of the kidneys. They vary considerably in shape, but are usually descri●ed as triangular or cresentric accessory depots of cortical tissue are frequently found in many mammalian species, most commonly in the perirenal fat and along the path of descent of the gonads.

During the third month of intrauterine life, the human adrenal reaches its maximum relative size and exceeds that of the kidney. This large size of the festal adrenal is due to the presence of a thick boundary zone between the definitive cortex.

Adrenal Gland

The suprarenal or adrenal glands are paired organs that are embedded in the retroperitonealpose tissue superior poleach kidney. These glands are roughly triangular and flattened in appearance; the approximate dimensions are 5 cm by 3 cm by 1 cm in thickness, and the total mass of both adrenals is around 15 gm in an adult. However, the size of the glands as well as their total weight may exhibit physiologic state of the individual.

The cut surface of a transected adrenal gland reveals two discrete regions within the thick, collagenous connective tissue

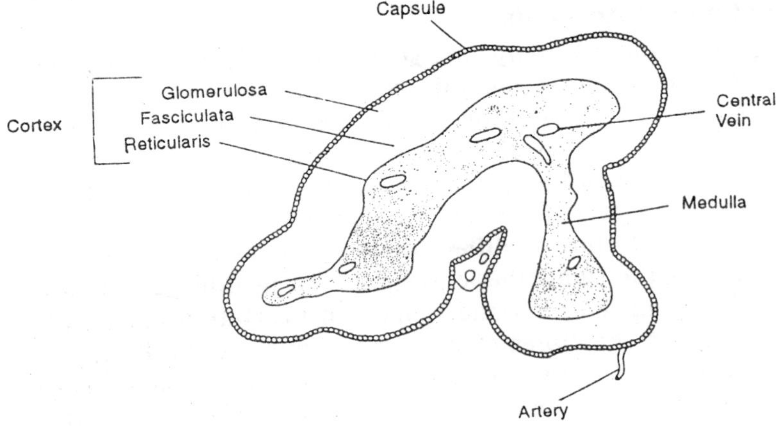

Fig. 7.1. T.S. of the adrenal gland.

capsule that surrounds that becomes reddish-brown in its innermost layer adjacent to the thin second inner layer, the greyish medulla. Consequently, the adrenal gland consists of two distinct endocrine organs that are located within a common structure. The cortex and medulla differ markedly in their embryologic origins, morphology, hormonal secretions, and functions.

The adrenal medulla synthesizes and secretes the two catecholammine hormones, epinephrine and norepinephrine. Although these hormones are not essential to survival (as are certain of the cortical hormones), they do enable the individual to adapt physiologically to stress or emergency situations.

The tissue of the adrenal cortex synthesizes and secretes all three of the general types of steroid hormones found in the body. First, glucocorticoids, which are essential to the normal metabolism of carbohydrate, protein, and fat by the body. Second, a mineralocorticoid which is critical in the regulation of normal electrolyte balance and maintenance of the proper volume of extracellular fluids in the body. Third, the adrenal cortical tissue also secretes small quantities of sex hormones, which exert minimal effects upon reproductive functions under physiologic conditions.

Functional Morphology of the Adrenal Cortex

Developmental Considerations

The adrenal cortex develops from the coelomic mesoderm that lies medially to the urogenital ridge. In human fetuses about 10 mm in length, mesothelial cells near the cranial end of each mesonephros multiply extensively and invade the adjacent highly vascular mesenchyme. Eventually these cells form the fetal cortex, which ultimately comprises roughly 80 per cent of the total mass of cortical tissue in the embryo. In the 14 mm embryo, further proliferation of the mesothelial cells forms the permanent cortex.

Following parturition, the fetal cortex degenerates rapidly, whereas the permanent cortex enlarges; the net effect of these processes is that the adrenals lose about 50 per cent of their total mass during the first weeks of postnatal life. The embryonic adrenal cortex is physiologically active, and already it is under the regulatory influence of corticotropin from the adenohypophysis.

Histology

The cortex forms the major portion of the total mass of tissue found in the adrenal gland, and in adults three concentric zones are discernible within this organ: (1) a thin outermost layer, the zona glomerulosa lies just beneath the capsule; (2) a thick middle region, the zona fasciculata which lies beneath the zona glomerulosa; (3) a relatively thick inner layer, the zona reticular is which lies adjacent to the medulla. The transition between one cortical zone and the next is gradual rather than sharply delineated.

Zona glomerulosa

The zona glomerulosa comprises about 15 per cent of the total cortical mass. The cells of the zona glomerulosa are arranged in tight groups and pillars consisting of large, columnar epithelial cells that are continuous with the cells of the zona fasciculata. The individual pillars of glomerulosa cells are separated by vascular sinuses.

The spherical nuclei of the glomerulosa cells generally contain one or more nucleoli, and stain deeply; the Golgi apparatus is juxtanuclear. The cytoplasm of glomerulosa cells is more sparse than in the cells of the other two cortical zones, and is usually acidophilic, although clumps of basophilic material are often present. Filamentous mitochondria are abundant.

The cytoplasm also contains a well-developed smooth endoplasmic reticulum that is present as an anastomosing network of tubules, as well as abundant free ribosomes in the matrix. Some rough endoplasmic reticulum is also present in these cells.

The glucocorticoid hormone corticosterone is secreted by all three layers of the adrenal cortex. However, the enzymatic machinery necessary for the synthesis of the mineralocorticoid hormone aldosterone is restricted to cells of the zona glomerulosa. Furthermore, the cells of the glomerulosa that are located adjacent to, and just beneath, the capsule are able to provide cells for regeneration of all three cortical zones if necessary.

Zona fasciculata

The zona fasciculata normally comprises the bulk of the cortical tissue, thus comprising roughly 75 per cent of the total mass of the adrenal cortex. The cells in this zone are polyhedral and much larger than those present in the glomerulosa. They are aligned in long cords, and the cords are arranged radially with respect to the medulla. Usually the cords of cells that are present within the zona fasciculata are only one cell layer thick, and they are separated by

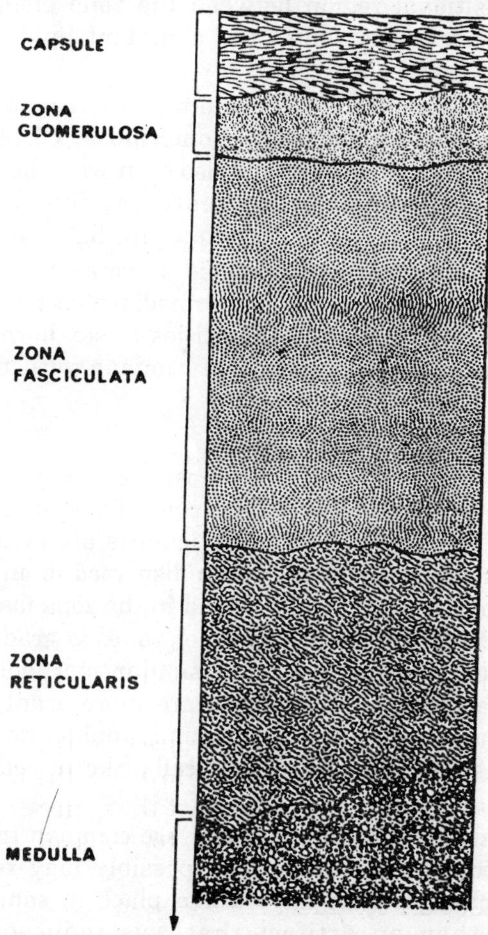

Fig. 7.2. Microscopic anatomy of the adrenal gland.

sinusoidal blood vessels. The cells of the fasciculata have one or two centrally located, spherical nuclei, the cytoplasm is weakly basophilic, and clumps of basophilic material are seen occasionally. The mitochondria are relatively more scarce and more variable in appearance than in the cells of the glomerulosa.

The cells of the fasciculata are filled with lipid droplets normally, but during routine histologic preparations, the lipid is extracted, so that the cells take on a vacuolated appearance. A narrow transitional region between the zona glomerulosa and zona fasciculata that lacks the abundant lipid droplets may also be present.

The smooth endoplasmic reticulum is even more highly developed than that in the zona glomerulosa. Cisternae of granular endoplasmic reticulum are also seen with the electron microscope, and these cisternae represent the clumps of basophilic material that can be observed with light microscopy.

As noted earlier, corticosterone is secreted by the zona fasciculata and the cells of this intermediate cortical region also secrete cortisol and small quantities of sex hormones, a functional property that is shared in common with the zona reticularis.

Zona reticularis

The zona reticularis is the innermost region of the adrenal cortex and comprises roughly 10 per cent of the total cortical tissue mass. The cells of the zona reticularis are arranged in an anastomosing network, rather than dispersed in an ordered parallel array of cords such as are found in the zona fasciculata. The transition between these two inner zones is gradual, and the cells in both regions are quite similar morphologically, save for the fact that lipid droplets are more scanty in the cytoplasm of the reticularis cells. The nongranular endoplasmic reticulum is abundant in reticularis cells like in cells of the fasciculata.

"Light" and "dark" staining nuclei are common in cells of the zona reticularis, and this effect possibly may represent degenerative changes that have taken place in some of the cells, rather than an artifact that was induced during preparation of the tissue for histologic examination. The cells of the zona reticularis normally secrete corticosterone and cortisols as well as minute quantities of sex hormones (androgens and estrogens).

Arterial blood supply, connective tissue, and lymphatics of the adrenal cortex

The adrenal gland is supplied abundantly with blood from several arteries that enter the capsule around its periphery. The major part of the blood supply is received via the superior suprarenal arteries which originate from the inferior phrenic artery. In addition, the middle suprarenal arteries (which arise from the aorta) and the inferior suprarenal arteries (which arise from the renal artery) also provide some blood to the gland. These arteries form a plexus of vessels within the capsule, and the cortical arteries originate from this plexus, thence blood is distributed to an anastomosing network of sinusoids which surround the masses and cords of cells within the three cortical regions. The sinusoids from local areas of the cortex then converge upon a collecting vein within the zona reticularis at the junction of cortex and medulla; consequently, the cortex has no venous system.

In certain regions of the adrenal gland, the suprarenal capsule extends into the substance of the cortex as thick pillars of connective tissue or trabeculae. In addition delicate reticular connective tissue fibers form a complex framework that supports the individual cells of the cortex as well as the medullary tissue. Lymphatic vessels are limited to the capsule and cortical trabeculae, as well as to the connective tissue surrounding the large veins.

A few random nerve fibers are present in the cortex. However, morphologically as well as functionally, the innervation of the adrenals is a topic that properly belongs to a discussion of the adrenal medulla, hence will be considered later.

Morphologic changes in the adrenal cortex induced by alterations in pituitary junction

As mentioned earlier, normal adrenal cortical functions and morphology are dependent upon an adequate supply of the hormone corticotropin from the adenohypophysis. Hypophysectomy or pituitary hypofunction causes marked histologic and functional alterations in the adrenal cortex. A deficiency of corticotropin, regardless of its cause, produces an almost complete atrophy of the zona fasciculata and zona reticularis, but little change is seen in the zona glomerulosa. This effect is prevented or reversed by injections of corticotropin.

Conversely, the administration of large quantities of the glucocorticoid hormone cortisol into intact animals or human patients can suppress the normal secretion of corticotropin by the pituitary so that the inner zones of the cortex atrophy.

Selective hypertrophy of the zona glomerulosa can result from inordinate demands placed upon the homeostatic mechanisms for regulating electrolyte balance. For example, this can be accomplished experimentally by placing an animal on a low sodium-high potassium diet. On the other hand, selective atrophy of the zona glomerulosa results from administration of the mineralocorticoid hormones aldosterone and deoxycorticosterone in large doses.

Possibly the zona glomerulosa is unaffected by corticotropin deficiency because other factors are involved in stimulating aldosterone biosynthesis, although persistent hypofunction of the adenohypophysis ultimately results in degeneration of this cortical region also.

Chemistry of the Adrenal Steroid Hormones

The hormones synthesized by the adrenal cortex are all steroids, and share in common the fact that they are chemical derivatives of the cyclopentanoperhydrophenanthrene nucleus.

About 50 steroids have been isolated in crystalline form from adrenal gland extracts. Almost all of these steroids have a total of either 19 or 21 carbon atoms, and thus are designated as C19 or C21 steroids.

Of the numerous steroid compounds that have been isolated chemically from adrenal cortical tissue in crystralline form, only a few of these are synthesized and secreted in physiologically significant quantities by the normal adrenal gland. These important hormones are classed as either glucocorticoid or mineralocorticoid, depending upon whether the major action of the hormone is on carbohydrate, fat, and protein metabolism or on electrolyte (sodium and potassium) excretion. The principal glucocorticoids are secreted by the adrenals are cortisol (synonyms: 17 α-hydroxycorticosterone, hydrocortisone, compound F) and corticosterone (synonym: compound B). The principal mineralocorticoid secreted by the adrenals is aldosterone.

These three adrenocortical hormones are C21 steroids, and belong to one structural type, chemically speaking. That is, each has a two-carbon side chain attached to the 17 position on the D ring, and thus each has a total of 21 carbon atoms in the molecule. The C21 steroids that have a hydroxyl (—OH) group at the 17 position in the molecule are often referred to as 17-hydroxycorticoids or 17-hydroxycorticosteroids.

A second chemical type of adrenocortical steroid hormone contains a keto (C = O) or hydroxyl (C—OH) group attached directly to the 17 position of the D ring; however only 19 carbon atoms are present in the entire molecule. The physiologically significant C19 steroids generally have a keto group at the 17 position, hence are called 17 ketosteroids. The C19 steroids exhibit androgenic activity, which effect is minimal in physiologically normal women. In adult men, however, this adrenal source of androgen apparently complements the normal testicular hormone output. The principal androgenic compound secreted by the adrenal cortex normally is dehydroepiandrosterone.

Before proceeding into a discussion of the biosynthesis of the adrenocortical hormones, a few additional remarks on the conventions of steroid nomenclature are necessary.

First, the individual rings, and the positions on the rings that may be occupied by various chemical groups, are indicated by letters and numbers respectively.

Second, the chemical groups that lie above the stereochemical plane formed by the steroid ring are designated by the Greek letter (beta) β and a solid line (—R). On the other hand, the chemical groups that lie below this stereochemical plane are designated by the Greek letter alpha (α), and indicated by a dashed line (—R) in the structural formula.

Third the Greek letter delta (Δ) denotes a double bond.

Fourth, the methyl group (—CH_3) in positions 18 and 19 of the steroid molecule usually is indicated only by short vertical lines in the structural formulae.

In the majority of the naturally occurring adrenal steroids, the 17-hydroxy groups occupy the α configuration, whereas the 3-, 11-, and 21-hydroxy groups are present in the β configuration.

The topic of stereoisomerism was discussed briefly in Chapter 1, as it is of considerable importance to the biochemical activity of various metabolites in the body. In this regard, naturally occurring aldosterone is in the D-isomeric form owing to the 18 aldehyde configuration and a-aldosterone is inactive physiologically.

In the presentation to follow, the names in most common usage of the various steroids will be employed, and the synonyms for these compounds will be included as appropriate.

Table 7.1. Steroid Nomenclature

Prefix	*Suffix*	*Signature*
allo-	—	*Trans* as opposed to *cis:* configuration of A and B rings
cis-	—	Arrangement of two groups in same plane
trans-	—	Arrangement of two groups in opposing planes
dehydro-	—	Conversion of —C—OH to —C=O by loss of two hydrogen atoms
dihydro-	—	Addition of two hydrogen atoms
α-	—	A group trans to the 19 methyl radical
β-	—	A group cis to the 19 methyl.
nor-	—	One less carbon in side chain compared to parent molecule
oxy-	-one, -dione	Ketones
hydroxy-dihydroxy	-ol, diol	Alcohols
—	-ane	A saturated carbon atom
—	-ene	A single double bond in the ring structure

Comparative Morphology

Adrenal tissues are present in all vertebrates from cyclostomes to mammals, but profound differences are encountered in the arrangement of the functional components, *i.e.,* steroid-producing cells and catecholamine-producing cells. Histologic studies indicate that the two kinds of adrenal tissue coexist in cyclostomes: cells, presumed to be steroidogenic,

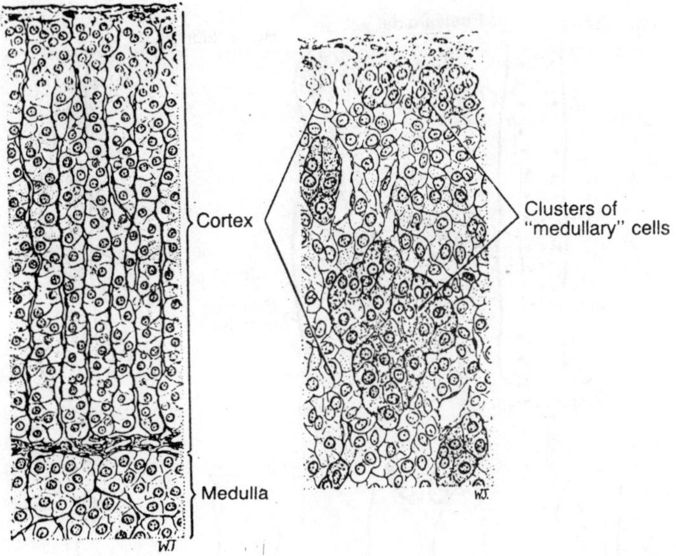

Fig. 7.3. A section through part of the adrenal gland in a mammal (rat), with division into cortical and medullary layers, and a reptile (Heloderma), in which the two tissues are intermingled.

are scattered along the walls of the postcardinal veins and in the mesonephric kidneys; small clusters of chromaffin cells are found in the same areas, and these occasionally come in contact with the steroidogenic cells. In elasmobranchs, the steroidogenic tissue is condensed into several well-formed bodies lying between the caudal ends of the kidneys; these are the *interrenal glands*. Paired aggregations of chromaffin cells are present between the kidneys, the more posterior ones being embedded in the kidneys. The two components of the chondrichthyean adrenal are typically separated, though small islets of chromaffin cells have been described in the interrenal glands of the ray *Raja clavata*. It should be noted that the chondrichthyes are the only vertebrates having an adrenal component which is accurately described as an "interrenal" gland, meaning located between the kidneys, and these fishes are not in the main evolutionary line leading to tetrapods.

Great variation is found among actinopterygian fishes with respect to the condensation and dispersal of the adrenal

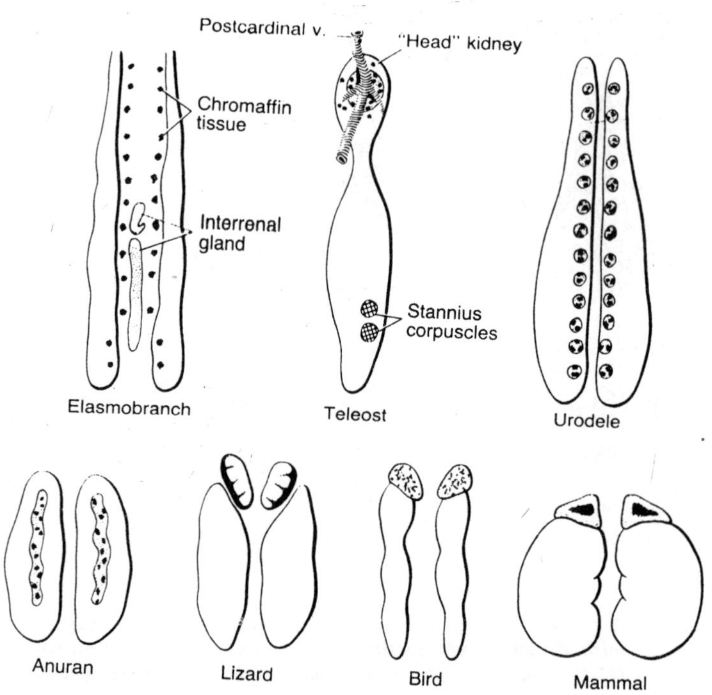

Fig. 7.4. Comparative morphology of the adrenal tissues of vertebrates.

tissues. These are generally located within or just anterior to the "cephalic kidneys," and occur around the postcardinal veins and their branches. In many species, the so-called "cephalic kidneys" consist largely of lymphoid or myeloid tissue, or both, and hence are blood-forming organs. The two types of adrenal cells are entirely separated in certain species (*Salmo*), but in others they are intermingled (*Cottus*). Branches of the postcardinal veins may be lined with chromaffin cells which are surrounded by steroidogenic cells.

The corpuscles of Stannius are bodies within the mesonephric kidneys of some teleosts and are thought to arise as proliferations from the urinary ducts. Some workers have detected adrenocortical steroids in these corpuscles and regard them as holocrine glands which playa physiologic role comparable to the adrenal cortices of mammals. Other workers have not been able to identify steroids in these

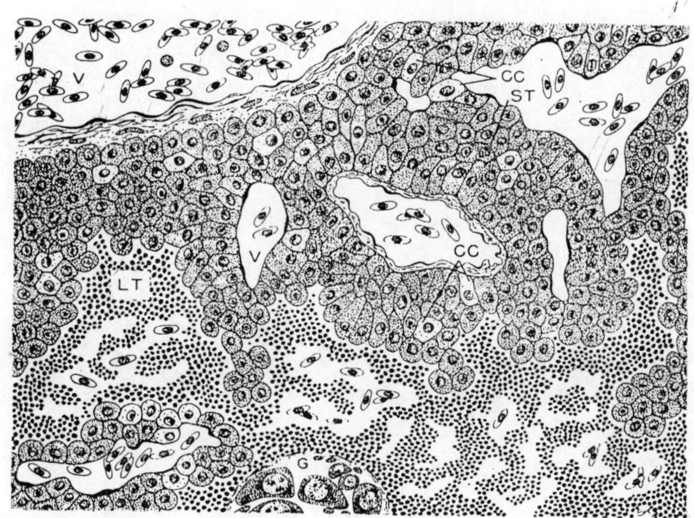

Fig. 7.5. Histology of the carp adrenal gland. CC—chromaffin cells; G—ganglion; LT—lymphoid tissue of "head" kidney; ST—steroidogenic tissue; V—vein containing red corpuscles.

bodies. The fact that they differ in embryonic origin and mode of secretion from established steroidogenic tissue in teleosts has led many investigators to abandon the view that they are comparable to adrenal cortices.

The amphibian adrenal consists of rather discrete bodies along the ventral surfaces of the kidneys, and the chromaffin and steroidogenic cells are interspersed. At certain points, the adrenal tissue may be completely buried within the kidney. The steroidogenic cells are organized into cords which are separated by conspicuous blood sinuses; the chromaffin cells occur singly or in small clusters and many neurons are present throughout the gland. A very characteristic feature of the adrenals of aquatic anurans is the presence of many "Stilling cells" scattered throughout the organs. These cells resemble mast cells and are PAS-positive and eosinophilic. The functional role of the Stilling cells is unknown, but they are said to be absent in three genera of terrestrial toads.

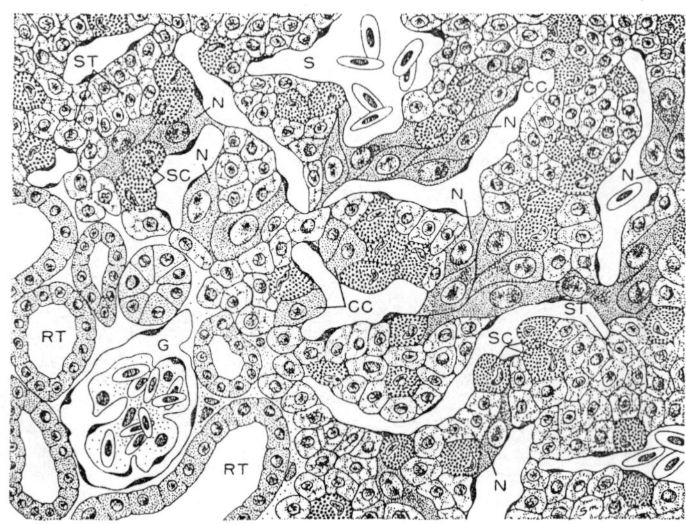

Fig. 7.6. Histology of the adrenal gland of Rana pipiens. CC—chromaffin cells; G—glomerulus of kidney; N—neuron; RT—renal tubule; S—sinusoid containing red corpuscles; SC—Stilling cells; ST—steroidogenic tissue.

The adrenal glands of reptiles and birds are more compact than in lower forms, and the two types of tissue are intermingled. The adrenals of turtles are discrete bodies located on the anterior ventral surfaces of the kidneys; in snakes, the adrenals are some distance anterior to the kidneys. In certain lizards and snakes, the chromaffin cells aggregate to form a distinct band of tissue at the periphery of the organ, partly surrounding the central mass of steroidogenic tissue. The avian adrenals are located near the anterior poles of the kidneys and consist of cords of steroidogenic tissue and clusters of chromaffin cells. It is only in mammals that distinct cortices and medullae are present, and even here, there may be considerable interdigitation and intermingling of the two kinds of tissue.

Studies on the adrenal gland of the rat suggest that the production of epinephrine by the medullary cells is regulated

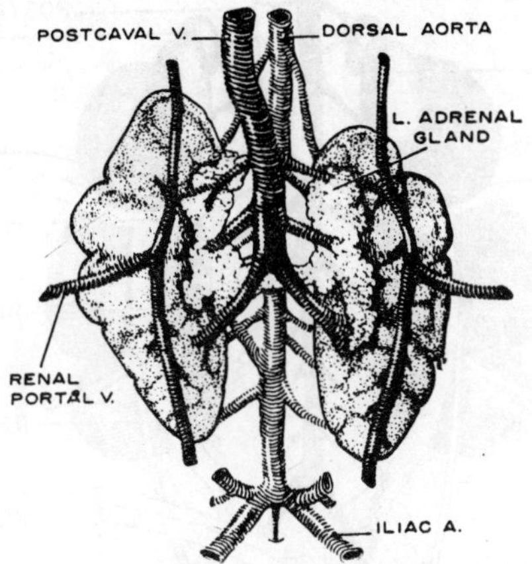

Fig. 7.7. Ventral view of the adrenal glands of the turtle (Pseudemys scripta), showing the main blood vessels of the kidney region.

by the pituitary-adrenocortical system. The conversion of norepinephrine to epinephrine requires the transfer of a methyl group, and this process is catalyzed by the enzyme phenylethanolamine-N-methyl transferase (PNMT). The activity of this medullary enzyme is markedly reduced after hypophysectomy, but may be restored by administering either ACTH or glucocorticoids. The ACTH produces its effect by

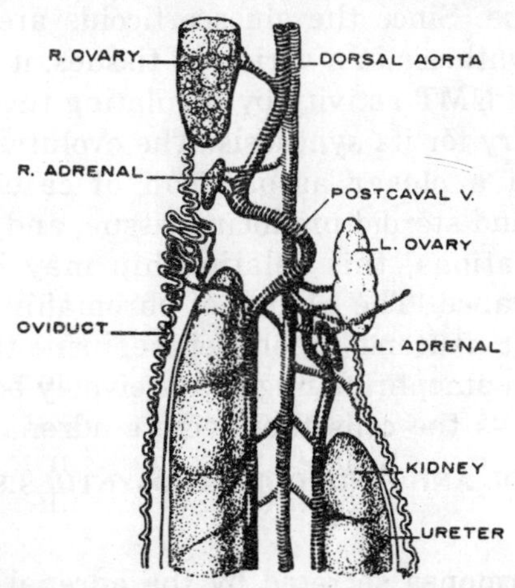

Fig. 7.8. Adrenal glands and adnexa of the snake (Natrix), from ventral view.

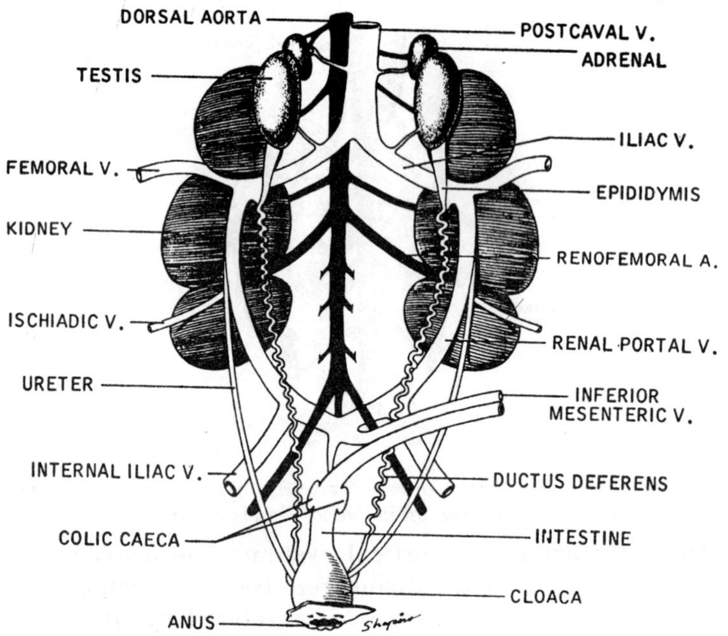

Fig. 7.9. Urnogenital organs of the male pigeon (Columba livia). Both testes are retracted laterally to expose the adrenal glands.

increasing the availability of glucocorticoids from the steroidogenic tissue. Since the glucocorticoids are known to promote protein synthesis in a variety of tissues, it is probable that they elevate PNMT activity by regulating the conditions which are necessary for its synthesis. The evolutionary trend has been toward a closer association of catecholamine producing tissue and steroid-producing tissue, and, in light of the above observations, this relationship may have some functional significance. The ability of chromaffin cells, and possibly adrenergic neurons, to form epinephrine through the methylation of norepinephrine might conceivably be enhanced by their nearness to the cells that secrete adrenal steroids.

Cortisol and Androgen Biosynthesis

Steroidogenesis

The major hormones secreted by the adrenal cortex are cortisol, the androgens, and aldosterone. The carbon atoms in the steroid molecule are numbered.

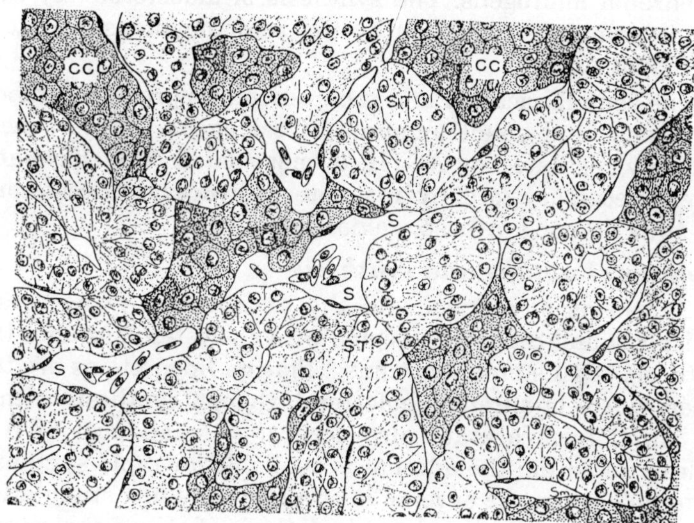

Fig. 7.10. Adrenal glands and adnexa of the snake (Natrix), from ventral view.

Zones and steroidogenesis

Because of enzymatic differences between the zona glomerulosa and the inner 2 zones, the adrenal cortex functions as 2 separate units, with differing regulation and secretory products. Thus, the zona glomerulosa, which produces aldosterone, lacks 17 α-hydroxylase activity and

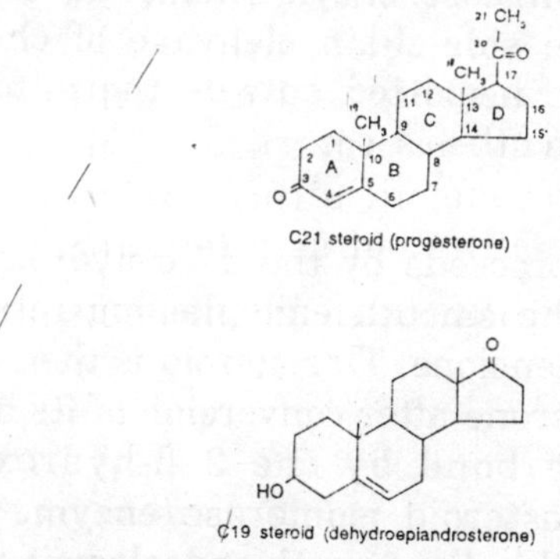

Fig. 7.11. Structure of adrenocortical steroids.

cannot synthesize 17 α-hydroxypregnenolone and 17 α-hydroxyprogesterone, which are the precursors of cortisol and the adrenal androgens. The synthesis of aldosterone by this zone is primarily regulated by the reninangiotensin system and by potassium.

The zona fasciculata and zona reticularis produce cortisol, androgens, and small amounts of estrogens. These zones, primarily regulated by ACTH, do not contain the enzymatic system necessary to dehydrogenate IS-hydroxycorticosterone and thus do not synthesize aldosterone.

Cholesterol uptake and synthesis

Synthesis of cortisol and the androgens by the zonae fasciculata and reticularis begins with cholesterol, as does the synthesis of all steroid hormones. Plasma lipoproteins are the major source of adrenal cholesterol, though synthesis within the gland from acetate also occurs. A small pool of free cholesterol within the adrenal is available for rapid synthesis of steroids when the adrenal is stimulated. When stimulation occurs, there is also increased hydrolysis of stored cholesterol esters to free cholesterol, increased uptake from plasma lipoproteins, and increased cholesterol synthesis within the gland.

Cholesterol metabolism

The conversion of cholesterol to pregnenolone is the rate-limiting step in adrenal steroidogenesis and the major site of ACTH action on the adrenal. This step occurs in the mitochondria and is mediated by cholesterol 20, 22-hydroxylase: 20, 22-desmolase enzyme complex, involving 2 hydroxylations and the side-chain cleavage of cholesterol. Pregnenolone is then transported outside the mitochondria before further steroid synthesis occurs.

Synthesis of cortisol

Cortisol synthesis proceeds by the 17 α-hydroxylation of pregnenolone within the smooth endoplasmic reticulum to form 17 α-hydroxypregnenolone. This steroid is then converted to 17 a-hydroxyprogesterone after conversion of its 5,6 double bond to a 4,5 double bond by the 3 β-hydroxysteroid dehydrogenase: Δ^5- oxosteroid isomerase enzyme complex, which is also located within the smooth endoplasmic reticulum.

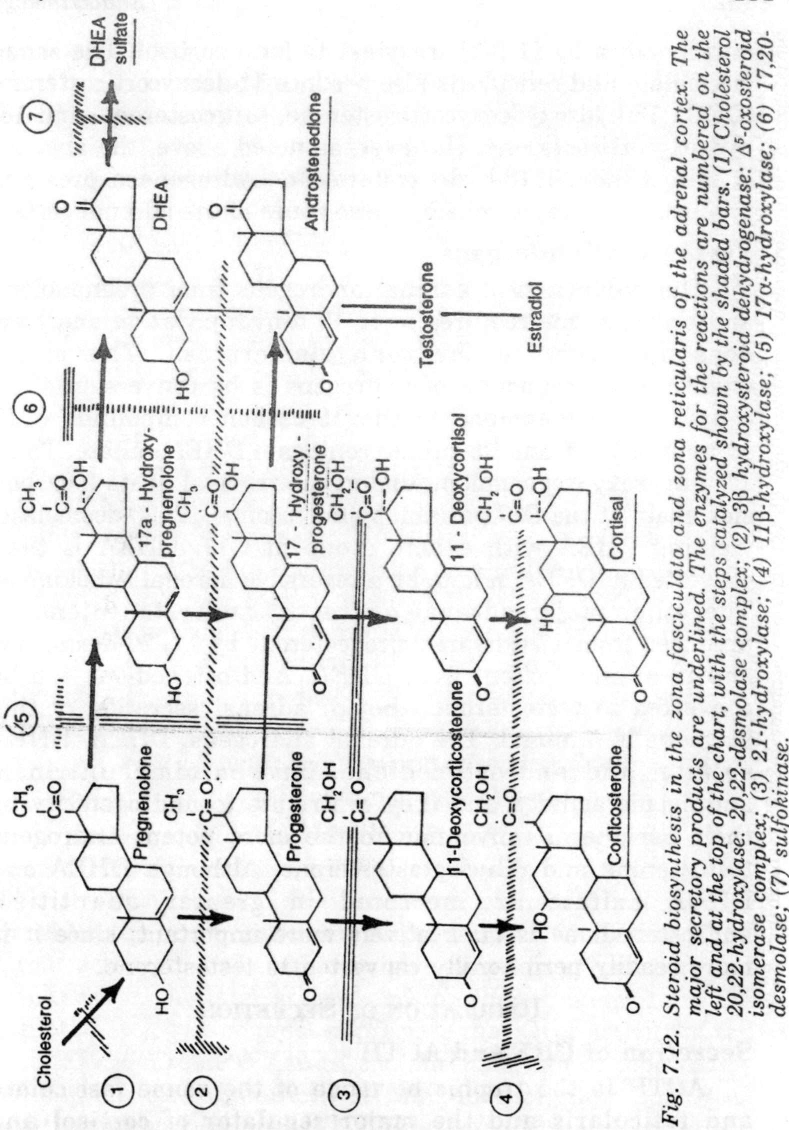

Fig. 7.12. Steroid biosynthesis in the zona fasciculata and zona reticularis of the adrenal cortex. The major secretory products are underlined. The enzymes for the reactions are numbered on the left and at the top of the chart, with the steps catalyzed shown by the shaded bars. (1) Cholesterol 20,22-hydroxylase: 20,22-desmolase complex; (2) 3β hydroxysteroid dehydrogenase: Δ^5-oxosteroid isomerase complex; (3) 21-hydroxylase; (4) 11β-hydroxylase; (5) 17α-hydroxylase; (6) 17,20-desmolase; (7) sulfokinase.

An alternative but apparently less important pathway in the zonae fasciculata and reticularis is from pregnenolone →progesterone → 17 α-hydroxyprogesterone.

The next step, which is again microsomal, involves the 21-hydroxylation of 17 α-hydroxyprogesterone to form 11-deoxycortisol; this compound is further hydroxylated within mitochondria by 11 β-hydroxylase to form cortisol. The zonae fasciculata and reticularis also produce 11-deoxycorticosterone (DOC), 18-hydroxydeoxycorticosterone, corticosterone, and 18-hydroxycorticosterone. However, as noted above, the absence of mitochondrial 18-hydroxysteroid dehydrogenase prevents production of aldosterone by these zones of the adrenal cortex.

Synthesis of androgens

The production of adrenal androgens from pregnenolone and progesterone requires prior 17 α-hydroxylation and thus does not occur in the zona glomerulosa. The major quantitative production of androgens is by conversion of 17 α-hydroxypregnenolone to the 19-carbon compounds (C19 steroids) DHEA and its sulfate conjugate DHEA sulfate. Thus, 17 a-hydroxypregnenolone undergoes removal of its 2-carbon side chain at the C17 position by microsomal 17, 20-desmolase, yielding DHEA with a keto group at C17. DHEA is then converted to DHEA sulfate by a reversible adrenal sulfokinase. The other major adrenal androgen, androstenedione, is produced from 17 α-hydroxyprogesterone by 17, 20-desmolase and to a lesser extent from DHEA. Androstenedione can be converted to testosterone, though adrenal secretion of this hormone is minimal. The adrenal androgens, DHEA, DHEA sulfate, and androstenedione, have minimal intrinsic androgenic activity, and they contribute to androgenicity by their peripheral conversion to the more potent androgens testosterone and dihydrotestosterone. Although DHEA and DHEA sulfate are secreted in greater quantities, androstenedione is qualitatively more important, since it is more readily peripherally converted to testosterone.

Regulation of Secretion

Secretion of CRF and ACTH

ACTH is the trophic hormone of the zonae fasciculata and reticularis and the major regulator of cortisol and androgen production by the adrenal cortex. ACTH in turn is regulated by the hypothalamus and central nervous system via neurotransmitters and corticotropin releasing factor (CRF). The neuroendocrine control of CRF and ACTH secretion is discussed in Chapters 3 and 4 and involves mechanisms to be discussed below.

ACTH effects on the Adrenal Cortex

The delivery of ACTH to the adrenal cortex leads to the rapid synthesis and secretion of steroids; plasma levels of these hormones rise within minutes after ACTH administration. Chronic stimulation increases protein, RNA, and DNA synthesis, leading to adrenocortical hyperplasia and hypertrophy; conversely, ACTH deficiency results in decreased steroidogenesis and is accompanied by adrenocortical atrophy, decreased gland weight, and decreased protein and nucleic acid content.

ACTH and Steroidogenesis

ACTH binds to high-affinity plasma membrane receptors of adrenocortical cells, thereby activating adenylate cyclase and increasing cAMP, which in turn activates intracellular phosphoprotein kinases. This process stimulates the rate-limiting step of cholesterol to Δ^5-pregnenolone conversion and initiates steroidogenesis. ACTH may also have other actions, although these are less well documented. The exact mechanisms of ACTH stimulation of the desmolase enzyme complex are unknown, as are their relative importance; however, ACTH has a number of effects, including increased free cholesterol formation as a consequence of increased cholesterol esterase activity and decreased cholesterol ester synthetase; increased lipoprotein uptake by the adrenal cortex; increased content of certain phospholipids, which may increase cholesterol sidechain cleavage; and increased binding of cholesterol to the cytochrome P-450 cholesterol cleavage enzyme in mitochondria.

Neuroendocrine Control

Cortisol secretion is closely regulated by ACTH, and plasma cortisol levels parallel those of ACTH. There are 3 mechanisms of neuroendocrine control: (1) episodic secretion and the circadian rhythm of ACTH, (2) stress responsiveness of the hypothalamic-pituitary adrenal axis, and (3) feedback inhibition by cortisol of ACTH secretion.

Circadian rhythm

Circadian rhythm is superimposed on episodic secretion; it is the result of central nervous system events that regulate both the number and magnitude of CRF and ACTH secretory

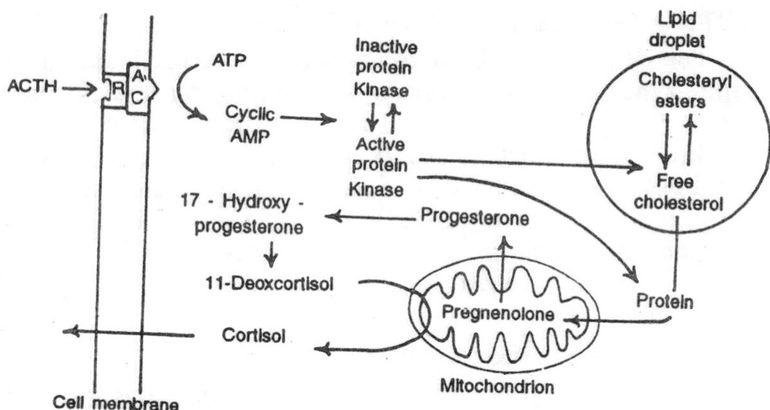

Fig. 7.13. Mechanism of action of ACTH on cortisol-secretion cells of the adrenal cortex. AC—adenylate cyclase; R—receptor.

episodes. Cortisol secretion is low in the late evening and continues to decline in the first several hours of sleep, at which time plasma cortisol levels may be undetectable. During the third and fifth hours of sleep, there is an increase in the secretion; but the major secretory episodes begin in the sixth to eighth hours of sleep and then begin to decline as wakefulness occurs. About half of the total daily cortisol output is secreted during this period. Cortisol secretion then gradually

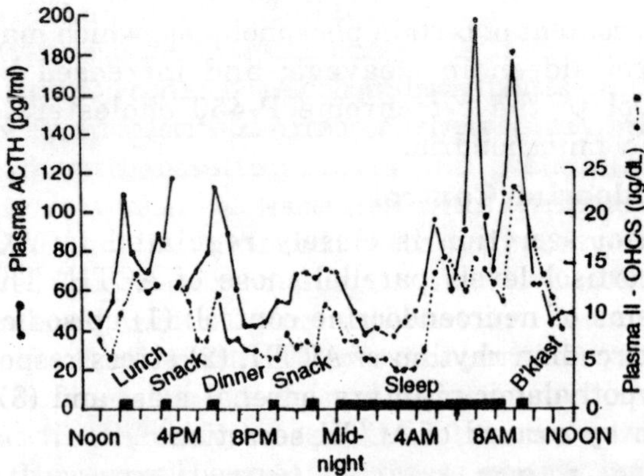

Fig. 7.14. Fluctuations in plasma ACTH and glucocorticoids (11 QHCS) throughout the day. Note the greater ACTH and glucocorticoid rises in the morning before awakening.

declines during the day, with fewer secretory episodes of decreased magnitude, however, there is increased cortisol secretion in response to eating.

Alterations in the circadian rhythm

Although this general pattern is consistent, there is considerable intra- and inter individual variability, and the circadian rhythm may be altered by changes in sleep pattern, light-dark exposure, and feeding times. The rhythm is also changed by (1) physical stresses such as major illness, surgery or trauma, or starvation; (2) psychologic stress, including severe anxiety, endogenous depression, and the manic phase of manic-depressive psychosis; (3) central nervous system and pituitary disorders; (4) Cushing's syndrome; (5) liver disease and other conditions that affect cortisol metabolism; (6) chronic renal failure; and (7) alcoholism. Cyproheptadine inhibits the circadian rhythm, possibly by its antiserotoninergic effects, whereas other drugs usually have no effect.

Stress responsiveness

Plasma ACTH and cortisol secretion are also characteristically responsive to physical stress. Thus, plasma ACTH and cortisol are secreted within minutes following the onset of stresses such as surgery and hypoglycemia, and these responses abolish circadian periodicity if the stress is prolonged. Stress responses originate in the central nervous system and increase hypothalamic CRF and thus pituitary

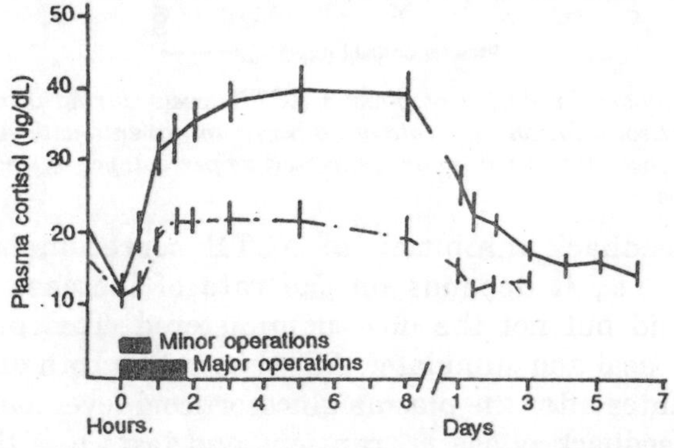

Fig. 7.15. Plasma cortisol responses to major surgery (continuous line) and minor surgery (broken line) in normal subjects. Mean values and standard errors for 20 patients are shown in each case.

ACTH secretion. Stress responsiveness of plasma ACTH and cortisol is abolished by prior high-dose glucocorticoid administration and in spontaneous Cushing's syndrome; conversely, the responsiveness of ACTH secretion is enhanced following adrenalectomy.

Feedback inhibition

The third major regulator of ACTH and cortisol secretion is that of feedback inhibition by glucocorticoids of CRF, ACTH, and cortisol secretion. Glucocorticoid feedback inhibition occurs at both the pituitary and hypothalamus and involves 2 distinct mechanisms—fast and delayed feedback inhibition.

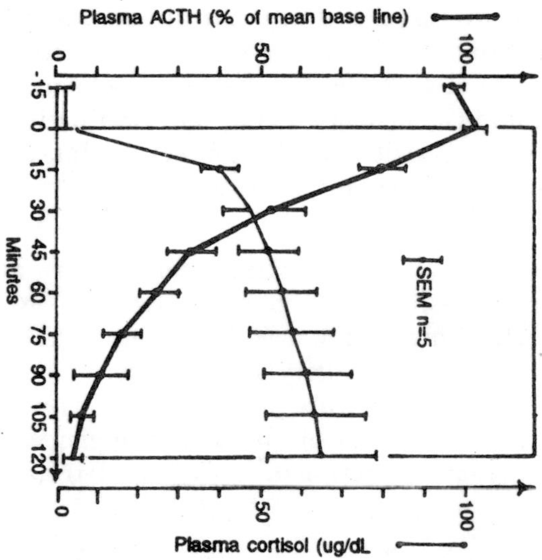

Fig. 7.16. Feedback inhibition of plasma ACTH levels during intravenous cortisol infusion at a rate of 50 mg/h in patients with Addison's disease. ACTH values are expressed as percentages of mean basal levels.

Fast feedback inhibition of ACTH secretion is rate-dependent, *i.e.,* it depends on the rate of increase of the glucocorticoid but not the dose administered. This phase is rapid, and basal and stimulated ACTH secretion both diminish within minutes after the plasma glucocorticoid level increases. This fast feedback phase is transient and lasts less than 10 minutes, suggesting that this effect is not mediated via cytosolic glucocorticoid receptors but rather via actions on the cell membrane.

Delayed feedback inhibition after the initial rate-dependent effects of glucocorticoids further suppresses CRF and ACTH secretion by mechanisms that are both time- and dose-dependent. Thus, with continued glucocorticoid administration, ACTH levels continue to decrease and become unresponsive to stimulation. The ultimate effect of prolonged glucocorticoid administration is suppression of CRF and ACTH release and atrophy of the zonae fasciculata and reticularis as a consequence of ACTH deficiency. The suppressed hypothalamic-pituitary-adrenal axis fails to respond to stress and stimulation. Delayed feedback appears to act via the classic glucocorticoid receptor (see below), thus reducing synthesis of the messenger RNA for the precursor to ACTH.

Regulation of androgen production

Adrenal androgen production in adults is also regulated by ACTH; both DHEA and androstenedione exhibit circadian periodicity in concert with ACTH and cortisol. In addition, plasma concentrations of DHEA and androstenedione increase rapidly with ACTH administration and are suppressed by glucocorticoid administration, confirming the role of endogenous ACTH in their secretion. DHEA sulfate, because of its prolonged metabolic clearance rate, does not exhibit a diurnal rhythm. Thus, adrenal androgen secretion is regulated by ACTH, and, in general, the secretion of these hormones parallels that of cortisol.

Circulation of Control and Adrenal Androgens

Cortisol and the adrenal androgens circulate bound to plasma proteins. The plasma half-life of cortisol is 70-90 minutes; this is determined by the extent of plasma binding and by the rate of metabolic inactivation.

Plasma Binding Proteins

Cortisol and adrenal androgens are secreted in an unbound state; however, these hormones bind to plasma proteins upon entering the circulation. Cortisol binds mainly to corticosteroid-binding globulin (CBG, transcortin) and to a lesser extent to albumin, whereas the androgens bind chiefly to albumin. The precise physiologic role of plasma protein binding is unclear, since bound steroids are biologically inactive and the active fraction is that which circulates unbound. Furthermore, CBG is not required for cortisol transport to target tissues nor for full biologic activity of the steroid. The plasma proteins may provide a pool of circulating cortisol by delaying metabolic clearance, thus preventing more marked fluctuations of plasma free cortisol levels during episodic secretion by the gland.

Free and Bound Cortisol

In basal conditions; about 10% of the circulating cortisol is free, about 75% is bound to CBG, and the remainder is bound to albumin. The plasma free cortisol level is approximately 1 µg/ dL, and it is this biologically active cortisol which is regulated by ACTH.

Corticosteroid-binding globulin (CBG)

CBG has a molecular weight of about 50,000, is produced by the liver, and binds cortisol with high affinity. The CBG in plasma has a cortisol-binding capacity of about 25 µg/dL. When total plasma cortisol concentrations rise above this level, the free concentration rapidly increases and exceeds its usual fraction of 10% of the total cortisol. Other endogenous steroids do not appreciably affect cortisol binding to CBG under usual circumstances; an exception is in late pregnancy, when progesterone may occupy about 25% of the binding sites on CBG. Synthetic steroids do not bind significantly to CBG—with the exception of prednisolone, which has a high affinity. CBG levels are increased in high-estrogen states (pregnancy; estrogen or oral contraceptive use), hyperthyroidism, diabetes, certain hematologic disorders, and on a genetic basis. CBG concentrations are decreased in familial CBG deficiency, hypothyroidism, and protein deficiency states such as severe liver disease or nephrotic syndrome.

Albumin

Albumin has a much greater capacity for cortisol binding but a lower affinity. It normally binds about 15% of the circulating cortisol and this proportion increases when the total cortisol concentration exceeds the CBG binding capacity. Synthetic glucocorticoids are extensively bound to albumin; e.g., 77% of dexamethasone in plasma is bound to albumin.

Androgen binding

Androstenedione, DHEA, and DHEA sulfate circulate weakly bound to albumin. However, testosterone is bound extensively to a specific globulin, sex hormone-binding globulin (SHBG), also called testosterone-estrogen-binding globulin (TEBG).

Metabolism of Adrenal Steroids

The metabolism of the steroids renders them inactive and increases their water solubility, as does their subsequent conjugation with glucuronide or sulfate groups; and these inactive conjugated metabolites are more readily excreted by the kidney. The liver is the major site of steroid catabolism and conjugation, and 90% of these metabolized steroids are excreted by the kidney.

Conversion and Excretion of Cortisol

Cortisol is modified extensively before excretion in urine; less than 1 % of secreted cortisol appears in the urine unchanged.

Hepatic conversion

Hepatic metabolism of cortisol involves a number of metabolic conversions of which the most important (quantitatively) is the irreversible inactivation of the steroid by Δ^4-reductases, which reduce the 4,5 double bond of the A ring. Dihydrocortisol, the product of this reaction, is then converted to tetrahydrocortisol by a 3-hydroxysteroid dehydrogenase. Cortisol is also converted extensively to cortisone, which is then metabolised by the enzymes described above to yield tetrahydrocortisone. Tetrahydrocortisol and tetrahydrocortisone can be further altered to form the carotic acids. These conversions result in the excretion of approximately equal amounts of cortisol and cortisone metabolites. Cortisol and cortisone are also metabolized to the cortols and cortolones and to a lesser extent by other pathways. Cortisol is also converted to 6 β-hydroxycortisol which is very water-soluble and excreted unchanged in the urine.

Hepatic conjugation

Over 95% of cortisol and cortisone metabolites are conjugated by the liver and then reenter the circulation to be excreted in the urine. Conjugation is mainly with glucuronic acid at the 3 α-hydroxyl position and to a lesser extent as the sulfate at the 21-hydroxyl group.

Variations in clearance and metabolism

The metabolism of cortisol is altered by a number of circumstances. It is decreased in infants and in the elderly. It is impaired in chronic liver disease, leading to decreased renal excretion of cortisol metabolites; however, the plasma cortisol level remains normal. Hypothyroidism decreases both metabolism and excretion; conversely, hyperthyroidism accelerates these processes. Cortisol clearance may be reduced in starvation and anorexia nervosa and is also decreased in pregnancy because of the elevated CBG levels. The metabolism of cortisol to 6 β-hydroxycortisol is increased in the neonate, in pregnancy, with estrogen therapy, and in patients with lever disease or severe chronic illness. Cortisol metabolism by this pathway is also increased by drugs that induce hepatic microsomal enzymes, including barbiturates, phenytoin, mitotane *(o,p'* -DDD), aminoglutethimide, and rifampin. These alterations generally are of minor physiologic importance, since secretion, plasma levels, and cortisol half-life are normal. However, they result in decreased excretion of the urinary metabolites of cortisol measured as 17-hydroxycorticosteroids. These conditions and drugs have a greater influence on the metabolism of synthetic glucocorticoids and may result in inadequate plasma levels of the administered glucocorticoid because of rapid clearance and metabolism.

Conversion and Excretion of Adrenal Androgens

Adrenal androgen metabolism results either in degradation and inactivation or the peripheral conversion of these weak androgens to their more potent derivatives testosterone and dihydrotestosterone. DHEA is readily converted within the adrenal to DHEA sulfate, the adrenal androgen secreted in greatest amount. DHEA secreted by the gland is also converted to DHEA sulfate by the liver and kidney, or it may be converted to Δ^4-androstenedione. DHEA sulfate may be excreted without further metabolism; however, both it and DHEA are also metabolized to 7α- and 16α-hydroxylated derivatives and by 17β- reduction to Δ^5-androstenediol and its sulfate. Androstenedione is converted either to testosterone or by reduction of its 4,5 double bond to etiocholanolone or androsterone, which may be further converted by 17β-reduction to etiocholanediol and androstanediol, respectively. Testosterone is converted to dihydrotestosterone in androgen-sensitive tissues by 5α-reduction, and it in turn is mainly metabolized by 3α-reduction to androstanediol. The metabolites of these androgens are conjugated either as glucuronides or sulfates and excreted in the urine.

Biological Effect—Glucocorticoids

General Considerations

Although glucocorticoids were originally so called because of their influence on glucose metabolism, they are currently defined as steroids that exert their effects by binding to specific cytosolic receptors which mediate the actions of these hormones. These glucocorticoid receptors are present in virtually all tissues, and glucocorticoid-receptor interaction is responsible for most of the known effects of these steroids. Alterations in the structure of the glucocorticoids have led to the development of synthetic compounds with greater glucocorticoid activity. The increased activity of these compounds is due to increased affinity for the glucocorticoid receptors and delayed plasma clearance, which increases tissue exposure. In addition, many of these synthetic glucocorticoids have negligible mineralocorticoid effects and thus do not result in sodium retention, hypertension, and hypokalemia. This secretion describes the molecular mechanisms of glucocorticoid action and the effects on individual metabolic functions and tissues.

Molecular Mechanisms

Glucocorticoid receptors

Glucocorticoid action is initiated by entry of the steroid into the cell and binding to the cytosolic glucocorticoid receptor proteins. After binding, activated hormone-receptor complexes enter the nucleus and interact with nuclear chromatin acceptor sites. These events cause the expression of specific genes and the transcription of specific mRNAs. The resulting proteins affect the glucocorticoid response, which may be inhibitory or stimulatory depending on the specific tissue affected. Although glucocorticoid receptors are similar in many tissues, the proteins synthesized vary widely and are the result of expression of specific genes in different cell types. The mechanisms of this specific regulation are unknown.

Other mechanisms

Although interaction of glucocorticoids with cytosolic receptors and their subsequent stimulation of gene expression are responsible for most glucocorticoid effects, other effects may occur through different mechanisms. The most significant example is that of glucocorticoid-induced fast feedback inhibition of ACTH secretion. This effect occurs within minutes of glucocorticoid administration, and its rapidity suggests that it is not due to RNA and protein synthesis but rather to glucocorticoid-induced changes in secretory function or cell membranes.

Glucocorticoid Agonists and Antagonists

The study of glucocorticoid receptors has led to the definition of glucocorticoid agonists and antagonists. These studies have also identified a number of steroids with mixed effects termed partial against, partial antagonist, or partial agonist-partial antagonist

Agonists

In humans, cortisol, synthetic glucocorticoids (*e.g., prednisolone, dexamethasone*), corticosterone, and aldosterone

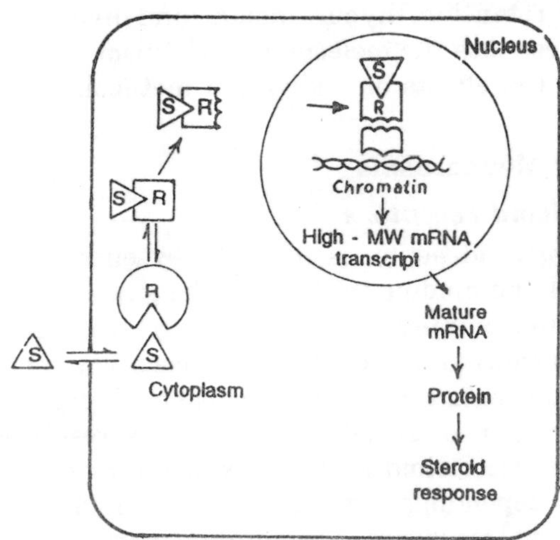

Fig. 7.17. Steps in steroid hormone action. Activation of the intracellular receptors by steroid hormones is followed by nuclear binding of the complex and stimulation of mRNA synthesis. R—receptor; S—steroid hormone.

are glucocorticoid agonists. The synthetic glucocorticoids have substantially higher affinity for the glucocorticoid receptor, and these have greater glucocorticoid activity than cortisol when present in equimolar concentrations. Corticosterone and aldosterone have substantial affinity for the glucocorticoid receptor; however, their plasma concentrations are normally much lower than that of cortisol, and thus these steroids do not have significant physiologic glucocorticoid effects.

Antagonists

Glucocorticoid antagonists bind to the glucocorticoid receptors but do not elicit the nuclear events required to cause a glucocorticoid response. These steroids compete with agonist steroids such as cortisol for the receptors and thus inhibit agonist responses. Other steroids have partial agonist activity when present alone, i.e., they elicit a partial glucocorticoid response. However, insufficient concentration, they compete with agonist steroids for the receptors and thus competitively inhibit agonist responses; i.e., these partial agonists may function as partial antagonists in the presence of more active glucocorticoids. Steroids such as progesterone, 11-deoxycortisol, DOC, testosterone, and 17 β-estradiol have antagonist or partial agonist-partial antagonist effects; however, the physiologic role of these hormones in glucocorticoid action is probably negligible, because they circulate in low concentrations.

Intermediary Metabolism

Glucocorticoid in general inhibit DNA synthesis. In addition, in most tissues they inhibit RNA and protein synthesis and accelerate protein catabolism. These actions provide substrate for intermediary metabolism; however, accelerated catabolism also accounts for the deleterious effects of glucocorticoids on muscle, bone, connective tissue, and lymphatic tissue. In contrast, RNA and protein synthesis in liver is stimulated.

Hepatic glucose metabolism

Glucocorticoids increase hepatic gluconeogenesis by stimulating the gluconeogenic enzymes phosphoenolpyruvate carboxykinase and glucose-6-phosphatase. They also have a permissive effect in that they increase hepatic responsiveness to the gluconeogenic hormones (glucagon, catecholamines), and they also increase the release of substrates from peripheral tissues, particularly muscle. This latter effect may be enhanced by the glucocorticoid-induced reduction in peripheral amino acid uptake and protein synthesis. Glucocorticoids also increase glycerol and free fatty acid release by lipolysis and increase muscle lactate release. These steroids also enhance hepatic glycogen synthesis and storage by stimulating glycogen synthetase activity and to a lesser extent by inhibiting glycogen breakdown. These effects are insulin-dependent.

Peripheral glucose metabolism

Glucocorticoids also alter carbohydrate metabolism by inhibiting peripheral glucose uptake in muscle and adipose tissue. This effect and the others described above may result in increased insulin secretion in states of chronic glucocorticoid excess.

Effects of adipose tissue

In adipose tissue, the predominant effect is increased lipolysis with release of glycerol and free fatty acids. This is partially due to direct stimulation of lipolysis by glucocorticoids, but it is also contributed to by decreased glucose uptake and enhancement by glucocorticoids of the effects of lipolytic hormones. Although glucocorticoids are lipolytic, increased fat deposition is a classic manifestation of glucocorticoid excess. This paradox may be explained by the increased appetite caused by high levels of these steroids and by the lipogenic effects of the hyperinsulinemia that occurs in this state. The reason for abnormal (central) fat deposition in states of cortisol excess is unknown.

Summary

The effects of the glucocorticoids on intermediary metabolism can be summarized as follows: (1) Effects are minimal in the fed state. However, during fasting, glucocorticoids contribute to the maintenance of plasma glucose levels by increasing gluconeogenesis, glycogen deposition, and the peripheral release of substrate. (2) Hepatic glucose production is enhanced, as is hepatic RNA and protein synthesis. (3) The effects on muscle are catabolic; i.e., decreased glucose uptake and metabolism, decreased protein synthesis, and increased release of amino acids. (4) In adipose tissue, lipolysis is stimulated. (5) In glucocorticoid deficiency, hypoglycemia may result, whereas in states of glucocorticoid excess there may be hyperglycemia, hyperinsulinemia, muscle wasting, and weight gain with abnormal fat distribution.

Effects on Other Tissues and Functions

Connective tissue

Glucocorticoids in excess inhibit fibroblasts, lead to loss of collagen and connective tissue, and thus result in thinning of the skin, easy bruising, stria formation, and poor wound healing.

Bone

The physiologic role of glucocorticoids in bone metabolism and calcium homeostasis is unknown; however, in excess, they have major deleterious effects. Glucocorticoids directly inhibit bone formation by decreasing cell proliferation and the synthesis of RNA, protein, collagen, and hyaluronate. Glucocorticoids also directly stimulate bone-resorbing cells, leading to osteolysis and increased urinary hydroxyproline excretion. In addition, they potentiate the actions of PTH and 1,25-dihydroxycholecalciferol ($1,25[OH]_2D_3$) on bone, and this may further contribute to net bone resorption.

Calcium metabolism

Glucocorticoids also have other major effects on mineral homeostasis. They markedly reduce intestinal calcium absorption, which tends to lower serum calcium. This results in a secondary increase in PTH secretion, which maintains serum calcium within the normal range by stimulating bone resorption. In addition, glucocorticoids may directly stimulate PTH release. The mechanism of decreased intestinal calcium absorption is unknown, though recent studies have shown that it is not due to decreased synthesis or decreased serum levels of the active vitamin D metabolites; in fact, $1,25(OH_2)D_3$ levels are normal or even increased in the presence of glucocorticoid excess. Increased $1,25(OH_2)D_3$ synthesis in this setting may result from decreased serum phosphcrus levels (see below), increased PTH levels, and direct stimulation by glucocorticoids of renal 1 a_hydroxylase. Glucocorticoids also increase urinary calcium excretion, and hypercalciuria is a consistent feature of cortisol excess. They also reduce the tubular reabsorption of phosphate, leading to phosphaturia and decreased serum phosphorus concentrations.

Thus, glucocorticoids in excess result in negative calcium balance, with decreased absorption and increased urinary excretion. Serum calcium levels are maintained, but at the expense of net bone resorption. Decreased bone formation and increased resorption ultimately result in the disabling osteopenia that can be a major complication of spontaneous and iatrogenic glucocorticoid excess.

Growth and development

Glucocorticoids accelerate the development of a number of systems and organs in fetal and differentiating tissues, although the mechanisms are unclear. As discussed above, glucocorticoids are generally inhibitory, and these stimulatory effects may be due to glucocorticoid interactions with other growth factors. Examples of these development-promoting effects are increased surfactant production in the fetal lung and the accelerated development of hepatic and gastrointestinal enzyme systems.

Glucocorticoids in excess inhibit growth in children, and this adverse effect is a major complication of therapy. This may be a direct effect on bone cells, although decreased growth hormone (GH) secretion and somatomedin generation may also contribute.

Blood cells and immunologic function

1. *Erythrocytes*—Glucocorticoids have little effect on erythropoiesis and hemoglobin concentration. Although mild polycythemia and anemia may be seen in Cushing's syndrome and Addison's disease, respectively, these alterations are more likely to be secondary to altered androgen metabolism.
2. *Leukocytes*—Glucocorticoids influence both leukocyte movement and function. Thus, glucocorticoid administration increases intravascular polymorphonuclear leukocytes by increasing release from bone marrow, by increasing circulating half-life of PMNs, and by decreasing movement out of the vascular compartment. Conversely, circulating lymphocytes, monocytes, and eosinophils are

reduced, mainly by increased movement out of the circulation. The converse—i.e., neutropenia, lymphocytosis, monocytosis, and eosinophilia-is seen in adrenal insufficiency. Glucocorticoids also decrease the migration of inflammatory cells (PMNs, monocytes, and lymphocytes) to sites of injury, and this is probably a major mechanism of the anti-inflammatory actions and increased susceptibility to infection that occur following chronic administration. Glucocorticoids also decrease lymphocyte production and the mediator and effector functions of these cells.

3. *Immunologic effects*—Glucocorticoids influence multiple aspects of immunologic and inflammatory responsiveness, including the mobilization and function of leukocytes, as discussed above. They also impair release of effector substances, antigen processing, antibody production and clearance, and other specific bone marrow-derived and thymus-derived lymphocyte function.

Cardiovascular function

Glucocorticoids may increase cardiac output, and they also increase peripheral vascular tone, possibly by augmenting the effects of other vasoconstrictors, e.g., the catecholamines. Thus, refractory shock may occur when the glucocorticoid-deficient individual is stressed.

Renal function

These steroids affect water and electrolyte balance by actions mediated either by mineralocorticoid receptors (sodium and water reteution, hypobalemia, and hypertension) or via glucocorticoid receptors (increased glomerular filtration rate due to increased cardiac output or due to a direct renal effect). Thus, corticosteroids such as betamethasone or dexamethasone that have little mineralocorticoid activity increase sodium and water excretion. Glucocorticoid-deficient subjects therefore have decreased glomerular filtration rates and are unable to excrete a water load. This may be contributed to by increased ADH secretion, which may occur in glucocorticoid deficiency.

Central nervous system

Glucocorticoids readily enter the brain, and although their physiologic role in central nervous system function is unknown, their excess or deficiency may profoundly alter behavior and cognitive function.

1. *Excessive glucocorticoids.* In excess, the glucocorticoids initially cause euphoria; however, with prolonged exposure, a variety of psychologic abnormalities Occur, including irritability, emotional ability, and depression. Hyperkinetic or manic behavior is less common; overt psychoses occur in a small number of patients. Many patients also note impairment in cognitive functions, most commonly memory and concentration. Other central effects include increased appetite, decreased libido, and insomnia, with decreased REM sleep and increased stage II sleep.
2. *Decreased glucocorticoids.* Patients with Addison's disease are apathetic and depressed and tend to be irritable, negativistic, and reclusive. They have decreased appetite but increased sensitivity of taste and smell mechanisms.

Effects on other hormones

1. *Thyroid function.* Glucocorticoids in excess affect thyroid function. Although basal TSH levels are usually normal, TSH responsiveness to thyrotropin releasing hormone (TRH) is frequently subnormal. Serum total thyroxin (T_4) concentrations are usually low normal; however, thyroxine-binding globulin is decreased, and free T_3 levels are normal. Total and free T_3 (triiodothyronine) concentrations may be low, since glucocorticoid excess decreases the conversion of T_4 to T_3 and increases conversion to reverse T_3. Despite these alterations, manifestations of hypothyroidism are not apparent.
2. *Gonadal function.* Glucocorticoids also affect gonadotropin and gonadal function. In males, they inhibit gonadotropin secretion, as evidenced by decreased responsiveness to administered gonadotropin-releasing hormone (GnRH) and subnormal plasma testosterone concentrations. In females, glucocorticoids also Suppress LH responsiveness to GnRH, resulting in suppression of estrogens and progestins with inhibition of ovulation and amenorrhea.

Miscellaneous effects

1. *Peptic ulcer.* The role of steroid excess in the production or reactivation of peptic ulcer disease is controversial. Ulcers in spontaneous Cushing's syndrome and with

modest exposure to glucocorticoid therapy are unusual, although current data suggest that steroid-treated patients with established ulcers and those on high-dose therapy may be at increased risk.

2. *Ophthalmologic effects*. Intraocular pressure varies with the level of circulating glucocorticoids and parallels the circadian variation of plasma cortisol levels. In addition, glucocorticoids in excess increase intraocular pressure in patients with open-angle glaucoma. Glucocorticoid therapy may also cause cataract formation.

ADRENAL ANDROGENS

The direct biologic activity of the adrenal androgens (androstenedione, DHEA, and DHEA sulfate) is minimal, and they function primarily as precursors for peripheral conversion to the active androgenic hormones testosterone and dihydrotestosterone. This section will deal with the adrenal contribution to androgenicity. Thus, DHEA sulfate secreted by the adrenal undergoes limited conversion to DHEA; this peripherally converted DHEA and that secreted by the adrenal cortex can be further converted in peripheral tissues to androstenedione, the immediate precursor of the active androgens.

Effects in Males

In males, conversion of androstenedione to testosterone accounts for less than 5% of the production rate of this hormone, and thus the physiologic effect is negligible. In adult males, excessive adrenal androgen secretion has no clinical consequences; however, in boys it causes premature penile enlargement and early development of secondary sexual characteristics.

Effects in Females

In females, ovarian androgen production is low; thus, the adrenal substantially contributes to total androgen production by the peripheral conversion of androstenedione. In the follicular phase of the menstrual cycle, adrenal precursors account for two thirds of testosterone production and one-half of dihydrotestosterone production. During midcycle, the ovarian contribution increases, and the adrenal precursors account for only 40% of testosterone production.

In females, abnormal adrenal function as seen in Cushing's syndrome, adrenal carcinoma, and congenital adrenal hyperplasia results in excessive secretion of adrenal androgens, and their peripheral conversion results in androgen excess, manifested by acne, hirsutism, and virilization.

Adrenal Medulla

The medulla of the adrenal gland is an entirely different endocrine organ from the adrenal cortex both from the standpoint of morphology as well as physiology. Furthermore, both of these structures are derived independently and from totally different tissues in the embryo.

For these reasons, the morphologic, biochemical, and physiologic properties of the adrenal medulla are discussed separately from those of the adrenal cortex, despite the fact that both of these organs are intimately related anatomically and also share a common vascular supply.

Unlike the adrenal cortex, the mammalian adrenal medulla is not essential to life, and thus may be extirpated without producing deleterious effects, insofar as survival is concerned.

Functional Morphology of the Adrenal Medulla

Gross anatomy

The adrenal medulla in adult humans and other mammals is invested closely by the cortical tissue (interrenal tissue). However, in other vertebrates the two structures either may be entirely separate or else interrelated in a number of ways. In a transacted fresh adrenal gland, the medullary tissue appears grayish in sharp contrast to the yellow cortex.

Histology

Under the light microscope, the boundary between the zona reticularis and the medulla generally is irregular and columns of cortical cells may extend for some distance into the medullary tissue.

The medulla is comprised of irregularly shaped epithelioid cells that are arranged in cords or clusters, and that are surrounded by capillaries and venules.

If adrenal medullary tissue is fixed in solutions containing bichromate ion, the medullary cells are seen to be filled with fine brown granules. This is the chromaffin reaction, and presumably this reaction is caused by the action of chromic ion on the catecholamines in the granules of the medullary cells so that oxidation and polymerization of the hormones epinephrine and norepinephrine within the cytoplasmic granules takes place. Similarly, the medulla is stained green following treatment with ferric chloride.

The chromaffin reaction is seen in other tissues of the body, principally in the argentaffin cells of the gastrointestinal tract and the mast cells. These cells give this reaction because of the 5-hydroxytryptamine and dopamine that they contain. The adrenal medullary chromaffin cells, however, originate from neural ectoderm in the embryo, secrete catecholamines, and are innervated by sympathetic preganglionic fibres.

Special histochemical techniques reveal that two different cell types are present in the adrenal medulla, one of which contains epinephrine, the other norepinephrine. Furthermore, electron microscopy of the adrenal medulla reveals that the cells contain abundant membrane-limited dense granules, having diameters ranging between 50 and 350 mm. The granules of certain cells exhibit a rather low electron density and are considered to contain epinephrine, whereas another cell population contains very high density granules that are believed to consist of norepinephrine.

The cisternae of the granular endoplasmic reticulum are arranged in parallel systems. The Golgi apparatus is located in a juxtanuclear position, and its cisternae contain electron dense material that is believed to represent a precursor of the hormone granules.

The secretory mechanism whereby the adrenal medullary hormones are released into the bloodstream is not understood fully. Some investigators feel that this process is similar to the secretion of exocrine products, whereas other workers believe that the chromaffin granules are secreted intact, and the catecholamine then diffuses out of these "packets" of hormone through the membrane. Alternatively, an active transport process may liberate the hormone from the granules into the cytoplasm and thence the hormone passes out of the cell.

Medullary blood supply and innervation

Certain features of the adrenal cortical blood supply may be reviewed here briefly together with a consideration of the medullary blood supply. As discussed previously, the rich blood supply to the adrenal gland is provided by a number of arteries that penetrate the capsule of the organ, and give rise to a capsular plexus. This plexus in turn gives rise to a system of cortical arteries, and these vessels are distributed to the anastomosing network of sinusoids that invest the cords of cortical parenchymal cells. The sinusoids within a specific region of the cortex then converge upon a collecting vein at the corticomedullary junction, and as noted earlier there is no venous drainage from the adrenal cortex.

Some branches of the adrenal arteries, in contrast to the cortical vessels, pass through the trabeculae and then traverse the cortical tissue with little or no branching until they reach the medulla. Within the medulla, these vessels now branch repeatedly and thus form the abundant capillary network that surrounds the cords and clumps of medullary chromaffin cells. Consequently the adrenal medulla has a double (or dual) blood supply. First, blood comes from the sinusoids of the cortex that anastomose with the medullary capillary bed at the corticomedullary junction, and second, blood also comes from the medullary arteries that pass directly from capsule to medulla.

The medullary capillaries drain into the same venous system, as do those of the cortex. Ultimately the numerous adrenal venules form large central veins within the medulla which anastomose and emerge from the capsule as the suprarenal vein.

The cells that line the medullary capillaries are composed of typical endothelial cells.

The abundant capsular nerve plexus contains some sympathetic ganglion cells; branches of these nerves pass through the cortex in the trabeculae and thence run directly, and almost exclusively, to the medulla. These preganglionic fibers terminate in multiple endings that surround individual medullary cells. In actual fact, therefore, the adrenal medulla may be considered from a functional standpoint to be a physiologic extension of the sympathetic nervous system.

Embryologic considerations

As noted earlier, the adrenal cortex develops from coelomic mesoderm located on the medial side of the urogenital ridge.

The adrenal medulla, on the other hand, arises from ectodermal tissue of the neural crest. This tissue also gives rise to the cells of the sympathetic ganglia. The sympathochromaffin cells of the neural crest that are destined to form the adrenal medulla migrate ventrally and penetrate the analgen of the adrenal cortex medially, so that ultimately they assume a central position within the primitive adrenal gland.

Chromaffin system (or paraganglia)

Within the body are a number of broadly scattered accumulations of cells that appear to share many common features with the adrenal medullary cells. These groups of cells collectively are termed the paraganglia or chromaffin system of the body. Included in this category are the organs of Zuckerkandl (which is located in the retroperitoneum) as well as similar cell groups found in the heart, liver, testis, and ovary. These paraganglia all contain "cells that exhibit a typical chromaffin reaction described above, and these cells also stain with ferric ion. Furthermore, the paraganglia contain cells that are usually arranged in cords, and also possess a rich vascular supply. However, a definite endocrine role has not been demonstrated as yet for this group of tissues, although they are known to synthesize catecholamines.

Hormones of the Adrenal Medulla

Chemistry

Two compounds having endocrine activity have been isolated from extracts of adrenal medullary tissue. These two hormones are the catecholamines epinephrine (adrenaline) and norepinephrine (or arterenol), and their structural formulas are given in figure. Both of these compounds are synthesized, stored, and secreted by the adrenal medullary tissue under physiologic conditions.

There is a considerable difference among various species of animal insofar as the relative proportions of these catecholamines in the adrenal medulla is concerned. For example, the adrenal gland of man and dog contain and secrete approximately 80 percent epinephrine and 20 percent norepinephrine, whereas cat adrenal secrete around 90 percent norepinephrine. On the other hand, the whale adrenal secretes norepinephrine exclusively. Norepinephrine may also enter the circulation from adrenergic nerve endings.

Biosynthesis of the adrenal medullary hormones

The biosynthetic pathway for the catecholamines by the adrenal medulla is summarized in Figure. The precursors of epinephrine and norepinephrine are the amino acids phenylalanine and tyrosine. The enzyme phenylalanine hydroxylase, which catalyzes the conversion of phenylalanine to tyrosine, is found in the liver but not in the adrenal gland.

The oxidation of tyrosine to 3, 4-dihydroxyphenylalanine (or DOPA) in the adrenal glands is catalyzed by tyrosine hydroxylase (2) and enzyme similar to, and which uses the same cofactors as, phenylalanine hydroxylase.

The aromatic L-amino acid decarboxylase that catalyzes the decarboxylation of 3,4dihydroxyphenylalneanial to 3,4-dihydroxyphenylethylamine (hydroxytyramine or dopamine, (3) requires pyridoxal phosphate, and is found not only in adrenal medullary tissue, but also in kidney and liver. Dopamine is converted into norepinephrine by 3,4-dihydroxy-phenyl-ethylamine-β-hydroxylase (4). This nonspecific oxidase enzyme requires copper and catalyzes the direct addition of oxygen to the β-carbon atom. The methylation of norepinephrine to produce epinephrine is catalyzed by phenylethanolamine-N-methyl transferase (5). S-adenosylmethionine may (or may not) be the methylating agent in vivo, and the activity of this transferase is regulated indirectly by the pituitaryadreno-cortical mechanism in that the excess adrenocortical steroid

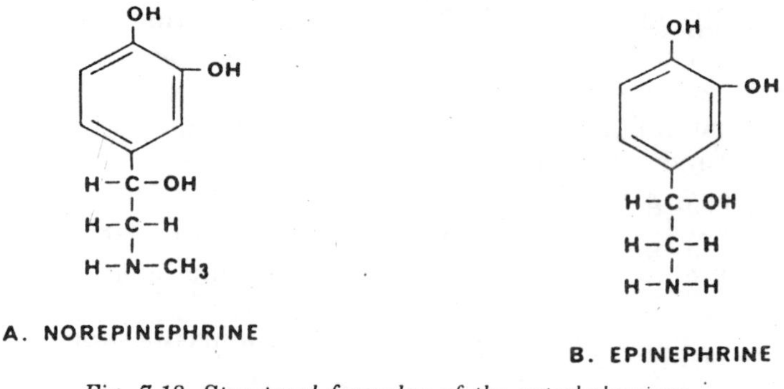

OH
OH
H−C−OH
H−C−H
H−N−CH₃

A. NOREPINEPHRINE

OH
OH
H−C−OH
H−C−H
H−N−H

B. EPINEPHRINE

Fig. 7.18. Structural formulas of the catecholamines.

production stimulated by corticotropin depresses the activity of this enzyme. Consequently, the adrenal cortical steroids may exert some of their physiologic effects via their action upon the adrenal medulla by altering the enzymatic methylation of norepinephrine to epinephrine.

The hydroxylation of tyrosine to 3,4-dihydroxy-phenylalanine (DOPA) is the rate-limiting step in the biosynthesis of the adrenal medullary hormones.

This pathway for the production of norepinephrine in the adrenal medulla is quite similar to the pathway for synthesis of this compound that is found in adrenergic nerve endings. Tyrosine is transported into, hence is concentrated within, the nerve endings by an active mechanism. At this site tyrosine is converted into DOPA by tyrosine decarboxylase, and then into dopamine by an aromatic L-amino acid decarboxylase. These enzymes are located within the neuronal cytoplasm. The dopamine now enters the granular vesicles and there is converted into norepinephrine. However, in distinct contrast to the adrenal medulla, the adrenergic nerve endings do not contain appreciable quantities of the enzyme phenylethanolamine-N-methyl transferase, so that only in the adrenal medulla are significant quantities of epinephrine synthesized from norepinephrine as indicated in step 5.

In a manner similar to the adrenals, the rate limiting step in the adrenergic nerve endings for the synthesis of norepinephrine lies in the conversion of tyrosine into dihydroxyphenylalanine (DOPA).

Adrenergic nerve endings also take up norepinephrine as well as small quantities of epinephrine from the blood, and these catecholamines, whether from endogenous synthesis or an exogenous source such as the adrenal medulla then are concentrated in the granulated vesicles of the neurons by an active transport mechanism and stored there until released by nerve impulses.

Secretion and Transport of the Adrenal Catecholamines

Several possible mechanisms whereby the adrenal catecholamines may be secreted by the medullary cells were mentioned above. The free hormones, epinephrine and norepinephrine, are found in the plasma in extremely low concentration (i.e., at micrograms per liter levels), and no special transport systems are present to convey these compounds in the bloodstream. Despite the fact that human medullary tissue contains between three and ten times more epinephrine than norepinephrine, the mean plasma concentration of epinephrine is only 0.06 μ-g/liter compared to around 0.3 μg/liter for norepinephrine. Adrenergic nerve endings also contribute some norepinephrine to the circulating catecholamine level, albeit the total quantity of norepinephrine that is released into the circulation from this source is negligible.

The epinephrine found in body tissues other than the adrenal medulla, and brain in general, is absorbed from the blood rather than synthesized by the tissues.

Metabolism and Excretion of the Catecholamines

The catecholamines have very short half-lives in the circulation. Hence these compounds are rapidly metabolized, principally in the liver, prior to excretion. The pathways for the catabolism of noreepinphrine and epinephrine are outlined in Figure. Each of these hormones is catabolized by three mechanisms: (1) o-methylation, a reaction catalyzed by catechol o-methyl transferase, (2) S-adenosylmethionine is used in this reaction; (2) oxidative deamination, a reaction catalyzed by monoamine oxidase (MAO, (3)); (3) conjugation reactions.

O-methylation is the chief metabolic pathway for epinephrine, and the principal metabolites of this hormone which are found in the urine are 3-methoxy-4hydroxymandelic acid (vanilmandelic acid [VMA]), 3-methoxyepinephrine (metanephrine), and 3-methoxy-4-hydroxyphenylglycol. The catechols 3,4-dihydroxymandelic acid, norepinephrine, and epinephrine are excreted in the urine only in minor quantities.

Roughly 50 percent of the total catecholamines secreted by the adrenal medulla are excreted in the urine as free or conjugated normetanephrine and metanephrine, and only trace quantities of free norepinephrine and epinephrine are excreted. The normal daily (24-hour) urinary output of catecholamines amounts to roughly 30 μg of norepinephrine, 6 μg of epinephrine, and 700 μg of VMA.

Neural Regulation of Adrenal Medullary Secretion

Essentially, the adrenal medulla is a functional extension of the sympathetic nervous system whose postganglionic neurons have no axons, and which have evolved into specialized secretory cells. The physiologic stimuli that induce secretion by this organ are transmitted as impulses that reach the medulla via the splanchnic nerves to enhance the output of catecholamines. Although the adrenal medullary hormones are not essential to life, and no clinical condition involving hyposecretion of these organs has been recorded, the secretion of epinephrine and norepinephrine is of importance in assisting the individual to adapt rapidly to emergency situations.

Basal catecholamine secretion is low in resting states, especially during sleep. However, an increased medullary secretion of epinephrine and norepinephrine is part of the overall regulatory mechanism of the individual for meeting emergency situations. The other component of this system is, of course, the sympathetic nervous system. Furthermore, there is experimental evidence indicating that the secretion of the adrenal medullary hormones is enhanced when the cholinergic sympathetic vasodilatory system discharges at the onset of a bout of exercise. Under these circumstances, increased adrenal medullary epinephrine secretion augments the vasodilatation produced by discharge of the sympathetic vasodilatory fibers to skeletal muscle. Thus, the principal physiologic regulatory mechanism for increasing adrenal medullary secretion operates via the sympathetic nervous system. In addition, a number of drugs act directly upon the adrenal medulla to increase secretion of the catecholamine hormones by this organ.

It is important to realize that, in general, stimulation of adrenal medullary secretion does not result in an alteration of the proportions of epinephrine and norepinephrine that are elaborated by these glands. However, certain particular circumstances can alter the ratio of epinephrine to norepinephrine secreted. Hypoxia, asphyxia and emotional stress all increase the quantity of norepinephrine secreted relative to epinephrine, whereas hemorrhage selectively increases the quantity of epinephrine secreted relative to norepinephrine. One must not conclude from these facts that the adrenal glands arbitrarily secrete the "best" hormone for a given situation, but rather that there is a "selective" release of epinephrine or norepinephrine that depends in turn upon the specific type of stimulus involved.

Physiologic and Biochemical Actions of Norepinephrine and Epinephrine

The hormones secreted by the adrenal medulla exert a broad spectrum of physiologic and biochemical effects upon a wide variety of bodily processes. The several possible effects of a single catecholamine may result from the mode of administration of the hormone (i.e., endogenous or exogenous), as well as the dosage of the hormone to which the subject is exposed. These factors are important in determining the degree as well as the type of response elicited by the adrenal medullary hormones.

Table 7.2. Physiologic and Biochemical Effects of Norepinephrine and Epinephrine

Physiologic Effects	*Norepinephrine*	*Epinephrine*
Heart rate	+; –[b]	+
Cardiac output	0; –[b]	++++
Systolic blood pressure	++++	++
Diastolic blood pressure	++	+; 0; -
Total peripheral resistance (TPR)	++	-
Action on central nervous system	0; +	++
Eosinopenic response	0	+
Biochcmical Effects		
Release of free fatty acids	++++	+++
Blood glucose	0; +	+++
Blood lactic acid	0; +	+++
Oxygen consumption; body heat production (*i.e.*, metabolic rate)	0; ++	++

Physiologic effects of the catecholamine hormones

Norepinephrine and epinephrine not only mimic the effects of sympathetic nervous discharge or stimulation, but also these hormones themselves stimulate the nervous system.

The catecholamines directly augment both the rate and contractile force of the isolated heart as studied in vitro. The heart in situ, however, may respond quite differently, because the various reflexes present in the body may cause secondary alterations in heart rate because of shifts in blood pressure induced by these hormones. The catecholamines also directly increase myocardial irritability, so that extrasystoles or even fibrillation may develop under certain conditions subsequent to the administration of these compounds.

Both norepinephrine and epinephrine dilate the coronary vessels. However, norepinephrine causes vasoconstriction in other organs, whereas epinephrine dilates the vascular bed in skeletal muscles and viscera when administered in moderate doses. Thus, epinephrine when injected intravenously produces a marked rise in heart rate and cardiac output concomitantly with a sharp rise in blood pressure that is due to a generalized vasoconstriction.

As noted above, norepinephrine produces vasoconstriction in most organs other than the heart; hence it too produces a marked pressor response. However, since epinephrine causes vasodilatation of the vascular bed in skeletal muscles, this effect overcompensates for the vasoconstriction produced elsewhere; the net effect of an increased adrenal medullary secretion in vivo is a fall in the total peripheral resistance. The systolic blood pressure rises because of the direct effect of epinephrine in stimulating the heart rate increases the cardiac output sufficiently to overcome the concomitant vasodilatory effect of epinephrine in the skeletal muscles.

Infusion of norepinephrine into normal humans (or animals) results in an increase in both the systolic and diastolic blood pressures. The ensuing hypertension then stimulates the aortic and carotid pressoreceptors, with the result that reflex bradycardia occurs, and this reflex-induced bradycardia dominates the direct stimulatory effect of norepinephrine upon heart rate; hence this rate shows a net decrease and the cardiac output falls.

Epinephrine, on the other hand, causes an increased pulse pressure; however, the pressoreceptor stimulation evoked by this hormone is inadequate to obscure the direct effects of epinephrine upon the heart rate and cardiac output. 'The heart rate and cardiac output increase.

In addition to the effects exerted upon the nervous and cardiovascular systems by the adrenal medullary hormones, these compounds also directly affect the physiologic activity of smooth muscle. In this regard the effect of norepinephrine and epinephrine are quite variable, and their effects upon this kind of tissue depend upon the site in the body where the muscle is located.

Norepinephrine exerts a weakly inhibitory action upon the contactile activity of smooth muscle in the gastrointestinal tract, but does not relax the smooth musculature of the pulmonary bronchioles. As noted above, this hormone also elicits a marked pressor response because of its direct effect upon vascular smooth muscle, with an attendant rise in total peripheral resistance.

Epinephrine, on the other hand, markedly inhibits the contractile activity of the smooth musculature of the gastrointestinal tract, producing relaxation, while simultaneously stimulating contraction of the pyloric and ileocecal sphincters. Epinephrine also produces a strong dilatation of the brachial musculature.

Note that these effects of the adrenal catecholamines strongly resemble the effects of adrenergic nerve stimulation to the same organs.

Epinephrine decreases the rate of onset of fatigue in skeletal muscle and also increases the rate and depth of respiration. Presumably these effects are due to the enhanced glycogenolysis and lipolysis produced by this hormone as well as the calorigenic action of these amines.

In adrenalectomized animals suffering from an overall adrenal insufficiency, the vascular smooth muscle becomes unresponsive to the effects of norepinephrine and epinephrine. Capillary dilatation ensues and permeability of these vessels (e.g., to dyes) increases markedly in the terminal stages. Lack of response to adrenergic stimulation (hence liberated norepinephrine) possibly hinders the compensation of the vascular bed to the reduced circulating blood volume (or hypovolemia) that attends adrenal insufficiency, and this in turn favours cardiovascular collapse. Vascular reactivity in such animals is restored by glucocorticoid administration.

Metabolic effects of the catecholamine hormones

Both epinephrine and norepinephrine exert certain effects upon the metabolism of carbohydrate and fats, and these actions of the catecholamines may be far more significant physiologically than their roles as endocrine extensions of the sympathetic nervous system.

Carbohydrate metabolism

Epinephrine stimulates glycogenolysis in both liver and muscle. This results in an elevation of the blood glucose level as well as an increased production of lactic acid in muscle. Epinephrine activates phosphorylase in both of these tissues and the plasma potassium level rises coincidentally with the increased glycogenolysis. Aside from their inherent physiologic significance, the mechanisms of these reactions provide examples of the first elucidation of the action of a hormone on a fundamental cellular process at the molecular level.

An increased oxygen consumption accompanies these glycogenolytic effects; this amounts to about 20 to 40 percent above the resting level in humans. The carbon dioxide production concomitantly rises to an even greater extent than does the increase in oxygen consumption, so that the respiratory quotient also increases.

Norepinephrine, on the other hand, exerts very little effect upon carbohydrate metabolism and oxygen consumption, a fact that provides the major exception to the generalization that the metabolic actions of the two adrenal catecholamines are similar.

Lipid metabolism

Epinephrine as well as norepinephrine are almost equally potent in their effects upon lipid metabolism. Thus, both of these catecholamines exert a profound lipid-mobilizing action, and they increase the blood level of nonesterified fatty acids by stimulating the release of free fatty acids and glycerol from adipose tissue. These effect results directly from an elevated rate of lipolysis that is produced in adipose cells by these hormones, and an increased oxygen consumption (or calorigenic action) also attends these processes.

In normal subjects, the administration of epinephrine also elevates serum cholesterol and phospholipids levels. The turnover rate of the cholesterol and phospholipids, also is elevated in cardiac muscle by exogenous epinephrine.

Epinephrine directly inhibits insulin release from the pancreas, and thus serves as an emergency hormone by exerting the following metabolic actions. First, epinephrine rapidly provides a supply of fatty acids for use as fuel by skeletal muscle. Second, epinephrine mobilizes glucose by stimulating hepatic glycogenolysis as well as gluconeogenesis directly. Third, by reducing the endogenous secretion of insulin, epinephrine decreases glucose uptake by the peripheral tissues,

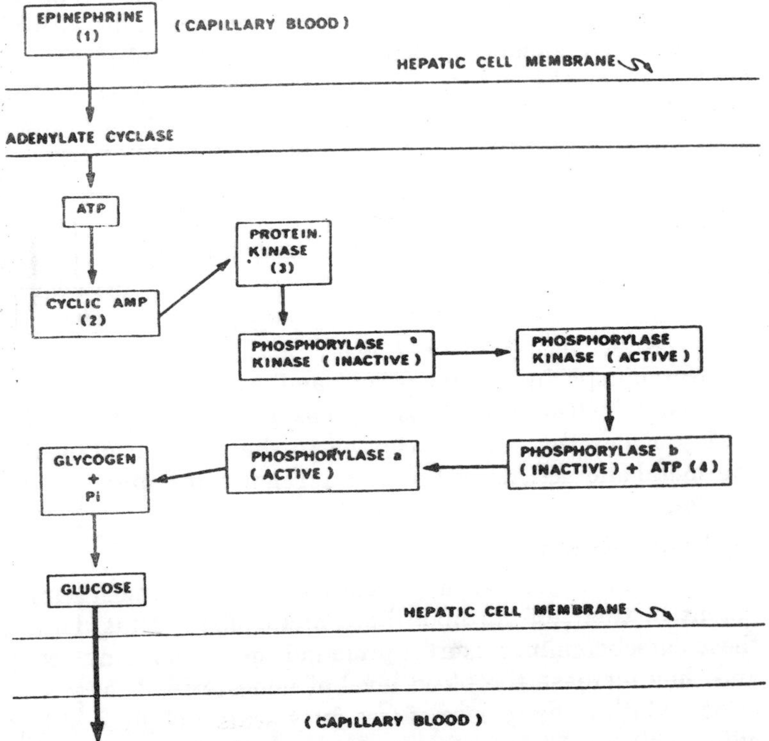

Fig. 7.19. Mechanism underlying the release of blood glucose from a hepatic cell by glycogenolysis. (1) Epinephrine, the first messenger, activates adenylate cyclase within the cell membrane so that some cytoplasmic ATP is converted into cyclic AMP, (2) which is the second messenger. (3) A protein kinase then is activated by the cyclic AMP and this in turn activates a second cytoplasmic kinase.

thereby increasing the supply of this substrate that is available to the central nervous system.

Mechanism and Action of the Adrenal Medullary Hormones

The general effects of norepinephrine and epinephrine have been divided into two categories depending upon their sensitivity to certain drugs. These effects depend in turn upon the actions of these catecholamines upon postulated α and β receptors located within the effectors organs. Drugs which inhibit a receptors (such as phentolamine) prevent such actions as the pressor effects of the catecholamines, whereas inhibition of β receptors by other compounds (e.g., isoproterenol or propranolol) prevents the inotropic and chronotropic responses of the heart to the catecholamines as well as the ability of these hormones to stimulate glycogenolysis in muscle and liver and lipolysis in adipos tissue. Experimental evidence has shown that most of the β responses are due to stimulation of adenyl cyclase activity, so that an increase in the synthesis of cyclic 3', 5'-AMP results. Epinephrine, for example, produces this effect in a number of tissues, including liver, skeletal muscle, adipose tissue, and cardiac tissue. The adenyl cyclase system in turn has been shown to be localized within the cell membranes of hepatic and other tissues as discussed in the introduction to this chapter. Furthermore, the effects of norepinephrine and epinephrine upon glycogen metabolism have been shown to be mediated by a marked increase in the activity of the adenyl cyclase system for epinephrine.

On the other hand, certain a responses have been demonstrated to reside in an inhibition of cyclic AMP synthesis.

It is important to realize that in certain tissues where epinephrine can stimulate both α and β types of response, the net effect of the amine depends upon the relative quantity and/or sensitivity of each type of receptor that is present in the tissue. For example, in the pancreas the α-adrenergic response to epinephrine is predominant; therefore, cyclic AMP concentration decreases, inhibiting insulin secretion. If, however, an α adrenergic blocking agent such as phentolamine (Regitine) is present, the β effect becomes predominant, and epinephrine now stimulates cyclic AMP synthesis, and insulin secretion is stimulated rather than inhibited.

8

Diseases of Suprarenal Gland

Primary adrenocortical hypofunction, or Addison's disease, is an uncommon disease which affects six in every 100 000 of the population. Some cases of Addison's disease are due to destruction of the adrenal cortex by tuberculosis, other granulomatous diseases or carcinoma (primary or secondary) (Table 8.1): the remaining cases, called idiopathic Addison's disease, are autoimmune in origin.

Table 8.1. Historical comparison of causes of Addison's Disease

	Years studied	
	1938-1958	*1962-1973*
Idiopathic	30%	78%
Tuberculosis	60%	20%
Others	10%	2%

Case

A 12-year-old girl presented with vague abdominal discomfort for 6 months. She had noticed occasional diarrhoea but had not passed any blood. She admitted to weight loss (6 kg) and anorexia. Her mother had thyrotoxicosis. On examination, she was obviously pigmented, although she thought this was sun-induced; however, her buccal mucosa and gums were also brown. There were no other physical signs.

She had a low cortisol level and her response to ACTH in a Synacthen test was poor. A diagnosis of adrenal cortical failure was made. X-ray of her abdomen showed no calcified areas in either adrenal gland, and her serum contained antibodies to adrenal cortex, consistent with a diagnosis of *autoimmune adrenalitis*. Her serum also contained antibodies to pancreatic islet cells and thyroid microsomes. In view of the young age at presentation and these serum antibodies, she will be followed at yearly intervals to see if she develops other autoimmune endocrinopathies.

Autoimmune Adrenal Disease

Three-quarters of all cases of Addison's disease are due to an *autoimmune adrenalitis* which affects the adrenal cortex but spares the medulla. Like other autoimmune endocrinopathies, it is more common in women and reaches a maximum incidence between 40 and 50 years of age. Patients' sera should be tested for all organ-specific autoantibodies as 40% of patients have at least one other autoimmune endocrinopathy. The presence of autoantibodies may predict future onset of the disease, so that replacement therapy (or other relevant treatment) can be started promptly.

Table 8.2. Association of 'Idiopathic' Addison's disease with other endocrine diseases

Associated autoimmune disease	*Percentage of 'idiopathic' Addison's disease patients with other organ involvement*
Thyroid diseases	19%
Diabetes mellitus	15%
Ovarian failure	8%
Hypoparathyroidism	4%
Pernicious anaemia	2%

Evidence for immune involvement in idiopathic Addison's disease is shown in Table 8.3. The histological evidence is, of course, only suggestive; *an inflammatory cell infiltrate may be the result of the damage rather than its cause.* The presence of antibodies to cytoplasmic adrenal cortex antigens in this disease may also be secondary to damage, although fewer than 5% of patients with adrenal damage due to tuberculosis have this antibody. Similarly, cell-mediated immunity to adrenal tissue can be demonstrated in about 60% of patients with 'idiopathic' Addison's disease and is probably secondary. Recently, circulating antibodies have been detected which block ACTH-induced adrenal cell growth in vitro; such antibodies are probably primary, pathogenic autoantibodies.

Table 8.3. Evidence for immune involvement in idiopathic Addison's disease

1. Association with other autoimmune diseases.
2. Presence of autoantibody to adrenal cortex and high incidence of other organ specific autoantibodies.
3. Diffuse lymphocytic infiltration of adrenal cortex.
4. Evidence of cell-mediated immunity to adrenal cortex antigens.
5. Adrenal failure produced experimentally in animals by immunization with adrenal tissue with transfer by lymph node cells.

The finding of antibodies to cytoplasmic adrenal antigens in some patients with Cushing's syndrome due to nodular adrenocortical dysplasia has led to a search for *stimulating adrenocortical antibodies*. Positive preliminary data are now being followed up with large scale studies.

Adrenogenital Syndrome

This term was first used (as i.e. *syndrome genito-surrenal*) by the French physician Gallais in 1912, and must be considered as a clinical label. It can be defined as a condition of amenorrhoea, or oligomenorrhoea, produced by an excess of adrenal cortex androgenic hormones, as indicated clinically by an associated development of hirsutism. By definition, the disorder as thus described, is limited to females and cannot be recognized as this syndrome before chronological puberty. The term therefore does not include congenital virilism dating from infancy, and it is to be regretted that the term is sometimes used less discriminatingly. The term also cannot include the similar clinical picture occasionally produced by an androgen-secreting ovarian tumour, since the ovary and not the adrenals in such cases is the source of the androgens.

Pathology

The lesion is usually bilateral hyperplasia of the adrenal cortex but a single adenoma or adenocarcinoma is not infrequent. The ovaries are involuted or atrophic, and multiple small cysts are sometimes present, the appearance resembling that produced by injecting animals with large doses of testosterone. The excess of adrenal androgens inhibits the normal pituitary secretion of gonadotrophins and there is thus failure of follicle development and of ovulation.

Clinical features

Amenorrhoea (or oligomenorrhoea) combined with the development or increase of hirsutism are two essential features. Chronologically there are two main types, (1) puberty, (2) adult. The puberty type becomes manifest at the age of chronological puberty, e.g. 14, when the hirsutism is noted and the patient in the course of the next few years fails to menstruate, or menstruates scantily and at long intervals. As it is now recognized that the adrenals develop increase activity (adrenarche) at the time of puberty, it is not surprising that one type of adrenogenital syndrome shows itself at this phase. Since the adrenarche may also occur earlier, *e.g.* age 7, partial manifestations may present earlier than puberty time, but the label adrenogenital syndrome cannot be applied before chronological puberty indicates the evolution, or lack of evolution. Failure of the breasts to develop at 14 or so, is another indication of suppression of normal ovarian evolution by the adrenal androgens. There is also absence of widening of the pelvis so that the relative measurements of shoulders and pelvis approximate to that of the average male.

The adult type of adrenogenital syndrome usually manifests itself in the third or fourth decade, but sometimes is met with in the later teens, after a puberty which is normal or nearly normal. It is difficult to understand, in the absence of an adrenal neoplasm, why the adrenals develop excessive androgenic activity at varying phases of adult life in different individuals (apart from puberty, pregnancy, and the climacteric), but that they do so is undoubted and not infrequent: and with gradually increasing androgen manifestations over a number of years.

Acne is another manifestation of excessive androgenic activity in all types of adrenogenital syndrome and this may be very severe, affecting both body and face. Enlargement of the clitoris is similarly explained and adds strength to the diagnosis, but it is by no means met with in all patients as it depends both upon the intensity of the androgenic stimulus and upon the sensitivity or responsiveness of the clitoric tissues.

Hirsutism

The responsiveness of the hair follicles partly determines the degree of hirsutism met with but, in addition, the distribution of hair follicles varies in different individuals. In some women the hirsutism may be almost limited to the face, in milder degrees affecting the upper sides of the face, the upper lip, and chin, as in the adolescent man, but in the more severe it may cover the face to the extent found in the normal adult male. Body hair may be limited to the appearance, or increase, of hair along the linea alba, and a few hairs round the nipple, but it may spread all over the abdomen, be present or increase over the lumbosacral area, and spread from the spine over the shoulders and scapular areas. It is of interest that in the hirsute male, the lumbar area may be free of hair, whereas it is a sensitive hirsute area in the virile female. The pubic hair increases in amount, and hair spreads downwards from the vulva on to the thighs as seen posteriorly. In severe hirsutism, hair spreads over the buttocks, and even in the absence of hair, the skin over the buttocks loses its smoothness and is covered by raised hair follicles and papules. From this it will be seen that the appearance of the patient's limnar area, buttocks, and posterior thighs, as seen from the posterolateral angle is of diagnostic significance. Hair appears, or increase on legs, forearms, backs of the hands, and fingers. Thick eyebrows, perhaps joined by a less intense middle hirsute ridge, and long eyelashes are also manifestations of heightened androgenic activity. The scalp hair becomes thick and coarse and tends to encroach upon the forehead and nape of neck, and greasy seborrhea is often present. Later scalp hair is lost from the top of the head and frontotemporal areas as in the normal male.

Amenorrhoea may be absolute, or scanty menstruation may occur at long intervals. Occasionally long but irregular intervals of menstruation are associated with menorrhagia and anovular menstruation.

Alternations in body fat

In the characteristic, or classical, variety of the varilizing adrenogenital syndrome, fat is lost, particularly from the breasts, abdomen, buttocks, thighs and upper arms, and face —areas in which the normal deposition of fat tends to produce the normal 'feminine' types of contours—and simultaneously muscular tissue tends to hypertrophy, both these effects being an expression of the protein anabolic effects of the excess of androgens. There is frequently an increase in muscular strength. In so far as these processes are reversed by successful removal of an adrenal androgen secreting tumour, there seems no doubt as to the cause, although the metabolic mechanism may be more involved than that of protein anabolism associated with fat catabolism.

In addition to the above, an adipose type of adrenogenital syndrome must be recognized in which not only is there no loss of fat but fat is deposited, especially in the sites mentioned above, with a resulting general increase in weight- sometimes considerable. In so far as some adrenal gluco-corticoids influence fat deposition and fat distribution, it seems probably that the adipose type of adrenogenital syndrome is influenced by a simultaneous excessive secretion of adrenal glucocorticoids, although this is not consistently confirmed by hormone assays. Be this as it may, the clinical adipose type must be recognized as existing, and in fact as being the more commonly met with. Its relationship to Cushing's syndrome and adipose gynism will be discussed later. In this type enlargement of the clitoris is relatively rare. The supposition that the adrenals are the cause of the adiposity receives some support from the loss of adiposity that follows subtotal adrenalectomy, or removal of an adrenal tumour, as such loss is not necessarily associated with any evidence of adrenal insufficiency. The relative excess of adrenal androgens and glucocorticoids determine the clinical picture, as these hormones are metabolic antagonists.

Diagnosis

Clinical diagnosis depends upon definition and is not difficult, although the differentiation between adrenal hyperplasia and neoplasia requires accessory aids. Clinically, the following varieties of hirsutism must be differentiated from the adrenogenital syndrome.

Androgenic ovarian tumour

The clinical picture resembles the virile rather than the adipose type. The raised urinary 17-ketosteroids are not diminished by the prednisone suppression test. When in doubt, examination under anaesthesia, and even lower abdominal laparotomy, is prudent.

Stein-Leventhalsyndrome. When adiposity and/or hirsutısm are prominent features of this syndrome, the writer considers that the primary condition is adrenal and that there is a confusion of nomenclature. In other words, many patients described under this heading are examples of the adrenogenital syndrome. This is supported by the clinical fact that bilateral ovarian wedge resection does not influence adiposity and hirsutism appreciably, except perhaps with the relatively rare condition of superluteinization of ovaries, luteal tissue having some functional kinship with adrenal cortex tissue.

Pituitary type of adrenogenital syndrome

Hirsutism, acne, and amenorrhoea are recognized complications of acromegaly but it is less generally appreciated that they may be features of even mild giantism, with oligomenorrhoea or anovular menstruation instead of amenorrhoea. Apart from the skeletal structure, this type can be identical with the adrenogenital syndrome and perhaps should be included as a third group in this syndrome. The pituitary fossa may be enlarged but a pituitary neoplams is rare. The patients are usually muscular and virile but a degree of adiposity may also be met with.

Adipose gynism

Although this clinical entity may have a spontaneous favourable outcome, the history in retrospect of some patients with the adipose type of adrenogenital syndrome suggests that this disorder is a continuation of the entity called adipose gynism or a relapse from a remission from this disorder, which starts in childhood. The diagnosis, however, in this adult phase is still correctly given as adrenogenital syndrome.

Familial hirsutism

This condition is sometimes termed 'essential' hirsutism or 'simple' hirsutism or 'genetic' hirsutism or 'racial' hirsutism, it being more common among Celtic and Mediterranean races. The nomenclature implies that the hirsutism is not associated with other features of adrenogenital syndrome such as amenorrhoea, enlarged clitoris, or adiposity. Theoretically this is a proper differentiation, but in practice it is not an absolute one, especially when one remembers that the adrenogenital syndrome is in itself often familial, genetic, and racial, and that when severe acne is associated with hirsutism it is difficult not to regard both as of androgenic origin; and further, that hirsutism by itself might not become severe until pregnancy, or the climacteric, when adrenal androgenic activity increases. The absence of raised urinary 17-ketosteroids cannot be an absolute criterion because these values are often not raised in the adrenogenital syndrome, and constitute a poor measure of adrenal androgenic activity. Nevertheless, it appears valid to say that the distribution of hair follicles is more dense and more widely distributed in some women than in others, and that they may be more sensitive to a androgenic stimulus. To state these facts does not solve the problem arising from the fact that most women will not accept facial hirsutism as being normal.

Treatment

When an adrenal or ovarian androgenic tumour is the cause, the tumour must obviously be removed surgically. Unlike adrenal glucocorticoids tumours, adrenal androgenic tumours do not usually lead to atrophy of the opposite adrenal gland so that it is not essential to give large doses of cortisone in the post-operative period, as in Cushing's syndrome. Nevertheless, it might be thought prudent to do so.

In the absence of a tumour, bilateral hypertrophy and hyperplasia of the adrenal cortex is the probable lesion to be dealt with. Unilateral adrenalectomy is of little use except in those rare cases in which one adrenal only is considerably hypertrophic. Subtotal adrenalectomy is best carried out by removing the whole of one adrenal gland and some seven-eighths of the other. Since the remnant left behind may be sufficient to maintain essential adrenal functions until it has been given time to undergo some hypertrophy, replacement therapy with cortisone or prednisone is advisable and may have to be continued.

Apart from surgery, adrenal androgenic hyperfunction, as measured by urinary 17 ketosteroids, may be suppressed by cortisone by mouth, e.g., 25 mg. three times a day, or by prednisone, 5 mg. three times a day, although not as effectively as in genetic virilism.

The denial of hirsutism in childhood does not necessarily exclude genetic metabolic virilism because the patient may have amnesia for early life. In one such patient, aged 50, a mild operative interference, without protective cortisone therapy, led to acute and fatal adrenal insufficiency. Every patient therefore should be investigated completely before any operative procedure is undertaken.

Prognosis

Removal of an adrenal tumour usually gives an excellent clinical result, although hirsutism of long standing may persist in the absence of raised 17-ketosteroids. Subtotal adrenalectomy is always effective in regard to menstruation and body-weight, but hirsutism is not always alleviated. Chronic suppression therapy with cortisone or prednisone should not be undertaken in the absence of raised 17-ketosteroids and in any case is more likely to benefit amenorrhoea and acne than hirsutism. Facial hirsutism and coincident loss of scalp hair may remain a provocative therapeutic problem.

Cushing's Syndrome

A syndrome due to adrenal cortex hyperfunction, hyperplasia, or neoplasia, associated with excessive secretion of adrenal glucocorticoids, and a resulting negative nitrogen balance, and manifested clinically by a type of plethoric adiposity, hypertension, purple lineae distensae, patchy cyanosis, amenorrhoea or impotence, some hypertrichosis, diabetes, osteoporosis, weakness, and psychoneurotic features. The condition can occur at any age and in either sex.

Cushing's original description deserves quotation: "The disorder is characterized by a rapidly acquired plethoric adiposity affecting the face, neck, and the trunk, the extremities being spared. It is associated in women with hypertrichosis and amenorrhoea. Other characteristic features are vascular hypertension, purplish striae distensae of the abdomen, and acrocyanosis with cutis marmorata of the extremities. It is often accompanied by hyperglycaemia, and a peculiar softening of the bones of the skeleton has been commonly found at autopsy. In its extreme forms, the malady has often been encountered in young adults, and the average duration of life in the fatal cases has been something over five years."

In defining the syndrome I have intentionally excluded a mixed syndrome in which the clinical picture is a combination of adrenogenital virilism and classical Cushing's syndrome, because the hyperfunction is associated with an excessive secretion both of adrenal androgens and of adrenal glucocorticoids. I have described this mixed syndrome, as a separate entity, under the heading of mixed adrenal hyperfunction, but it is recognized that this classification of convenience is not absolute. It is necessary, however, since androgens are antagonistic to glucocorticoids e.g. they are protein anabolic and prevent esteoperosis. The inclusion of hypertrichosis in classical Cushing's syndrome is partly justified by Cushing's original description and partly by the fact that some metabolic products of adrenal glucocorticoids are androgenic, but severe generalized hirsutism and other signs of virilism are not compatible with a diagnosis of classical Cushing's syndrome.

History

There is no doubt whatsoever that the clinical picture of Cushing's syndrome has been known to physicians for some fifty years or more: and has been frequently reported under a variety of clinical names, including 'fat bearded diabetic woman': Ercheim (1903), Launois (1911), Turney (1913), Anderson (1915). Achard and Thiers (1921), and Parkes Weber (1926). It was, however, the keen clinical and pathological observations of Cushing, as well as his intensive study of the literature, which resulted in the crystallization and separation of a distinct clinical entity, so ably and dramatically presented in his classical paper of 1932; and to which, in the same year, Bishop and Close appropriately appended his name. Cushing chose an arresting title for his paper, namely, *The Basophil Adenomas of the Pituitary Body and their Clinical Manifestation (Pituitary Basophilism).* We shall see in the pathology section that this theory of origin could not be sustained.

Pathology

The more immediate cause of Cushing's syndrome is hyperfunction of the adrenal cortex, hyperplasia being more frequent than neoplasia. Hyperplasia, particularly in children, may be simple or associated with multiple adenomata, macroscopic or microscopic. The zona fasciculata is the zone principally involved. It must be recognized, however, that in some well-developed and characteristic cases, the adrenal glands weigh no more than the normal, and the appearance, macroscopic and microscopic, gives no indication of the hyperfunction and hypersecretion which is obvious clinically, by hormone assays, and by the results following subtotal adrenalectomy.

When due to a neoplasm, the whole gland may be replaced by an adenoma or adenocarcinoma, the latter being more common in children; or a narrow flattened streak of adrenal gland may be attached to the tumour. The opposite adrenal gland is frequently atrophied, owing to the suppression of pituitary ACTH secretion by the high concentration of blood glucocorticoids.

Cushing attached more significance to the finding of a basophil adenoma in some patients with his syndrome than did Erdheim before him, and he thought that diligent search, with serial sections, would show an actiologically significant basophil adenoma in most cases. This did not prove to be so and this theory of the primary cause of the syndrome is now discarded.

Crooke (1935) found constant changes in the basophil cells of the pituitary gland in Cushing's syndrome, whether the adrenal lesion was hyperplasia or neoplasia, namely degranulization, hyalinization, and vacuolization of the basophil cells. He recorded: the normal cytoplasm, charged with ripe basophil granules, is replaced by a dense homogenous cytoplasm. This may be complete or partial, and vacuolization may occur. His observations were confirmed by Rasmussen (1938). Since the clinical syndrome is cured by removal of an adrenal adenoma, Simpson (1948) concluded that these basophil changes were secondary to the adrenal hyperfunction and reversible and this was experimentally proved by producing such basophil changes in animals by injections on cortisone or ACTH.

Nevertheless, quite apart from the fact that Cushing's syndrome is occasionally associated with an enlarged pituitary fossa and even with acromegaly, bilateral hyperplasia of the adrenal cortex is probably caused by an excessive secretion of pituitary ACTH, and further back in the link by an initial hypothalamic stimulus, experimental or in man, from third ventricle tumours; and an excessive concentration of plasma ACTH has been found in some patients with Cushing's syndrome. The cellular source of the ACTH is probably of basophil type but Cushing's syndrome has been found with the following pituitary lesions: a basophil adenoma, a basophil carcinoma, a chromophobe adenoma (Fuller and Russell, 1936), and an eosinophil adenoma. If there is a primary hypersecretion of ACTH in Cushing's syndrome, the resulting excessive secretion of hydrocortisone would tend to reduce this. It is also necessary to explain why the adrenal response is mainly or entirely glucocorticoid and not equally androgenic. For this, and other more complex reasons, there are some who regard the essential lesion as a primary adrenal cellular abnormality and hypersensitivity to a normal pituitary ACTH stimulus. The innate difficulty of this qualitative conception is the fact that the onset may occur at any age.

It would appear that a primary adrenal adenoma or adenocarcinoma is independent of any abnormal pituitary stimulus: although, as has been considered in the adreno-genital syndrome, it is possible that hyperplasia proceeds to multiple adenomata and that one of these may become large and autonomous, with resulting involution of the others.

Ovary

Whatever the primary cause of Cushing's syndrome may be, the ovaries usually show secondary changes of atrophy or involution. Thus Cushing (1933) recorded senile, fibrotic, or atretic ovaries; atretic follicles with no corpora lutea; numerous follicular cysts of variable size and normal germinal epithelium, but no primordial ova; follicular development without ovulation and luteinization. Such changes are characteristic, but occasionally normal ovaries and normal corpora lutea are found. Exceptionally an ovarian tumour, developing from an adrenal 'rest' in the ovary, may in itself produce Cushing's syndrome. Other ovarian tumours may produce virilism but not Cushing's syndrome.

Testes

Atrophy of the spermatogenous epithelium, and of the interstitial cells, with fibrosis, is the usual finding, but normal testes have been recorded.

Thymus

In two males, Cushing's syndrome was associated with a carcinoma of the thymus, but the adrenals were enlarged, and the pituitary basophil cells showed hyalinization. In the majority of cases, however, the thymus gland has been atrophic or completely replaced by fat. In Some it has been normal, and in one boy of 19 the thymus was found to be enormous, the adrenals long, thin, and hypoplastic, and the testes atrophic. A basophil adenoma was present in the pituitary. The significance of the thymus as an aetiological factor in Cushing's syndrome remains doubtful. The involution or atrophy of the thymus usually found is in keeping with animal experiments following cortisone administration.

Thyroid

The thyroid gland is often slightly enlarged, but may be small and fibrotic. The epithelium is flattened and colloid is present in the vesicle, and the appearance is that of a resting gland, in keeping with the known inhibition of thyroid function by cortisone.

Parathyroids

These may be normal, or fibrotic, or infiltrated with fat. Occasionally an adenoma may be present, but there was no evidence of hyperplasia or neoplasm in one case showing a marked increase in calcium excretion, and a negative calcium balance.

Pancreas

Although no constant findings have been recorded, and although fatty degeneration of the islets may occur, the pancreas is usually normal, even in the presence of gross diabetes. This is in keeping with the initial steroid character of the diabetes and the later secondary effects on the pancreatic islets in untreated cases. In children, hyperplasia of the pancreatic islet cells has been found.

Kidneys

The kidneys may be normal or may show histological evidence of chronic nephritis; and sometimes the cardiovascular and renal lesions are identical with those of malignant nephrosclerosis. Minute deposits of calcium may be found in the kidney substance and in the calyces—as in hyperparathyroidism. With or without renal lesions, cardiac hypertrophy and arteriosclerosis are not infrequent findings, as might be expected from the fact that hypertension is a common symptom.

Clinical features

The disorder is four times as frequent in women as in men, and pregnancy immediately preceded the onset in five of the eleven women described in Cushing's monograph. It is commonest in the second and third decades of life, but there is a group in which the syndrome commences at the time of chronological puberty, and it may start in earlier childhood, although this is unusual in the uncomplicated form. A climacteric form of incomplete type has also been described.

In view of the multiple manifestations of excess of cortisone, or hydrocortisone, the disorder may present in a variety of ways, but most characteristically as a change of appearance, with development of the characteristic moon-like fat face, and fatigability, which draws attention to the wider symptomatology. However the writer has met with cases which presented primarily as diabetes, severe pain in the back from spontaneous fractures of osteoporotic vertebrae, hypertension; polycythaemia, spontaneous subcutaneous haemorrhages, general adiposity, facial hirsutism, amenorrhoea, malignant nephrosclerosis, depressive psychosis, and polyuria diagnosed as diabetes insipidus. It is therefore a condition which needs to be though of in many branches of medicine. Since massive cortisone therapy has been used in the treatment of collagen and allergic disorders, most of these symptoms, or manifestations, have been produced inadvertently by excessive therapy, demonstrating unequivocally their aetiological basis in Cushing's syndrome. Confirmatory proof has been given by animal experiments.

Adiposity

Clinically, fat is deposited in the face and neck, over the breasts, abdomen, buttocks, supraclavicular and lower cervical areas. In children it is more generally deposited—including even forearms, and legs—but in some adults these parts and even the upper arms and thighs may lose fat. Fat deposition is not always a prominent feature, and even when it is obvious in the face and trunk, an absence of increase in weight suggests a transference of fat rather than a general increase.

In some patients, however, there is a considerable and rapid increase in weight and fat. The endocrine basis of adiposity has been discussed in the physiology section and here it will be sufficient to mention some associated indications:

1. The high blood cholesterol, not explained by the mild degree of thyroid suppression that may be present, but reduced to normal by subtotal adrenalectomy, or adrenal tumour removal.
2. Fatty infiltration of the liver, found at autopsy, with concentrations of hepatic fatty acids five times the normal.
3. Hypergluconeogenesis from protein with conversion to fat and resulting fat deposition.
4. Loss of fat after subtotal adrenalectomy, even when unassociated with adrenal insufficiency.

Lineae distensae

These are frequently, but not invariably, found, the sites being the lower abdomen and inguinal regions, the loins, the lumbar regions, along the iliac crests, and even the breasts. They may be short narrow lines or long broad zones, bright red or dusky violet. They are rightly regarded as being the result of the stretching and giving way of the skin by accumulation of fat in the underlying subcutaneous areas, and may, therefore, be met with in other conditions in which fat is deposited. However, the rapidity of fat deposition is one factor that tends to their production, the thinness of the skin in Cushing's syndrome is another, and noteworthy is the deep red or purple colour that is associated with the plethora and cyanosis met with in this syndrome, with or without polycythaemia. Similar coloured lineae are met with in pregnancy and are regarded as physiological, but it is also known that in pregnancy there is a hypersecretion of adrenal glucocorticoids; and because of this and the fact that pregnancy may be a precursor of Cushing's syndrome, pregnancy has been referred to as transient physiological Cushing's syndrome. Of course, the striae of pregnancy eventually lose their red or purple colour, and become pallid. In so far as the red colour depends upon the visibility of the underlying capillary-venous network through the stretched and 'broken' skin—and this vascular network is active in an attempted repair process—it is probable that all white lineae distensae have at one time been red. Nevertheless, the degree of redness and particularly of purple coloration seems to depend upon the concentration of adrenal glucocorticoids, and the rapidity of the process. Adiposity does not seem to be the only factor, since such coloured lineae are sometimes found in Cushing's syndrome when the adiposity is minimal. Following successful operation the red-violet lineae become pale.

It is recognized that red lineae appear in a variety of infections, particularly tuberculosis, mainly in children or adolescents, and in a number of normal children at puberty, and in adipose gynism or gynandrism, but a hypersecretion of adrenal glucocorticoids is a probable factor in all these conditions or phases (Simpson, 1945).

Other cutaneous features

Apart from the coloured lineae distensae, other cutaneous features are thinness of the skin, patchy cyanosis, plethora, subcutaneous haemorrhages, spider-like telangiectases, hirsutism, and acne. The thinning and delicacy of the skin is believed to be due to a general breakdown of tissue protein associated with protein catabolism and a negative nitrogen balance, all due to the excess of adrenal glucocorticoids. If, however, there is an associated increase in adrenal androgens which are protein anabolic and antagonistic to glucocorticoids, this thinning is minimal and is not found in the mixed syndrome. Cyanosis is met with in patchy form in the face, over the breast, abdomen, buttocks, thighs, and legs and upper arms. Over the thighs and legs, because of its resemblance to a marble pattern, it was called cutis marmorata. It is not unrelated to underlying fat deposition, but there is probably another vascular factor which is not well understood. The cyanosis may be very intense. The face has a congested plethoric appearance, aggravated by a tendency to cyanosis and telangiectases, although in mild cases it could be referred to as a ruddy and robust facies. Spontaneous subcutaneous haemorrhages, especially on the upper limbs, give the appearance of heavy bruising, often excessive and of deep blue-black colour, but aetiological trauma is absent or trivial. Capillary friability and a decrease in fibrinogen levels have been found. Purpura, ecchymoses and bruising are also complications of massive cortisone therapy in non endocrine disorders.

Hirsutism is not an essential feature and is usually limited to the face. It may develop on the body but this suggests an associated increase in adrenal androgens, even when not revealed by 'normal' 17-ketosteroid assay values. When cortisone is given in high dosage for non-endocrine cases, any hirsutism that results is usually limited to the face and chin. Acne may also be a feature of Cushing's syndrome, and it is also a complication of cortisone therapy. Both hirsutism and acne may be severe in the mixed type of syndrome, where there is also a high concentration of androgens. In the male, facial and bodily hair may decrease because of the associated hypogonadism and in the adolescent it may not develop, but in the mixed type of syndrome with increased androgens, the facial and bodily hair growth may become more intense. Crops of boils or pustules may be a troublesome feature of Cushing's syndrome, being part of a general susceptibility to infections, particularly staphylococcal.

Hypogonadism

Amenorrhoea is the rule in women. Occasionally, menstruation is present but scanty, and very rarely menstruation may be normal. Swan and Stephenson (1935) recorded normal menstruation in a woman of 31, after which there was a period of amenorrhoea, followed by a resumption of normal menstruation on the giving of thyroid. In this case the ovaries were normal at autopsy; more usually the ovaries are found to be atrophic or involuted, or show cystic degeneration. The uterus is usually infantile, and the endometrium thin and atrophic.

In the male, involution or failure of development of the sex organs, and an absence of potency, is the rule. In Josephson's case (1936) of a boy of 17, the failure of sexual development led to a initial erroneous diagnosis of Frohlich's syndrome. Hypogonadism in both sexes is due to the suppression of pituitary, gonadotrophins by the high blood concentrations of adrenal glucocorticoids, as is also evident from massive cortisone therapy and animal experiments.

Blood

Polycythaemia is inconstant but nevertheless part of the syndrome. Just as in Addison's disease there is a relative lymphocytosis and eosinophilia, so in the opposite condition of Cushing's syndrome there is an excess of polymorphonuclear leucocytes, a lymphopenia, and a diminution or complete disappearance of eosinophils. The sedimentation rate is usually normal, but may be raised in the absence of obvious infection. An alkalosis, with low blood potassium, and raised blood sodium is occasionally found.

Skeletal changes

The bones tend to become rarefied and soft. Clinically this is shown by kyphosis and bent shoulders, with resulting lessening of height. Cushing particularly observed that 'nearly all the females appear to have been definitely undersized'. Radiographically the vertebrae are seen to be compressed, and all the bones of the body rarened, as compared with controls. Backache in the lumbosacral region is not uncommon and may be due to compression or spontaneous fracture of vertebrae—as also occurs in ribs, clavicles, and long bones. This clinical picture is quite comparable to that of hyperparathyroidism, but although an increased calcium excretion in the urine has been demonstrated (Albright, 1938) values for serum calcium, phosphorus, and even phosphatase are usually normal. Further, the parathyroid glands in most cases are histologically normal, but fatty infiltration, slight hyperplasia, and, in one case, adenoma, have been reported. A further analogy with hyperparathyroidism are the fine deposits of calcium which sometimes are found throughout the kidneys, or minute brown calculi in the calyes, and secondary chronic nephritis, with sclerosed glomeruli and calcium deposits, as in Case P. 4211 of Cushing (1933). In spite of these similarities it is probable that the hormone mechanism is unrelated to the parathyroids and that it is all secondary to the negative nitrogen balance, determined by the excess of adrenal glucocorticoids, which results in loss of bone matrix so that normal calcium concentrations lack adequate osteoid sites for normal deposition and are therefore in relative excess in the soft tissues and urine.

Hypertension and the kidneys

High blood pressure, systolic and diastolic (e.g. 220 and 140 respectively), is a frequent finding in Cushing's syndrome. It may occur in the absence of clinical or pathological renal disease or in association with chronic nephritis or malignant nephrosclerosis. Hypertension, probably of adrenal origin, may be regarded as the opposite of the hypotension of Addison's disease. Since an excess of aldosterone is a rarity in Cushing's syndrome, the hypertension is more appropriately correlated with the excess of glucocorticoids and changes in the intima of the blood vessels with associated hypercholesterolaemia. There is no excess of adrenaline or noradrenaline in Cushing's syndrome. Basophilia of the anterior pituitary has been correlated with hypertension in Cushing's syndrome, as also in eclampsia, and Selye (1947) described hypertrophy and hyperplasia of the pituitary basophils in experimentally induced stress. Selye also produced hypertension and nephrosclerosis with deoxycortone in rats. Since cortisone, in massive therapy, can produce hypertension in man, reversible by ceasing the therapy, and endogenous excess of cortisone or hydrocortisone is more compatible with clinical facts and hormone assays as the cause of the hypertension.

The kidneys are not infrequently involved secondary to vascular endarteritic changes, although this explanation is not adequate in the rare complication of malignant nephrosclerosis. In some patients the renal changes are due to minute depositions of calcium throughout the kidney substance, and in the tubules, as may occur in hyperparathyroidism. Evidence of renal disease may not be present in life, or may only be found on special examinations, *e.g.* granular casts in the urine, and poor renal functions test. In some cases, however, as in that of Close (1935), there may be frequency, haematuria, blurring of vision, vomiting, cramps, dyspnoea, drowsiness, and anuria, as well as albuminuric retinitis and intractable headaches. Experimental evidence indicates a more direct relationship between hypersecretion of the adrenal cortex and renal pathology.

Psychoneurotic symptoms

Depression, with suicidal tendencies in severe untreated cases, has been the writer's experience, often completely reversible with successful surgery. Personality changes are also met with, including mental and emotional aberrations, in some 50 per cent or more of patients. Although depressive states are more usual, maniacal agitation and violence are met with. Personality changes and behaviour patterns, as well as the psychoneuroses and psychoses, have been aetiologically associated with a variety of hormone disorders, although the basic genetic personality influences the type of response.

Thyroid function

This may be normal or decreased, the latter in keeping with the known depressant effect of cortisone on thyroid function and with the histological picture of a resting thyroid gland. The raised blood cholesterol, characteristic of Cushing's syndrome, is found even when thyroid function is normal and its degree is not paralleled by the mildness of thyroid insufficiency. The thyroid gland is diffusely enlarged to a moderate degree in some 50 percent of patients, in the presence of normal or decreased thyroid function. Patients with Cushing's syndrome rarely complain of sensitivity to cold or other features of myxoedema.

Exophthalmos

This is not infrequent in moderate degree and suggests a pituitary exophthalmic hormone dissociated from the pituitary thyrotrophic hormones.

Steroid diabetes

Experimental and clinical evidence leaves us in no doubt as to the potency of the adrenal glucocorticoids as diabetogenic hormones. Nearly all patients with Cushing's syndrome showed a delayed fall of blood sugar values after glucose, some 50 per cent give clinical and biochemical evidence of mild diabetes mellitus without ketosis, and a small percentage present as diabetes, with or without ketosis, or with a complication of diabetes such as neuritis or retinitis. The diabetes usually has the main characteristics of steroid diabetes, being relatively mild, slowly progressive, and not difficult to control with diet only. Insulin is required by some patients and these are not necessarily insensitive to insulin. Insulin-sensitivity curves indicate incomplete refractoriness to intravenous insulin. The severity of the diabetes usually fluctuates with the severity of the general symptomatology and its degree of control by therapy. In some patient with an adrenal tumour, and high concentrations of hydrocortisone in the urine, the only manifestation was severe diabetes mellitus, which disappeared on surgical removal of the tumour. If a steroid diabetes is allowed to continue untreated, a secondary pancreatic islet-cell degeneration results in a severe pancreatic diabetes. In most patients who have come to autopsy, the islet cells have been found to be normal, but in a few, especially children, the islet cells were hyperplastic, which should be regarded as a homeostatic reaction to the steroid hyperglycaemia, and an explanation of why diabetes is not found in all cases. The potency of cortisone in aggravating diabetes and precipitating ketosis is dramatically demonstrated by its use in patients with combined Addison's disease and diabetes; as also by the fact that cortisone in massive doses has been observed to produce diabetes in nonendocrine diseases.

Polyuria and polydipsia

These symptoms may be caused by glycosuria or by an excessive calcium excretion. Clinically speaking, they are not infrequent in Cushing's syndrome.

Pigmentation

The pigmentation of the orbits, face, and neck that is sometimes found in Cushing's syndrome, in the absence of any evidence of adrenal insufficiency, must be attributed to an excess of pituitary ACTH or the closely associated pituitary melanocytic hormone. The same must be true of the pigmentation that sometimes follows successful subtotal adrenalectomy, in the absence of any other indication of adrenal insufficiency.

Fatigability

The explanation of this symptom is perhaps not too obvious as an excess of cortisone suggests the opposite of Addison's disease, where the weakness is immediately remedied by cortisone replacement. However, the excess is associated with protein catabolism and a negative nitrogen balance and clinically there is wasting and flabbiness of the musculature. Weakness may be extreme and possibly other less obvious metabolic factors are involved, e.g. the loss of potassium in the urine and low tissue concentrations of potassium in some cases. In the mixed syndrome, where there is also an excess of adrenal androgens, weakness is not a feature.

Climacteric type of Cushing's Syndrome

This condition cannot really be included as a variety of Cushing's syndrome, since amenorrhoea is a normal feature of the climacteric, but nevertheless, many of the other features of Cushing's syndrome can develop at this time (Simpson, 1936), and it may be presumed that development of excessive pituitary gonadotrophic secretion becomes extended to pituitary ACTH secretion and this may be transitory in some, but persistent in other. A patient described by Achard and Thiers (1921) presented the more dramatic features, spontaneous leg fracture, diabetes, and hypertension at the age of 71, but, as in many other patients, other features were present in mild form much earlier and in this patient hair on the chin at the age of 9 suggested adrenal pathology. It might be queried whether a slowly progressive disorder occurring over a period of fifty or more years could justifiably be included under the definition of Cushing's syndrome. It must be regarded as the opposite of characteristic and yet cannot be ignored.

Diagnosis

Amenorrhoea, a change of facies in regard to plethora and rounding of cheeks and chin, and a deposition or transfer of subcutaneous fat, are prominent indications of the syndrome. When the development is progressive following pregnancy, the diagnosis is even more probable. It is obvious from the clinical description that the syndrome may first present itself to the gynaecologist, the orthopaedic surgeon, the psychiatrist, the vascular renal clinic, or as the plethoric type of diabetic.

Simple urinary assay or chromatography will reveal high values for glucocorticoids and normal values for 17-ketosteroids. Blood glucocorticoids are raised and following intravenous ACTH, 15 Units in 100 ml. of saline, will show a sharp rise within four hours, the adrenals in Cushing's syndrome showing a supernormal hypersensitive response in this respect.

The differentiation between hyperplasia and neoplasia depends upon radiography and hormone suppression tests, remembering clinically that hyperplasia is four times as frequent as neoplasia. Intravenous pyelogram may indicate depression of the renal calyces below an adrenal tumour, so that they lie at right angles to the vertical. Perirenal insufflation of air has given way to the safer procedure of episacral insufflation of air. When successful, the adrenal area is outlined by the surrounding air and an enlarged or neoplastic adrenal is indicated by the shape and the outline of the adrenal shadow.

When the cause of the excessive secretion of adrenal glucocorticoids is bilateral hyperplasia, the giving of cortisone, e.g. 25 mg. by mouth four-hourly or 100 mg. by injection once daily, should result in a significant fall in 17-hydroxycorticoids in the second twenty four hours' collection of urine. However, cortisone itself is partly excreted as 17-hydroxycorticoids and it has been found more satisfactory to use the more powerful prednisone, 5 mg., six-hourly by mouth: or the still more potent fludrocortisone (9-a-fluorohydrocortisone), 2 mg. six-hourly by mouth. Both prednisone and fluorohydrocortisone are excreted in insignificant amounts as urinary 17-hydroxycorticoids. In the case of adrenal neoplasm, there is no suppression of 17-hydroxycorticoids in contrast with hyperplasia. However, anomalous results occasionally occur both with neoplasm and hyperplasia, so that the tests are not absolute and must be interpreted in conjunction with other evidence (Nobano, Moxham, and Walker, 1958).

Albright (1938) considers that low values for 17-ketosteroids are an indication of an adrenal adenoma, since the high glucocorticoid secretion results in suppression of pituitary ACTH arid consequently of adrenal androgenic secretion; whereas high values for 17-ketosteroids and 17-hydroxycorticoids (glucocorticoids) are suggestive of adrenal carcinoma.

Treatment

When an adrenal tumour is present surgical removal is the obvious course. As the opposite adrenal is usually atrophied the patient is given cortisone (e.g. 100 g. by injection daily) or prednisone, (e.g. 25 mg. by injection daily) for two days prior to operation, on the day of operation, and for one week afterwards, after which the dose is gradually reduced and changed to oral therapy; and may be tentatively omitted after a further three weeks if the clinical condition of the patient suggests that the remaining adrenal gland is functioning adequately.

In the unusual circumstances of the pituitary fossa being enlarged, or where the technique of pituitary radiation has been highly developed, it appears to some to be justifiable to attempt radiation therapy before surgery, in the absence of an adrenal neoplasm, and amelioration or cure is recorded in a varying proportion of patients: 5-40 per cent. Permanent loss of hair over the radiation sites is a hazard even ii the best centres. The insertion of radon seeds or cauterization of the pituitary by open operation are in the opinion of the writer more hazardous than bilateral subtotal adrenalectomy.

Unilateral adrenalectomy is as useless for Cushing's syndrome, with bilateral hyperplasia, as was hemithyroidectomy for thyrotoxicosis. In fact, it is sometimes followed by exacerbation of symptoms because of excessive compensatory hyperactivity of the remaining gland. Subtotal adrenalectomy, comparable to subtotal thyroidectomy for thyrotoxicosis, is the logical procedure. It was first attempted at the Mayo Clinic, and particulars of their first series of 29 cases were published by Priestley et al. in 1951. Six of the 29 patients died in the post-operative period, but the last 10 patients (and a further 9 referred to in a footnote) were operated upon without any mortality, the safety depending Upon cortisone becoming available. Of 20 patients followed up, 1 was not improved, and 19 obtained excellent remissions, although in 3 of them there was subsequent recurrence of major symptoms, three, two, and one year after operation. Another 3 of the 19 had chronic adrenal insufficiency, and required substitution therapy with cortisone. The adrenals were removed at separate operations. Two days before the second operation, one day before, on the day of operation, and for three or more days after operation 200 mg. of cortisone acetate were injected intramuscularly, after which the dose was gradually reduced and finally omitted in the absence of adrenal insufficiency. In the case of bilateral hyperplasia, there is really no need to give cortisone for the first adrenalectomy. Neither is cortisone required until the actual day of the second adrenalectomy, but giving the massive injections in oily solution for two days prior to this provides a depot from which absorption takes place gradually during the day of operation—at least, that is the rationalization of the pre-operative therapy while blood cortisone is already high.

In the absence of cortisone therapy, or with inadequate therapy, acute adrenal insufficiency may prove fatal within a few days of operation or a delayed reaction may occur in the second or third week. Manifestations are anorexia, nausea, vomiting, weakness, fever, tachycardia, abdominal pain, muscular and articular pains, hypotension, and pigmentation. Smaller series of patients treated by subtotal adrenalectomy have been published subsequently by several observers, with broadly similar results, including the Guy's group, who advocate simultaneous radiotherapy of the pituitary gland to avoid recurrence. Others have recommended total bilateral adrenalectomy, preferring constant post-operative therapy with cortisone to the risk of recurrence of Cushing's syndrome. At St. Mary's Hospital, Simpson, Robb, and Wynn have not found recurrence to be such a major problem as to warrant the intentional production of a state of severe Addison's disease by bilateral total adrenalectomy. One whole gland has been removed and seven-eighths of another, so that if clinical recurrence is met with and does not respond to pituitary and adrenal radiation, it is known on which side to explore for compensatory growth. In one patient without any evidence of adrenal insufficiency, and with no chronic cortisone therapy, there were no signs of relapse even after two successful pregnancies. Unless unusual difficulties are experienced with the first adrenalectomy, the second is carried out at the same time, using two loin incisions. Cortisone is, at first, given by intramuscular injection but within a few days of operation it may be given by mouth, e.g. 25 mg. four times a day, then twice a day, with further gradual reductions until, in the absence of signs of adrenal insufficiency, it is discontinued. None of the patients have developed permanent adrenal insufficiency, and the occasional pigmentation could be ascribed to excessive secretion of pituitary ACTH that follows the operation. A 0.4 per cent. solution of noradrenaline in normal saline intravenously is rarely needed during operation, or for the more immediate post-operative period. Cortisone has been replaced or reinforced by hydrocortisone (intravenously), hydrocortisone succinate (intravenously), prednisone and fluorohydrocortisone. Peeling of the skin is an interesting post-operative feature that may go on for months, and one patient developed Raynaud's phenomena in the fingers in cold weather. There have been no deaths. Results have been uniformly good, and subtotal bilateral adrenalectomy would appear to be the operation of choice. The more immediate change in the contour of the facies and the prominence of the abdomen suggests that factors other than fat are involved.

Mason has summarized the results of his own series, and communications from others, and concludes that subtotal adrenalectomy is a satisfactory procedure but that the margin of safety between adrenal insufficiency and danger of recurrence is small for patients under 25.

Mixed type of Adrenal Hyperfunction

Adrenal hyperfunction involving both the adrenal glucocorticoids and adrenal androgens and producing a mixed clinical picture, with features both of Cushing's syndrome and of the adrenogenital syndrome. It may be caused by adrenal tumour or adrenal hyperplasia. The title is admittedly a clumsy one, even when reduced to the letters M.T.A.H. for brevity, but it is difficult to omit recognition of such a condition under this heading, without excluding from consideration endocrine syndromes which are district variants of Cushing's syndrome and cannot come within the classical definition of the latter, without permitting considerable latitude and qualification.

Further justification for the burden of additional classification will be forthcoming in a consideration of the many contrasting and antagonistic actions of the adrenal glucocorticoids and androgens, and in a consideration of the symptomatology, with illustrated cases, as well as the response to surgical therapy.

Hormone basic

Although the adrenals secrete many hormones, the two most important groups in this syndrome are the glucocorticoids and the androgens.

In the table 8.4. are seen ten contrasting or antagonistic actions. Most of these are related to the fact that glucocorticoids cause a breakdown of protein and protein tissues, with a negative nitrogen balance, and gluconeogenesis from protein. We have seen that this is also the most likely explanation of the osteoporosis of Cushing's syndrome. Osteoporosis does not appear if there is a coincident increase in adrenal androgens which are nitrogen anabolic. The potassium loss from the tissues is partly associated with the protein breakdown but cortisone does produce potassium loss as well as sodium and water retention when given in massive doses, although its analogue, prednisone, is relatively free from these electrolyte effects. Rarely in Cushing's syndrome there is also an associated increased secretion of aldosterone, which is chemically closely related to cortisone.

Table 8.4. Contrasting actions of Glucocorticoids and Androgens

	Glucocorticoids	*Androgens*
1.	Protein catabolic	Protein anabolic
2.	Negative nitrogen balance	Positive nitrogen balance
3.	Osteoporosis caused	Osteoporosis prevented
4.	Diabetogenic	Nil
5.	Flabby musculature	Strong musculature
6.	Somatic growth inhibited	Somatic growth increased
7.	Skin and hair thin and dry	Skin and hair thick and greasy
8.	Fat anabolism preponderant	Fat catabolism preponderant
9.	Potassium depletion	Potassium retention
10.	Thyroid function depressed	Thyroid function increased

If hormone assays were a true reflection of the degree of secretion and activity of hormones, it would be relatively easy to predict or interpret the clinical picture from such assays. Unfortunately this is not the case, particularly if one is limited to the urine and to standard methods, of assay. In particular, the 17-ketosteroid values are frequently 'normal' where there is undoubted clinical evidence of excessive androgenic activity. In regard to hormone assays in general, it would be prudency to say that 'normal' values, as for sedimentation rates, do not necessarily exclude disorders but raised values are always significant.

Clinical picture

Hirsutism in classical Cushing's syndrome, particularly in the absence of an adrenal tumour, is usually slight and limited to the face. In so far as mild hirsutism may be produced by massive cortisone therapy for non-endocrine conditions, cortisone has a mild androgenic activity due to weak androgenic by-products of its metabolism. When, however, hirsutism of the body has become extensive, it is due directly to adrenal androgens. Similar considerations apply to acne, especially when the skin is thick and greasy, because of well-developed sebaceous glands, rather than thin and atrophic. Although the initial action of androgens causes the scalp hair to thicken, become greasy, and grow more densely and rapidly, there is subsequent retraction of hair line in the frontal and temporal areas and loss of hair from the vertex. In some patients, with an excess of both androgens and glucocorticoids, the latter appear to inhibit the androgenic activity of the former, in that facial and body hair is not excessive and in some constitutional mixed endocrine types may even be scarce. In animal experiments, cortisone inhibits the growth of hair follicles.

Hormone influence on somatic growth is most obvious in the period of skeletal growth. There is no doubt that cortisone inhibits skeletal growth (including height increment) and this is a characteristic feature of Cushing's syndrome in childhood and adolescence. On the other hand, and excess of adrenal androgens in virilism or macrogenitosomia praecox is found to cause an increased rate of growth, as well as skeletal maturation, as is also the case when testosterone is used therapeutically in hypopituitarism or hypogonadism in childhood. Therefore in a mixed syndrome in childhood or adolescence the androgens counteract the glucocorticoid inhibition of growth, and the resultant may be a 'normal' rate of growth.

Fat metabolism is complicated and has been discussed more fully under adiposity. Briefly, it is in children that one sees more clearly an obvious general excess of fat deposition from an excess of glucocorticoids, whereas in adults there is often only a shift of fat from extremities to face and trunk. Virilizing androgenic tumours appear to cause a loss of fat so that muscular contours are clearly visible, but both glucocorticoids and androgens produce an increased appetite and food intake. The weakness of Cushing's syndrome is due both to muscle breakdown and atrophy and to potassium loss, in contrast with the excessive muscular strength found with adrenal androgenic tumours, particularly in childhood. In the combined syndrome therefore, muscular weakness may not be a feature. The total clinical picture will also vary according to whether the adrenal glucocorticoids or the adrenal androgens preponderate, or whether they are equally balanced.

Diagnosis

Diagnosis depends upon definition, and diagnostic tests are described in the section on Cushing's syndrome and on the adrenogenital syndrome.

Course and prognosis

This condition is often surprisingly slow in its evolution and this may be the case even when an adrenal tumour is the cause. In the absence of an adrenal tumour, it would appear justifiable to be content with symptomatic treatment in mild cases but the fact that an adrenal adenoma might become an adenocarcinoma and give rise to metastases after some twelve or more years should arouse caution and point the need for thorough and repeated radiological and hormone investigations. An increasing proportion of dehydroiso-androsterone, indicates neoplastic changes. The endocrine results of surgery are good but hirsutism may persist in some degree.

Treatment

An adrenal tumour should be removed and the results are usually excellent. The same precautions as in uncomplicated Cushing's syndrome should be taken and cortisone therapy is required. The hirsutism may be persistent in some degree even when the 17-ketosteroids are reduced to low subnormal values by removal of an adrenal tumour and this is even more likely when subtotal adrenalectomy is carried out in the absence of an adrenal tumour. For this reason subtotal adrenalectomy must not be undertaken in this mixed syndrome with promise of complete cure, Nevertheless, in the writer's opinion it is a logical and justifiable procedure in severe cases, and patients are appreciative even when hirsutism is not abolished.

The persistence of hirsutism in some patients, when hormone assays indicate subnormal concentrations of androgens, suggests that once the condition has been produced by androgens, the change may be irreversible, particularly if it has been present for a long time.

The cure or improvement of hirsutism may also be merely temporary, if hyperplasia recurs, as it may do in some patients. The treatment of hirsutism by cortisone or prednisone suppression of adrenal androgens is only mildly effective in these mixed syndromes, as might be expected, although acne tends to be improved and menstruation frequently returns.

Case histories

Case 1. A married woman of 29 complained of the following feature of three years' duration arising during or within six months of pregnancy: Redness and plethora of face, hairiness of face and body, greasiness and thinning of scalp hair, acne of face and body, fatness of face and body, bruising, persistent red striae, and amenorrhoea. The hair of the head became thick and greasy; eyebrows very thick and thick black hair grew over the shoulders and spine: B.P. 190/120. It is of interest that this mixed syndrome arose in pregnancy as does Cushing's syndrome. Urinary 17-ketosteroids were high, 33 mg. per 100 ml as were also the plasma 17-hydroxysteroids, 35 μg. per 100 ml Bilateral subtotal adrenalectomy was followed by good clinical result, including loss of plethora, acne, hirsutism and adiposity, and return of menstruation.

Case 2. Married woman of 40 with no children, who for twelve years had amenorrhoea, severe hirsutism of face and body, will loss of scalp hair, acne, easy bruising extremely plethoric facies, polyuria, hypertension, 200/140 mm.; polycythaemia, 6.3 million red cells per c.mm. and raised haemoglobin, 19 g. per 100 ml. blood; raised urinary 17-ketosteroids, 250 mg., raised urinary glucocorticoids, expressed as 20-ketosteroids, 6 mg. (normal 2 mg.), raised urinary oestradiol, 600 i.u. (normal 25 to 350). Removal of adrenal adenocarcinoma led to disappearance of all features except hirsutism. Subsequent relapse and death due to hormone-secreting metastases.

Adrenal Hyperfunction in Children

Bulloch and *Sequeira* (1905) first drew attention to the fact that the adrenals are related to normal puberty, as well as to pathological pseudo-puberty. In modern phraseology, at the age of 14 (11-16) years, the adrenals as well as the gonads develop an increased activity, because of the simultaneous secretion of pituitary ACTH as well as pituitary gonadotrophins; and, perhaps, because of increased sensitivity of the end-organs to the stimuli. This adrenal activity is shown by the fact that the urinary 17-ketosteroids are increased in females as well as males at puberty. Albright (1941) applied the term adrenarche to this conception. Simpson (1951) extended the idea in two directions: (1) the adrenarche to include adrenal glucocorticoid as well as adrenal androgenic activity; and (2) recognition that in some normal children a preliminary adrenarche starts at the age of 7 (approximately) and proceeds gradually to maximum intensity at 14, or there is a relatively stationary period between the pre-adrenarche at 7 and the puberty adrenarche at 14. These conceptions are based upon a study of large numbers of normal children and children with adrenal disturbances in childhood, and will be found useful in considering major endocrinopathies in childhood. Both in normal and mildly abnormal children, some of the clinical criteria of adrenal androgenic activity are the degree of hair on the face, body, and limbs, greasiness of the skin and scalp hair, with seborrhoea and scurf, thick eyebrows and long eyelashes, darkening of fair scalp hair, increased rate of growth and weight increment, and increased muscular development and strength. It is obvious that these features are accompaniments of normal puberty. However, if the hair on the sides of the face, upper lip, and forehead is excessive, or if hair is present over the shoulders and scapulae, and extensive over the lumbar area, or more than well marked on the limbs, then the indefinite border region between accepted normality and endocrinopathy may have been crossed; or if these androgenic features develop at 7 instead of 14 the query of abnormality is raised, particularly if vulval hair also starts growing. Evolution to adult 'normality' is usual, but in some there persists a tendency to hirsutism with oligomenorrhoea as a post-puberty feature, and in others a more severe androgenic endocrinopathy develops in the second or third decade or following pregnancy or the climacteric.

When the adrenarche is associated with preponderant glucocorticoid hypersecretion, the patients tend to be fat and plethoric.

Adrenal Tumours in Children

Bulloch and Sequeira (1935) drew attention to adrenal tumours in childhood causing pseudo-sexual precocity in boys, with enormous strength, due to the high secretion of adrenal androgens (Hercules type), to hirsutism and other androgenic features in girls, and also to a plethoric type of adiposity, resembling a condition later termed Cushing's syndrome, but usually combined with an associated excessive androgenic secretion, giving a mixed clinical picture. Lawson Wilkins (1948) reviewed the literature and found that up to 1948, there were records of 70 children (53 girls and 17 boys) who had developed an adrenal cortex tumour before the age of 12 years. In three of these cases, the neoplasm occurred in aberrant adrenal tissue in or near the ovaries. It is probable that the published literature on adrenal tumour in children only reflects a percentage of the total number of cases. Of 53 girls, 31 showed virilism, and 22 showed symptoms suggestive of Cushing's syndrome. All the latter were obese and most showed hirsutism and other evidence of excessive androgens, and it is of additional interest that 6 gave evidence of oestrogen secretion probably of adrenal origin. Although 12 showed obvious virilism as early as the first year of life, none showed a persistent urogenital sinus, as is found with congenital adrenal virilism. Of the 17 boys, 12 showed/ precocious pseudosexual development accelerated growth and osseous development and muscularity rather than obesity. The testes, unlike the penis, were not precociously developed, with two exceptions difficult to explain. Two showed Cushing type obesity. Malignancy and metastases were frequent.

Wilkins' first case was of special interest being an adrenal tumour in a boy of 5 years of age producing gynaecomastia. The breasts began to enlarge at 6 months of age and reached then maximum size at 2 years. Otherwise clinically he was normal except that the bone age was advanced, namely that of 10 years of age. The areolae of the breasts were well developed but not pigmented and superficial veins were prominent. The penis and testes were normal for the age but the prostate was' definitely enlarged, and thought to be hard and fibrous. The 17-ketosteroids per twenty-four hours were 4.1 mg., a figure considered by Wilkins only slightly greater than average for the age, and the oestrogens (5 rat units per twenty-four hours) were considered to be only a questionable slight increase for the age. Histological examination showed hyperplasia of the duct epithelium and very marked peridectal hyperplasia of the connective tissue. Removal of an adrenal encapsulated adenoma led to a return to normal, observed for a period of four years. No further assays are given. The total evidence is interpreted as indicating an oestrogenic feminizing adrenal tumour.

Wilkins' second case was a malignant but encapsulated adrenal tumour in a girl of 5, producing enlarged clitoris, deep voice, pubic, anal, and upper lip hari, seborrhoea, acne, increased height, bone age, and musculature: all these features noticed at the age of 5. The 17-ketosteroids per twenty-four hours were greatly raised, 22 mg., falling to 2 mg. after removal of tumour. Clinical improvement was followed by relapse and death two years later, as a result of abdominal metastases, with 17-ketosteroid excretion rising to 124 mg. per twenty-four hours. There was no enlargement of the breasts and no features of Cushing's syndrome. No oestrogen assays were recorded but the vaginal smear showed no oestrinization. Since many adrenal tumours in children are malignant, the β-fraction (dehydroisoandrosterone) of the urinary androgens is usually considerably raised but in this case it was only 1 per cent, from which a good prognosis might have been erroneously made: the histology, however, suggested possible malignancy. Prognosis should always be cautious with adrenal tumours in childhood, even when capsulated, and early diagnosis and removal is obviously important.

Other aspects of adrenal tumours in childhood are considered under sexual precocity and pseudohermaphroditism.

Virilism in Children

When the appearance at birth is normal and the virilism commences in childhood before puberty, the cause is nearly always an adrenal tumour, and the incidence of malignancy and metastases is high. Apart from hirsutism, acne, and seborrhoea, the clitoris is considerably enlarged, the voice deep, skeletal growth and maturation (radiographic bone age) much advanced, muscular development and strength much above normal, and the 17-ketosteroids greatly increased. In the female the urethral and vaginal orifices are separate in contrast to the common urogenital sinus found in genetic virilism and this clinical differentiation is particularly important when a tumour occurs in the first year or so of life.

Urinary 17-ketosteroids are not depressed by giving cortisone to a patient with an adrenal tumour although there are occasional anomalies. In the male the penis is large and the testes small.

When virilism starts between the ages of 7 and 14, and particularly when it is not severe or rapidly progressive, adrenal hyperplasia, rather than neoplasia may be the cause. Further investigation of a large series of cases by modem methods will give us a better indication than we have at present as to what percentage of such patients are simple qualitative hyperplasia and what percentage have genetic metabolic errors comparable to those to be described under the heading of genetic virilism but not so potent as to become manifest until the adrenarche comes into play. This consideration is not meant to conflict with the main differential diagnosis between severe virilism that appears in childhood and is caused by an adrenal tumour and genetic virilism already apparent at birth. The differential diagnosis between adrenal tumour and hyperplasia is dealt with also in the previous adult section.

The difficulties of diagnosis in exceptional cases is shown by the fact that general hairiness at birth may be due to an adrenal tumour, which does not really become manifest until some years later, as in the girl of 5 years of age described by Kolf and Tjiook (1950). Whether this is an example of hyperplasia passing on to multiple adenomata and then a single dominating adenocarcinoma must remain conjectural.

Genetic Metabolic Virilism

This is due to an innate inability of the adrenals to convert 17-hydroxyprogesterone, a normal adrenal intermediate metabolite, to adrenal glucocorticoids (cortisone and hydrocortisone). This innate defect of adrenal metabolism results in a deficiency of glucocorticoids, and an excess of adrenal androgens, as measured by total urinary 17- ketosteroids, and by assay of the urinary end-products of 17-hydroxyprogesterone, especially pregnanetriol. Since pituitary ACTH secretion is controlled by the blood level of hydrocortisone, the deficiency of the latter results in an excessive secretion of ACTH, and consequent hyperplasia of the adrenals, which then produce an additional secretion of normal and abnormal adrenal androgens. This theory was invoked by Bartter *et al.* (1951), Bongiovanni *et al.* (1953, 1954), and

Jailer (1953) to explain the clinical and biochemical facts in genetic virilism in females and pseudo-sexual precocity (macrogenitosomia praecox) in their male siblings.

Clinical features

1. The condition is present from birth, although it is not always obvious to observers until a year or so has elapsed.
2. The hirsutism and muscularity, together with a much enlarged clitoris, are clinical signs of a considerable excess of androgens.
3. The patients are apt to go into Addisonian adrenal insufficiency crisis, either spontaneously, or following extra stress such as intercurrent infections or minor operations. This is mainly manifest in regard to salt and water metabolism some patients even showing a chronic salt craving. Addisonian type pigmentation and hypoglycaemia are rare, and before their nature was recognized and cortisone became available adrenal crises often proved fatal. Other manifestations are vomiting, anorexia, and diarrhoea, and vomiting may be the presenting symptom. The inherent biochemical defect probably extends to the synthesis of aldosterone (electrocortin) but this has not yet been suggested or demonstrated. Some patients reach middle age without any adrenal insufficiency showing itself and yet go into severe crisis following a minor operative procedure.

Pathology

The adrenal glands in fatal cases were found at autopsy to be considerably enlarged often three times the normal, and sometimes nodular with multiple adenomata, as well as bilateral hyperplasia. The hyperplasia consisted of polyhedral eosinophil granular cells resembling those of the reticular zone.

Hormone assays

The blood adrenal glucocorticoids when measured as plasma 17-hydroxysteroids are low, and so are the urinary 17-hydroxysteroids if measured by appropriate methods, e.g. chromatography or Reddy's method. (Confusion was initially caused by failure to realize that normal or high values are given by other methods, e.g. assays of ketogenic steroids, which include pregnanetriol in their total figure.)

Response to ACTH

The blood and urinary 17-hydroxysteroids fail to increase after the injections of ACTH, since the synthetic defect prevents this, but there is an increase in pregnanetriol excretion. Although the excess of pregnanetriol is a metabolic abnormality characteristic of genetic virilism, pregnanetriol is found occasionally in small quantities in normal women and has been found in excess in Cushing's syndrome due to adrenal hyperplasia.

Response to cortisone

Cortisone in high dosage, or prednisone, results in a decrease of total 17-ketosteroids and of pregnanetriol in the urine, explained by the resulting inhibition of the raised pituitary ACTH secretion. This is accompanied by dramatic clinical improvement, e.g. breast development, maturation of testes, slowing of the rate of maturation and ossification of bones, decreased hirsutism, & c. Bergstrand and Gemzell (1957) found an excess of pregnanediol, as well as pregnanetriol in genetic virilism and a fall to normal values after cortisone.

Response to 17-hydroxyprogesterone

Hormone Assays. The administration of 17-hydroxy-progesterone results in an increased secretion of pregnanetriol (3α-17 α-20 α-pregnanetriol) in genetic virilism but not in normal patients.

The clinical manifestations of genetic virilism (pseudo-hermaphroditism) and macrogenitosomia praecox due to the same genetic defect are described in the gonad section.)

Cushing's Syndrome in Children

This disorder is rare in childhood, apart from the puberty type, but it may occur at any age, even in infancy; and it is usually, but not invariably, due to an adrenal tumour which is frequently malignant. The symptoms differ from those in adults in the following respects, but must nevertheless be interpreted and determined by the relative and absolute excess of adrenal glucocorticoids over adrenal androgens.

1. Skeletal growth is inhibited by the excess of glucocorticoids.
2. The adiposity is usually more extensive than in adults and may involve the extremities, including even the hands

and wrists; the appearance may almost resemble that of a monstrosity.

3. Androgenic features such as acne, seborrhoea, and hirsutism are more prominent, because of the frequency of high androgen secretion by the adrenal neoplasm.
4. Hypertension may be extreme, e.g. an infant of 6 months of age, with a blood pressure of 245/145, followed by death from cardiac failure (Marks, Thomas and Waskany, 1940).
5. Diabetes is rare, and the absence of disturbed carbohydrate metabolism can be correlated with a greater tendency to compensatory pancreatic islet-cell hyperplasia.

Series of cases in children have been reviewed by Marks *et al,* (1940), Chute *et al,* (1949) and Melicow and Cahill (1950). The manifestations in some children with adrenal neoplasm is a mixture of Cushing's syndrome and virilism, as might well be expected.

Adipose Gynism and Gynandrism

A-G. Syndrome-Simpson's Syndrome-Simpson-Pende Syndrome

This syndrome was first described by Simpson in 1948 and adipose gynandrism in males and as childhood adiposity with some features of Cushing's syndrome in females; and more specifically under the title of adipose gynism in girls and adipose gynandrism in boys respectively in 1950. The term gynism was used for girls because their general configuration is intensely female, whereas the term gynandrism was used for boys to indicate a mixture of female and male characteristics. Pende (1951), in Italy, described the syndrome in girls only, apparently independently. The respective eponyms were used by Maranon (1953) in Spain, and by Berardinelli (1955) in Brazil, in confirmatory clinical accounts.

Definition

A constitutional familial pituitary-adrenal hyperfunction manifested by a plethoric adiposity, height above normal, delayed puberty in boys, early puberty in girls, elements of feminization in boys and advanced or intense womanly characteristics in girls, characteristic behaviour patterns and attitudes, with a demonstrable increase in the secretion of adrenal glucocorticoids and a probable increase in the secretion of growth hormone.

Clinical features

The fact that similar clinical description of the syndrom were made at first independently, by observers in England and Italy, and later by other clinicians, who had seen these illustrated accounts, seems to establish the conditions as a clinical entity. Marquand (1951) regarded the condition as of pituitary origin, because of the incidence of familial giantism and diabetes among the relatives, and Prunty (1956) in spite of some assay problems described below, concluded that we are inclined to agree with Simpson that there is an underlying abnormality of adrenal cortical function. The time of onset usually corresponds with early puberty or with the pre adrenarche, e.g. at 7 or 8 years, but it may occur at any age, and is sometimes precipitated by an acute infection, or an operation, e.g. tonsillectomy.

In both sexes the patients are taller than average; the adiposity effects the face, chin, chest breast area, abdomen, buttocks, and thighs in contrast with thin, distal extremities; the face is red or cyanotic, often with multiple spider telengiectases which are also found in the cerbical area and sometimes on the trunk. The general facial appearance is similar to, if not identical with, that of Cushing's syndrome in childhood and this similarity is heightened by the general type of adiposity, the tendency to a short, thick neck and the forward stoop of the head and neck. The buttocks often resemble the face in colour and may be deeply cyanotic in some patients particularly in colder weather, and very cold to the touch; cyanosis is also present over the breast and thighs in patchy form, and the marble pattern on the thighs and legs justifies the term cutis marmorata, which latter is frequently found in Cushing's syndrome.

The skin is thin and delicate and in milder cases this contributes to a healthy and beautiful complexion with rosy cheeks. Whatever the cause may be, the eyes also are often beautiful giving the appearance of a lovely face perhaps babyish, attached to a massive body. There is a tendency to easy bruising associated with the thin skin and capillary friability. Red lineae distensae of the trunk are frequent and are identical with those found in Cushings syndrome, showing similar variability in intensity, colour, and distribution. They are not always present and diagnosis may be made before they appear. In both sexes the scalp hair tends to be dry and to come well forward on the front and side of the forehead.

Osteoporosis is not found, partly because of the small excess of adrenal glucocorticoids, partly because adrenal androgens are compensatory. Nevertheless, the forward inclined head and round shoulders, together with a short, thick neck, gives many children an appearance and posture resembling that of Cushing's syndrome.

In both sexes the pelvis is broad in relation to the shoulders and gives a waddling gait, noticeably in the male, and explains in part the usual inability to run or to participate in sports requiring rapid movement. On the other hand, swimming is a favourite sport with high achievement, and the general configuration of both sexes is characteristic of champion swimmers. Muscular strength is variable but it may be considerable, probably when associated with a simultaneous excess of adrenal androgens. The height is some two inches above the average throughout the period of growth and usually above average when the epiphyses have joined, in spite of the fact that the epiphyses join prematurely in the female, associated with a tendency to an early menarche. The radiographic bone age is usually advanced in females and normal or retarded in males. Puberty is retarded in males and descent of the testes is not infrequently delayed. It is of interest that Cushing (1933) described a typical example of his syndrom in a girl of 15, with precocious adolescence and menarche at the age of 10, and three years regular menstruation followed by amenorrhoea; and that Cushing's syndrome with adrenal tumour, in the male, has been associated with delayed Sexual maturation and gross adiposity (Josephson, 1936).

The breasts in girls are not only prematurely developed, but to a greater extent than in many normal adults, often giving the appearance of a woman's breast during pregnancy. The areola is especially broad and prominent, with numerous Montgomery's tubercles surrounding the nipple, and tortuous superficial veins may be present. Gynaecomasitia is not infrequent in the male. In neither sex has it been possible to demonstrate an excess of urinary oestrogens, and pituitary or adrenal mammogens must be postulated. Such breasts are met with in cases of adrenal tumour in children. The patchy cyanosis of the enlarged glandular and fatty breast has been referred to as well as the red or purple lineae distensae in this area in some patients.

As regards the nature, mannerisms, and behaviour pattern, both sexes tend to be hypersensitive, emotional (sometimes going pale and sometimes blushing, even over the neck, with little provocation), exhibitionistic, actively social and conversational, artistic, musical, histrionic, sentimental, with close maternal association in males. They are good dancers despite their massive bulk, entertainers, and often very intelligent and gifted. In addition to these elements and the physical points mentioned previously, the gestures of the male, the tendency to high-pitched voices and scanty beard growth, their flair for clothes and cooking, all tend to justify the term gynandrism. Nevertheless, their gonadal evolution, although delayed, is normal (apart from the tendency to horizontal pubic hair), as is their potency and fertility. Further, the behaviour pattern and attitudes are considerably modified when there is an associated adrenal androgenic excess.

Diabetes is a rarity in the childhood or adolescent phase, but is frequently mentioned in the family history. Diabetes is also a rarity in Cushing's syndrome in childhood, because of the compensatory pancreatic islet-cell hypertrophy. In six cases very carefully studied by Martin and Simpson (unpublished) it was not possible to demonstrate unequivocal abnormalities of carbohydrated tolerance or insulin sensitivity. In another series using the technique of the glucose uptake of the isolated rat diaphragm, an excess of plasma insulin activity was demonstrated in three of twenty cases studied. However, elevated plasma insulin activity was not constantly found in these three patients, so that the significance of these results must be questioned at the present time. Polycythaemia, hypertension, renal casts, and albuminuria are rarities and cannot be considered as an essential part of the syndrome, although occasionally met with, and of interest in relation to Cushings' syndrome.

Development and family history

Some detailed case histories have been given in previous publications. In both sexes, subsequent development in adolescence may approximate to normality, and for this reason the condition has been interpreted as an intensification of normal puberty. This is a valid supposition, as far as it goes, and incompatible with the broader conception of the normal adrenarche: and study of the syndrome throws much light on normal childhood and puberty, and the wide variation met with. At the time study of individual cases over many years, as well as of family histories, shows that many stigmata remain, even when the general appearance becomes normal, that many patients do not evolve to normality, and that others suffer a relapse in adult life, sometimes without obvious cause but often precipitated by pregnancy or the climacteric. Adiposity, hypertension, diabetes, and gallstones, appear to be significantly frequent in family histories.

In males, in adolescence, there is often in late adrenal androgenic phase, in which general body and facial hairiness is rapidly acquired, together with increased muscularity and strength, a spurt in growth, a broadening of the shoulders, and a loss of fat.

Hormone assay

Initially J. Bornstein, using the biological method of hepatic glycogen deposition in the adrenalectomized rat, found raised values for urinary glucocorticoids in five patients: 71-106 units compared with the normal 32-57 units. R. Gardner found raised values for urinary glucocorticoids using the copper reduction-arsenomolybdate method of Talbot. Prunty found normal values by measurement of the ketogenic steroids and, contrary to Cushing's syndrome with adrenal hyperplasia, no augmentation following ACTH stimulation. Gray *et al.* (1956), using chromatographic methods, found that 'the excretion of compound F was significantly increased above that of normal children; and, in adipose gynandrism, the excretion of an unknown delta Δ^4-3-ketosteroid, X_4, was also significantly increased'. Later, this substance X_4 was identified as Reichstein's compound U (Δ^4-pregnene-17-α-20 β-21-triol-3-11-dione). Delphine Parrott (1951) found an increase in plasma ACTH in four out of five patients. Gardner found a variable increase in urinary 17-ketosteroids but Gray and others found no constant difference from normal controls.

Mixed androgenic types

Whether or not 17-ketosteroid assays confirm clinical findings, there seems no doubt that in some of these patients there is also an excess of adrenal androgens as indicated by body and facial hairiness, severe acne, thickness of eyebrows and very long eyelashes, great muscular strength, greasiness and seborrhoea of scalp hair rather than dryness, and rapid growth of scalp hair, and nails. The androgenic phase in adolescent males has been referred to. The clinical picture is best interpreted by reference to the previous section on the mixed syndrome in adults. The androgenic element explains the occurrence of especially strong fat men in the family history, e.g. circus weight-lifters and iron-bar benders, and why some of these patients are great athletes, e.g. Rugby players and boxers. In one heavyweight boxer of this type, not hirsute, the 17-ketosteroids were 34 mg., and in a very strong tall adult woman of this type, also not hirsute, they were 32 mg. The absence of hirsutism in these cases suggest an antagonism between androgens and glucocorticoids in regard to hair growth. Two strong athletic brothers, with this mixed syndrome both developed steroid diabetes, as had other older members of the family (Simpson and Oakley, unpublished).

Treatment

There is no specific treatment available. Reduced diet, with or without appetite restrainers, e.g. dexamphetamine, is effective in some degree, as in other endocrinopathies. Chorionic gonadotrophins will accelerate descent of the testicle and testicular maturation, but this will usually happen spontaneously although delayed.

Adrenal Feminization

The term is applied to a clinical condition in man, produced by an oestrogen-secreting adrenal tumour and characterized by gynaecomastia, testicular tatrophy, with loss of libido and potency, and deposition of fat, especially on the breast, abdomen, and hips. The condition is rare.

The adrenal-cortex tumour is an adenoma, or adenocarcinoma, and is not distinguishable histologically from other hormone-secreting adrenal tumours; but hormone assays have shown that such tumours in man secrete a gross excess of oestrogens, demonstrated for the first time in the case of Simpson and Joll (1938), and confirmed in a similar case by Roholm and Teilum (1942). The high level of blood oestrogens inhibits the secretion of pituitary gonadotrophins, so that involution and atrophy of the testes follow. Although the urinary androgens of adrenal origin may be normal or slightly raised, impotence and loss of libido is the rule in the presence of a gross excess of oestrogens. The latter also explains the gynaecomastia. Adiposity is not invariable and, when present, must be presumed to have the same adrenal hormone basis as has been considered with other adrenal tumours. A review of adrenal feminization is given by Armstrong and Simpson (1948).

Holl's two cases (1930) will be briefly described as illustrating that: (1) puberty gynaecomastia may be associated with an adrenal carcinoma, which fact throws some light on the probable physiological mechanism of puberty gynaecomastia, and (2) the adipose type in the adult male may be characterized by a feminine appearance and removal of the tumour result in a return to normality.

Case 1. A lad of 15 who for two months had observed progressive enlargement of the breasts, which projected in bud-like fashion, like those of a young girl. The nipples were deeply pigmented. The suprapubic hair was feminine type with horizontal border. Death followed removal of a malignant cortical tumour.

Case 2. A male of 44 years of age with two children. The breasts slowly enlarged, with some pain, and the nipples became large and pigmented. The testes and penis became smaller, libido was lost, and sexual intercourse ceased. The patient became fat and his face took on a soft feminine appearance. The interesting feature in this case however, is that removal of an adenomatous cortical tumour was followed by a return to complete normality.

Simpson and Joll's classical case (1938) will be briefly described because of the oestrogen assays before and after operation and following metastatic recurrence:

In a physician of 34, the presenting symptom was pain in the left scapula and hypochondrium. However in retrospect, it was seen that two years before there had been enlargement of the breasts, with general increase of fat deposition; and one year before, diminution in size of the penis and testes, with loss of libido and potency. The adiposity, however, only lasted some months, and was followed by loss of weight. A large malignant cortical tumour was removed from the left side and incomplete clinical improvement ensued, namely decrease in size of breasts and partial recovery of libido and potency. This was only temporary, and relapse followed the development of hepatic and visceral metastases from which the patient died.

The oestrogens, which were exceedingly high at the time of operation, fell to values below normal after operation, and subsequently increased to very high values (3,000 I.U., per litre) with the development of metastases. Burrows *et al.* (1936) stated that 'an attempt to isolate the oestrogenic hormone from the urine has not been successful, but the results are consistent with the view that the hormone is oestrone'. Urinary androgens were just above normal, as measured by comb growth assays.

An unusual case of an adrenal-cortex tumour in a male of 25, producing gynaecomastia and hypoglycaemia, was described by Staffieri, Cames, and Cid (1949). Two atypical cases have been described in adult males (McFadzean, 1946, and Chambers, 1949), in which urinary gonadotrophins were present in excess, apparently secreted by the adrenal tumour. This was evidenced in McFadzean's case by the disappearance of a positive Friedman test after successful removal of a adrenal adenocarcinoma. Chamber's patient died before removal of an adrenal carcinoma was possible and excreted 50,000 mouse units of gonadotrophin per 100 ml of urine; the Friedman test was positive. Gonadotrophins have also been found in excess in a woman with adrenal virilism and disappeared after removal of an adrenal adenocarcinoma. It seems extraordinary and yet established that an adrenal tumour secretes gonadotrophins.

Aldosteronism

A syndrome produced by excessive secretion of aldosterone, with resulting sodium retention, hypokalaemic alkalosis, and alkaline urine; and manifested clinically by thirst, polyuria, nocturia, polydipsia, hypertension, weakness, periodic paralysis, muscular cramps, and tetany, but usually with an absence of oedema. The condition is sometimes referred to as primary aldosteronism to differentiate it from secondary aldosteronism. Which is a compensatory mechanism in oedematous conditions (renal, cardiac, and hepatic) and in saltlosing nephritis. Some cases previously diagnosed as potassium losing nephritis were subsequently proved to be examples of primary aldosteronism.

Incidence

The condition may occur at any age, usually above 12, and in either sex. By June 1957 a total of 16 cases of primary aldosteronism had been recorded, of which 3 were diagnosed only at autopsy.

Pathology

Of the 16 recorded cases, 13 were due to adrenal adenoma. 1 to adrenal adenocarcinoma, and 2 to adrenal hyperplasia. The adenoma usually arises from cells of the zona glomerulosa, but in one patient (Camphell *et al.,* 1956), the adenoma was thought to consist mainly of zona fasciculata cells. In Buchem's (1956) patient with bilateral hyperplasia, each gland weighing 8 g., the fascicular zone was more hyperplastic than the glomerular. In Conn's case the adrenal contralateral to the adenoma showed atrophy of the zona fasciculata. In others the contralateral gland was apparently normal.

The kidneys show arteriosclerosis and a diffuse vacuolar change and hydropic degeneration in the tubular epithelium, proceeding in some areas to necrosis.

Clinical features

Conn's case was a 34-year-old housewife who complained that for seven years she had been having attacks off intense generalized muscular weakness, with on four occasions, complete paralysis of the lower limbs lasting a few days. She had been quite well between attacks, although polydipsia, polyuria, and nocturia had been present for some years, with a known hypertension of 190/100, and proteinuria. Chvostek's and Trousseau's signs were present. Chalmer's patient, a woman of 43, presented with periodic weakness of limbs and neck and inability to lift the arms or head. Buchem's patient was a boy of 17 who as well as retarded physical and genital development-had had polyuria and polydipsia since the age of 2, but who presented with hypertension (220/150), and papilloedema. The urine in twenty-four hours exceeded 4 litres and the specific gravity was 1.004. The patient of Holten and Petersen was a girl of 13, admitted with a pyrexial infection who, during subsequent observation, manifested attacks of tetany, convulsions, and unconsciousness, with hypertension (195/150) and papilloedema. Mader and Iseri's patient was a negro woman of 33, who had noticed spasma of the hands during pregnancy, and six months later muscle weakness, cramps, paraestesia, restlessness, and nervousness. Milne and Muehrche described two patients, one originally thought to be a potassium-losing nephritis but with a twelve-year history of attacks of hypokalaemic muscular paralysis lasting, on one occasion, as long as twenty-one days; and the other, almost asymptomatic, with mild transient headaches and moderate hypertension, in whom; an electrocardiogram that was characteristic of potassium depletion, led to the discovery and removal of a small aldosterone-secreting adrenal adenoma. Hewllett's patient was a man of 44 complaining of dryness of the throat for three years, but also suffering from period of weakness, polyuria, and headaches, with hypertension for many years. These selected clinical histories give a characteristic clinical picture of this syndrome in its various manifestations and presentations. Although the primary electrolyte action of aldosterone is on sodium retention, the potassium less and hypokalaemia explain many of the clinical manifestations, especially attacks or paresis or paralysis. The paraesthesia and tetanic spasms are caused by potassium depletion and metabolic alkalosis decreasing calcium ionization, the serum calcium concentration being normal. The tetany does not usually respond to intravenous calcium. The hypertension and associated arterial and renal changes are secondary to the sodium retention and are reproducible in animal experiments,. However, the renal tubular defect is a recognized complication of chronic hypokalaemia.

Diagnosis

An awareness of the possibility of aldosteronism is called for in a variety of conditions such as hypertension, nephritis, attacks of muscular weakness or paralysis, thirst, polyuria, and tetany. The serum sodium is only moderately raised, e.g. 145 m Eq. per litre, and may be normal, in spite of considerable sodium retention. The serum potassium is usually considerably lower than normal, e.g. 2.0 mEq. per litre. The serum chloride is normal or even low. The salivary sodium/potassium ratio is readily obtained by the flame photometer and may be 1: 3 as compared with the normal of 1: 1. An electrocardiogram is typical of potassium depletion. The alkali reserve of the blood is characteristically high, e.g. 34 m Eq. per litre. The urine is usually persistently alkaline but neutral urine has been recorded. The urine is of low specific gravity and there is striking impairment of urine-concentrating power, but renal excretion and urea concentration tests are normal and proteinuria slight.

Although the Chvostek and Trousseau signs are positive, the serum calcium and phosphorus are normal, and the signs of tetany are usually not influenced by intravenous calcium but are ameliorated by prolonged potassium therapy. The polyuria and polydipsia and low specific gravity of the urine simulate diabetes insipidus, but they are entirely uninfluenced by vasopressin. Familial periodic paralysis is associated with low serum potassium but not the other features of aldosteronism and analysis of muscle biopsy will show high or normal potassium concentration in the former but low potassium concentration in the latter.

The rare condition of a potassium-losing nephritis, sometimes associated with obvious pyelonephritis and multiple calculi, can be distinguished from aldosteronism by evidence of primary nephritis, a good clinical and biochemical response to high potassium intake (to which aldosteronism is refractory), renal biopsy and the absence of high values for aldosterone in blood and urine.

The final proof of aldosteronism is the demonstration of increased quantities in the blood and urine. The initial step is partition chromatography and then bio-assay of the purified fraction in the adrenalectomized rat, measuring the effect by radioactive sodium and potassium excretion. The normal values for aldosterone are less than 10 mg. per twenty-four hours in urine and less than 0.1μg. per 100 ml. in the blood; in aldosteronism these values are raised considerably, e.g. three times the normal.

In secondary aldosteronism, nephrosis, cardiac failure hepatic cirrhosis, oedema, or free fluid are primary features, and the hyperaldosteronism is an endeavour to maintain the extracellular fluid volume. Oedema does not occur in primary aldosteronism. In a case of idiopathic oedema due to capillary permeability to protein, hyperaldosteronism proved to be secondary. The rats condition of primary salt-losing nephritis may be compensated by secondary aldosteronism. Potassium depletion is not a prominent feature of secondary aldosteronism. Hyperaldoteronism, with urinary values ten times the normal, is met with in the third trimester of normal pregnancy but is not related to toxaemia of pregnancy. It is believed to be secondary, to maintain the extracellular fluid volume.

Tomography after episacral air insufflation may reveal an adrenal tumour, when such is the cause of primary aldosteronism.

Treatment

This consists of removal of an adrenal tumour or subtotal adrenalectomy in the rare cases of bilateral hyperplasia. Cortisone postoperatively is not essential as cortisone secretion is usually normal but as partial atrophy of the remaining adrenal has been reported, its immediate post-operative use might be considered prudent Apart from operation, or where operation has to be deferred, potassium in very heavy doses may be ameliorative, but is rarely really effective, e.g. 600 mEq. of potassium chloride by mouth or 200 mEq. per litre intravenously. These are heroic doses.

Adrenal Hypofunction Addison's Disease

A condition of adrenal cortical insufficiency, usually due to tuberculosis or atrophy of the adrenal cortex, and characterized by malaise, pigmentation of skin and mucous membranes, low blood pressure, weakness, loss of weight and fat, anorexia and gastrointestinal symptoms, abnormal loss of salt in the urine and saliva, a tendency to develop hypoglycaemia, and a liability to go into adrenal-insufficiency crisis spontaneously or as a result of infection or trauma.

History

Addison's disease was first described in 1849 by Thomas Addison, physician to Guy's Hospital, in an address to the South London Medical Society. The paper was entitled

'Anaemia: Disease of the Suprarenal Capsules', since he had not then differentiated the idiopathic anaemia, which was also subsequently named after him. In 1885, however, assisted by his junior colleague, Samuel Wilks, he produced his clinical monograph, *On the Constitutional and Local Effects of Disease of the Supra-renal Capsules.* William Hunter, in 1909, suggested the term Addison's disease, a very worthy eponym for an epoch-making discovery, that was first received, as is not infrequent, with scepticism and even scoffing, and later acknowledged as the clinico-pathological foundation of endocrinology.

Aetiology and pathology

The disease occurs in both sexes, is most frequent in the third and fourth decades, and is rare in children before puberty, as it is also rare in old age. Tuberculosis and idiopathic atrophy (necrosis) are the two common causes, and although tuberculosis is still the most frequent basis (60 per cent), its incidence is tending to lessen with the decrease in tuberculosis generally. Idiopathic necrosis is responsible for some 30 per cent. of cases. Other causes may be regarded as rare and exceptional, e.g. amyloidosis, bilateral carcinomatous metastases, bilateral primary adrenal carcinoma, mycosis fungoides, and histoplasmosis. Prolonged treatment of non-endocrine states with high doses of cortisone produce a chronic latent adrenal insufficiency, liable to become acute and fatal with operative procedure. Acute bilateral adrenal haemorrhage, with adrenal crisis, is associated with some severe infections, particularly in childhood (Waterhouse-Friderichsen syndrome). In genetic virilism severe adrenal insufficiency may be coexistent but generally a latent adrenal insufficiency exists, precipitated by trauma or infection.

In O'Donnell's series of 8 patients with tuberculous adrenal glands, 4 showed evidence of old healed pulmonary tuberculosis, 2 advanced active pulmonary tuberculosis with cavitation, 1 tuberculous osteomyelitis of the lumbar vertebrae, and 1 with no other recognized tuberculous focus. This indicates a rather higher percentage of associated active pulmonary tuberculosis than is generally found.

The tuberculous adrenals are often enlarged and both the cortex and medulla are almost totally involved in a fibrocaseous and granulomatous tuberculous destruction, with occasionally a few islets of regenerated cortical cells. In parts of the gland there may be found more proliferative lesions, with many tubercles, gian cells, endothelial cells, fibroblasts, and lymphoblasts. Areas of calcification may be found. It is often difficult or impossible to recognize the medulla, macroscopically or microscopically.

The cause of idiopathic atrophy is unknown, although bacterial and metabolic cytotoxic agents have been postulated and a comparison with hepatic necrosis made. The glands are very small and difficult, or occasionally impossible, to find. The pathological process is limited to the cortex, the medulla being usually intact, these findings giving pathological support to the fact that the adrenal-cortex steroids are the essential missing elements in Addison's disease, and that the absence of adrenaline and noradrenaline plays no part in determining its symptomatology. The cortical layer is very thin, being almost entirely replaced by distended capillaries lying in a sparse, delicate, connective tissue, infiltrated with lymphocytes, plasma cells, and histiocytes, Small areas of surviving cortex and large atypical cells, in the process of regeneration and compensatory hypertrophy may be detected. The histological picture bears a closer resemblance to a toxic necrosis, as, for example, occurs in the liver, than to a simple atrophy. Lymphatic hyperplasia, involving the lymphatic glands, the spleen, the thyroid, and the thymus is frequently met with, in keeping with animal adrenalectomy experiments. Although recognizing a tendency to thymic hypertrophy in man, Sloper (1955) was unable to substantiate this by weight measurement in his series. Experimentally cortisone cause involution of the thymus and lymphatic glands.

Sloper (1953) has made a special study of the thyroid gland in Addison's disease, and found that in nine of twenty-one cases the thyroid showed severe involutional changes. In two patients there was severe atrophy of the thyroid, and in the remainder the thyroid was normal. Lymphoid infiltration of the thyroid is frequent but Sloper noted that the lymphoid thyroids of Addison's disease were typically unenlargeds, on this point differing from the enlarged lymphoid goitres of thyrotoxicosis and Hashimoto's disease. The pathological findings suggested that 'in some instances the patient may be approaching a stage of subclinical myxoedema, averted in some by the presence of areas of hyperplasia, in others by the number of surviving normal acini'; Sloper (1953) was unable to confirm the previous observations of Hinerman (1951) that there is invariably an islet-cell hyperplasia of the pancreas, although in two cases he found small islet-cell adenomata. The pancreas was usually diminished in weight; and was atrophic in one case with coincident diabetes. The testes and ovaries may be normal or involuted. There was marked fall in the number of pituitary basophil cells and to a less extent of the eosinophil cells. It is perhaps difficult to correlate these pituitary findings with the excess of pituitary ACTH found in the plasma in Addison's disease, but they must be regarded as histologically factual.

The heart is small and shows brown atrophy, but these classical findings are rare in the modern cortisone-therapy era, The kidneys are usually normal, but may show tubular degeneration.

Clinical features

The first manifestation of Addison's disease may be an acute crisis, but more usually the onset is insidious. Sometimes tiredness and malaise precede the full clinical picture for many years and postmortem evidence indicates that some nine-tenths, or more, of the adrenal cortex must be destroyed by disease to produce the full syndrome. The classical description of Addison (1855) still constitutes an excellent clinical picture:

The patient, in most of the cases I have seen, has been observed gradually to fall off in general health; he becomes languid and weak, indisposed either to bodily or mental exertion; the appetite is impaired or entirely lost; the whites of the eyes become pearly; the pulse small and feeble, or perhaps somewhat large, but excessively soft and compressible; the body wastes, without however, presenting the dry and shrivelled skin and extreme emaciation usually attendant on protracted malignant disease; slight pain or uneasiness is from time to time referred to the region of the stomach, and there is occasionally actual vomiting which in one instance was both urgent and distressing, and it is by no means uncommon for the patient to manifest indications of disturbed cerebral circulation. We discover a most remarkable and, so far as I know, characteristic discoloration taking place in the skin-sufficiently marked indeed as generally to have attracted the attention of the patient himself or the patient's friends. This discoloration pervades the whole surface of the body, and is commonly most strongly manifested on the face, neck, superior extremities, penis, and scrotum, and in the flexures of the axillae and around the navel. It may be said to present a dingy or smoky appearance, or various tints or shades of deep amber or chestnut brown; and in one instance the skin was so universally and so deeply darkened that but for the features the patient might have been mistaken for a mulatto. In some cases the discoloration occurs in patches, or perhaps rather certain parts are so much darker than others as to impart to the surface a mottled or somewhat chequered appearance; and in one instance there were in the midst of the mottling, certain insular portions of the integument presenting a blanched or morbidly white appearance, either in consequence of these portions having remained altogether unaffected by the disease, and thereby contrasting strongly with the surrounding skin, or, as I believe, from an actual defect of colouring matter in these parts. Indeed, as will appear in the subsequent cases, this irregular distribution of pigment cells is by no means limited to the integument, but is occasionally also made manifest on some of the internal structures. We have seen it in the form of small black spots beneath the peritoneum of the mesentery and omentum—a form which in one instance presented itself on the skin of the abdomen. This singular discoloration usually increases with the advance of the disease; the anaemia, languor, failure of appetite, and feebleness of the heart become aggravated, a darkish streak usually appears upon the commissure of the lips; the body wastes, but without the emaciation and dry harsh condition of the surface so often observed in ordinary malignant disease; the pulse becomes smaller and weaker; and without any special complaint of pain or uneasiness, the patient at length sinks and expires.

Pigmentation

The pigmentation is an increase of the normal melanin pigment in the basal layers of epidermis. The precursor of melanin in vertebrates is tyrosine, oxidized by the enzyme tyrosinase to dihydroxyphenylalanine and through further intermediate steps to melanin (Lerner, 1950). The pigmentation is especially intense in areas exposed to light, pressure, and irritation. The palms of the hands and soles of the feet, however, escape pigmentation, except for the creases at the interphalangeal joints. Almost pathognomonic of the disease is the pigmentation of the mucous membrane of the mouth, involving the inside of the cheeks, the inner lips, gums, and posterior aspect of the palate. In some patients, however, the pigmentation is limited to the skin. Some degree of desquamation of the latter is not uncommon. As Addison himself pointed out, a leukodermic type of pigmentation is occasionally met with, and the correctness of the diagnosis with this pattern of pigmentation has been confirmed *post mortem*. Idiopathic leukodermia itself is not of adrenal origin but when coincident with Addison's disease, the pigmented areas become more pigmented, but the leukodermic areas remain white.

Although patient tend to become more pigmented in phases of relapse and less pigmented in periods of improvement, the degree of pigmentation is not necessarily a measure of the severity of the disease. In fact, deep pigmentation may be present for some time before any other major manifestation of the disorder. On the other hand, precipitate onset and an acute downhill course may be met with in patients with little or no pigmentation. Normally, fair people and those who go red rather than brown when exposed to the sun, have little or no melanin formation in the skin, and so fail to become pigmented, or deeply pigmented, if they should develop Addison's disease. Such a failure to pigment may increase the difficulties of diagnosis in such cases.

. Since pigmentation is slight or absent in Simmonds' disease, in contrast with the average case of Addison's disease, it has been suggested that pigmentation depends upon an excess of pituitary ACTH secretion, which is present in Addison's disease but absent in Simmonds' disease. This view is supported by the development of pigmentation after ACTH therapy for non-endocrine states. In frogs, darkening of the skin, produced by expansion of the melanophores and dispersion of pigment in them, results from injection of intermedin, especially in the pale hypophysectomized animal. Intermedin will produce darkening of the human skin with development of pigmented naevi (Lerner *et al.*, 1954).

Intermedin, so named because it is derived from the pars intermedia of animal pituitaries, is also called melanocyte-stimulating hormone (M.S.H.) or melanophore-expanding hormone, or chromatophoric hormone, and has been found in normal human blood and inhuman pituitaries; and in increased amounts in Addison's disease, after bilateral adrenalectomy, in Cushing's disease, and in pregnancy (Sulman, 1956). The anterior lobe of the pituitary is the probable source of chromatophoric hormone in man, and frog-test assays were found by Sulman to parallel assays of corticotrophin by the ascorbic acid depletion test in several conditions. He further found that the output of these hormones, or biochemical complexes was diminished by cortisone, desoxycorticosterone, and aldosterone, all of which diminish pigmentation in Addison's disease.

While ACTH and M.S.H. appear to be closely associated in pituitary extracts, they have been partially separated by chemical processes and M.S.H., unlike ACTH, is stable to alkali. Salassa and colleagues at the Mayo Clinic (1954) made the significant observation that an ACTH preparation, injected into a patient with primary pituitary insufficiency, produced stimulation of adrenal function but only slight cutaneous pigmentation, whereas an M.S.H. preparation produced little effect on adrenal function but intense pigmentation both of skin and mucous membranes. From a practical angle it seems justifiable to refer to an ACTH complex, containing M.S.H, being an important factor in determining increased pigmentation.

Gastro-intestinal system

Anorexia, especially for fatty foods, is usual, often being associated with nausea and vomiting. Anorexia is also an early sign of relapse. Constipation is usual, but diarrhoea, sometimes precipitated by a mild purgative, is met with and is serious as it may herald a crisis. The onset of diarrhoea should be regarded as a warning of adrenal insufficiency in a patient previously stabilized on appropriate treatment. Various types of abdominal pain may simulate intestinal colic, gastric ulcer, gastric perforation, gall-bladder disease, and renal disease. This last is suggested by the pain in the loin due to the underlying tuberculous adrenal; and tenderness in the costovertebral angle is often present. Irritation of the overlying diaphragm and adjacent pleura produces pain, referred to the shoulder (diaphragmatic pleurisy) and round the costal margin, the pain sometimes being very severe. Hypochlorhydria, or achlorhydria, are frequently found in the course of investigation, if looked for.

Respiratory system

Disorders of rhythm and sighing respiration are not uncommon in the more severe phases of the disease. There is also a susceptibility to respiratory infection, and the supervention of bronchitis or pneumonia may precipitate a crisis and prove fatal. Antemortem evidence of active pulmonary tuberculosis is infrequent, and the incidence of Addison's disease in tuberculosis clinics or sanatoria is small.

Muscular system

Muscular weakness and easy fatigability are characteristic. Asthenia affects all the muscles, and not one special group of muscles as in myasthenia gravis. There may be considerable muscular wasting and creatinutia. The latter, in contrast with the creatinuria of myasthenia graves, disappears on treatment with cortical extract. Cramps of the calf muscles are sometimes met with, and maybe due to hypochloraemia, although hypoparathyroidism is rarely a coincident condition. The lack of muscular strength is explicable on several grounds: (i) general illness; (ii) the decrease of adrenal androgens, as indicated by the 17-ketosteroid assays, particularly in women the adrenal are the only source of androgens; (iii) the disturbance of potassium metabolism; and (iv) the absence of hepatic glycogen reserves and available glucose for sustained effort.

Nervous system

Inertia, lassitude, and apathy are common, but there may be periods of restlessness, excitability, and insomnia. Sometimes the patients tend to lie curled up in bed, or to sink down beneath the covering. Involuntary cries and grimaces, and later, delirium, may precede a crisés. Negativism, contrariness, and pessimism may be features of the more chronic phases. Psychotic states, depressive and delusional, are also met with, but must be regarded as rare. The psychoneurotic manifestations, or abnormal behaviour pattern can be ascribed to (1) abnormal loss of salt, since McCance observed depression, apathy, and irritability in normal volunteers deprived of salt, (2) abnormal carbohydrate metabolism with phases of hypoglycaemia, and (3) absence of cortisone since cortisone will almost immediately produce a return to normal of the abnormal electroencephalogram often associated with severe Addison's disease.

Blood count

Leucopenia, relative lymphocytosis, and eosinophilia are characteristic, being the opposite of the changes found in Cushing's syndrome and of those produced by an excessive administration of cortisone. In the patient with Addison's disease, cortisone will reverse these abnormalities in white cell counts. Hypochromic microcytic anaemia is often found and may be due to achlorhydria but it is also corrected by cortisone. The anaemia may not be obvious initially if there is a degree of haemoconcentration when the patient is first seen.

Plasma volume and electrolytes

The basic conceptions are discussed in the physiological section. The absence of aldosterone is the most important factor in determining electrolyte and fluid changes in Addison's disease although the absence of hydrocortisone is an additional factor. The primary result of aldosterone deficiency is sodium depletion with excessive excretion of sodium in the urine and in the saliva, the latter indicating that the renal is not the only mechanism involved. The sodium loss results initially in a decrease in extracellular fluid and plasma volume in an endeavour to maintain plasma concentration of sodium at normal values, but if the sodium loss is severe this compensating mechanism fails and there is a fall in plasma sodium values. In addition to loss of extracellular fluid through the kidneys, some of the fluid passes from the extracellular to the intracellular spaces. This latter is important clinically since treatment with desoxycortone, or fluorohydrocortisone, or aldosterone, can restore a decreased plasma volume even without the addition of extraneous fluids, causing a redistribution of fluid from the intracellular fluid to the extracellular fluid, including plasma. Severe sodium depletion impairs renal circulation so that initial polyuria gives way to oliguria, and even anuria, particularly in crisis.

Sodium depletion leads to the clinical picture of dehydration, and the diminution in interstitial fluid is indicated by a loss of tissue elasticity, a loss of superficial tissue fullness, or turgor, and by a low intraocular tension. The deficiency of plasma fluid is indicated by a rise in the haematocrit and in protein concentration and by poor venous filling, the veins being collapsed and difficult to enter with a needle. Even when the vein is properly entered, it will collapse after withdrawing 1 or 2 ml. so that one must be patient and wait until it slowly refills before drawing up a total of 20 ml. The additional fall in blood pressure is also an expression of sodium depletion and a compensatory vasoconstriction leads to cold extremities. Muscle cramps are also an expression of sodium depletion, the serum calcium being normal or raised.

In mild Addison's disease the plasma sodium may be normal, even when the daily excretion of sodium is in excess, but in severe cases the plasma sodium is diminished. The plasma chloride is a much less reliable index of adrenal electrolytic function, as several factors determine chloride values, but a low chloride value is confirmatory. With the increased excretion of sodium there is a decreased excretion of potassium so that plasma potassium values may be normal or raised, the latter only being found in severe adrenal insufficiency.

Cortisone, or hydrocortisone, has a mild sodium an associated fluid retaining action so that its absence contributes to the electrolytic defects in adrenal insufficiency. In so far as many patients in severe adrenal insufficiency, and even in crisis, respond dramatically to cortisone only, its influence on sodium and fluid may be greater in this condition than experimental considerations would appear to permit and in particular it may influence considerably the transfer of fluid from the intracellular fluid to extracellular fluid. In Simmonds' disease we have seen that the absence of cortisone results in haemodilution with increase of plasma volume causing lower values for haeamoglobin and plasma proteins; and cortisone causes a diuresis with a correction of this. This phenomenon is found in investigation water-loading in Addison's disease (Kepler test) and is corrected by cortisone but it does not make itself manifest otherwise in Addison's disease and this is probably explained by the coincident absence of aldosterone, which latter is normally present in Simmonds' disease.

The excessive loss of the kation sodium in Addison's disease results in a metabolic acidosis and a lowering of the plasma bicarbonate. This is aggravated by the fact that there is defective ammonia formation by the kidneys in Addison's disease, as ammonia formation spares sodium loss.

Renal function

Although adrenalectomy may result in ultimate renal tubular degeneration (Simpson and Korenchevsky, 1935) in animals, and comparable changes may be found in fatal cases of Addison's disease, the disturbance of renal function in Addison's disease is functional rather than organic, except in the terminal phase. It is rare to find any abnormality of the urine except in a crisis, when oliguria is associated with albuminuria and granular casts. The blood urea is raised in severe adrenal insufficiency but this is a consequence of abnormal salt loss and can be corrected by the giving of salt, or by aldosterone. Sodium depletion causes a fall in cardiac output and a reduction in renal blood flow, and the extrarenal uraemia of sodium depletion may even simulated primary renal failure. Since aldosterone stimulates renal tubular absorption of sodium, the sodium depletion of Addison's disease is partly due to the failure of renal tubular function, determined by the absence of aldosterone. However, the abnormal loss of sodium in the saliva indicates that this is not the only factor in a negative sodium balance.

Basal metabolism and body temperature

The basal metabolism is usually normal, or slightly subnormal, in the more subacute phases, or in a crisis. The infrequency of a low basal metabolism in the more chronic phase is a little surprising in view of the frequency of involutional changes in the thyroid gland, described in the pathology section: but the absence of the inhibitory effect of cortisone on thyroid function may playa part. The temperature in Addison's disease is usually subnormal, but these patients are very susceptible to any infection and then easily run a high temperature. These periods of intermittent pyrexia may also occur in the apparent absence of a definite infection. This is also true of phases of increased sedimentation rate without apparent cause but corrected by cortisone. Patients are also very sensitive to cold, in keeping with animal experiments, and adrenalectomized animals die rapidly on exposure to severe cold. Cortisone is protective against low temperatures and the sensitivity to cold is related to the absence of hepatic glycogen stores for carbohydrate metabolism. Certainly cold is one of the 'stress' stimuli to increased cortisone secretion in the normal animal. The extremities of the Addisonian patient are frequently cold to the touch and the fingers may become white or blue when exposed to cold. The phenomenon has been observed by the writer to occur after bilateral adrenalectomy in the absence of major indications of adrenal insufficiency.

Genital system

In the more severe phases of the disease, impotence and amenorrhoea are not uncommon, but in some women menstruation may continue quite normally. It is probable that the adrenals influence the gonads via the pituitary. There is certainly a close relation between the adrenals and the gonads, the adrenals undergoing hypertrophy after castration and secreting oestrogens after bilateral ovariectomy. Pregnancy was rarely met with but it is becoming more frequent and successful since the availability of the cortisone. Two successful pregnancies in one patient, with live normal babies, have been recorded. Since there is a hypersecretion of cortisone in pregnancy in normal women, pre-existing Addison's disease should be aggravated by pregnancy. There may, however, be some initial improvement and this is explained by the fact that the placenta can secrete cortisone. Theoretically the adrenals of the foetus might playa part. Sometimes the extra demands of pregnancy may reveal latent Addison's disease and in other cases the pigmentation of pregnancy persisted until the more complete picture of adrenal insufficiency became clear some months or years later. In one patient the climacteric increased the intensity of adrenal insufficiency but subsequently there was a large measure of apparent recovery.

Carbohydrate metabolism

Porges (1910) was the first to point out that a low blood sugar occurred in Addison's disease. Wadi (1928) described hypoglycaemic attacks in a man of 24 a few days prior to death. His blood sugar concentration was 71 mg. per 100 ml. in the more chronic phase, but 43 mg. and 36 mg. per 100 ml. in hypoglycaemic attacks that were relieved by intravenous glucose. The diagnosis of Addison's disease was confirmed at autopsy, the suprarenal gland being almost completely destroyed by tuberculosis. Snell and Rowntree (1929) found that hypoglycaemia was present in the terminal phases. Simpson (1932) found blood sugar values to be normal or slightly subnormal in the more chronic phases of Addison's disease but low values in crisis. Before the cortisone era Simpson (1953) found the most frequent cause of sudden death to be hypoglycaemic coma. This was a very dangerous complication when treatment was limited to desoxycortone and fatal hypoglycaemia could occur with great suddenness and be fatal within a few hours. These observations are in keeping with experimental knowledge that adrenalectomized animals have little or no hepatic glycogen and will go into hypoglycaemia under the influence of fasting, stress, or infection.

As long ago as 1909, Eppinger, Falta, and Rudinger found flat carbohydrate tolerance curves, suggesting an increased tolerance for carbohydrates in Addison's disease. Simpson (1932) found similar curves in some patients, partly explained by the absence of the diabetogenic action of adrenal glucocorticoids, and partly by retarded intestinal absorption.

In contrast with the above, Addison's disease is sometimes complicated by diabetes mellitus. Simpson (1949) described three such cases under his personal care and reviewed the literature of a further fifteen cases, proved by autopsy. The diabetes may precede or follow the clinical evidence of Addison's disease or the two may coexist from the onset. *Post mortem* the pancreatic lesion is seen to be atrophy of the islets, or reduction in their size, with hyalinization; but the adrenal lesion may be atrophy or tuberculosis. Haemochromatosis involving the adrenals and the pancreas has also been described, and in this metabolic disorder it is known that haemosiderin can be deposited in the liver, spleen, pancreas, adrenals, pituitary, and testes, & c.

Simpson's three cases are worth a brief summary. The first was a boy of 16 who stopped growing at 13 when lassitude developed, and who appeared to be a case of infantilism until pigmentation at the age of 16 led to the correct diagnosis of Addison's disease. On admission to hospital blood sugar levels were low, 50-60 mg. per 100 ml, but in the course of the next few weeks they rose to 250 and 300 mg. per 100 ml., and 10 Units of insulin produced hypoglycaemic values and symptoms. Treatment was limited to cortical extract and the patient died in Addisonian crisis in 1931, autopsy, showing atrophy of the pancreas and adrenals. It is probable that both these lesions were present simultaneously but that the diabetes was initially masked by the Addison's disease. The case also illustrates the instability of the blood sugar in these cases, the extreme hypersensitivity to insulin and liability to hypoglycaemic reactions.

The second case was a male of 37 who developed diabetes mellitus at the age of 11 and Addison's disease twenty-one years later at the age of 32. He died in 1947, at the age of 37, in Addisonian crisis with autopsy findings of atrophy of the adrenals and of the pancreas. The development of Addisons' disease led to a reduction of insulin requirement from 72 Units to 6 Units daily and instability of blood sugar concentrations was indicated by values of 45,56,917, and 1,540 mg. per cent.

The third case was a male of 20 who developed Addison's disease and diabetes mellitus in 1947, at the age of 18, the latter obvious four months after the former, but probably present in latent form from the onset. The patient was initially treated by desoxycortone and the special interest of the case was the effect of changing over to cortisone (in 1952) on the diabetes. Cortisone 25 mg. daily produced diuresis, ketosis, and incipient diabetic coma within forty-eight hours and the insulin requirements increased from 8 to 34 Units daily. In this patient cortisone was so powerfully diabetogenic and ketogenic that even with increased insulin the dose of cortisone had to be limited to 12 mg. daily. Another feature in this case was a raised sedimentation rate of 20 mm. in one hour (Westegren) partially reduced to 8 mm. by cortisone but variable in the absence of obvious infection. The patient continued well for two years and then died in diabetic coma a few hours after admission, his family doctor having stopped insulin forty-eight hours previously because of hypoglycaemic features. The adrenals were atrophic and the pancreas showed atrophy with some 'quite reasonable islets but curious basophil cells scattered throughout'.

Crisis

The term 'crisis' in Addison's disease is applied to an acute phase of the disorder in which the patient is collapsed and sometimes even comatose. Death is imminent unless active specific therapy is available. The patient may pass from a chronic state of insufficiency into a crisis even in the absence of any obvious precipitating cause. More usually, the incidence of intercurrent infection, to which patients with adrenal insufficiency are especially susceptible, is an important factor. On the other hand, it should be remembered that patients inadequately treated, and therefore liable to enter upon a crisis, are also much more liable to contract an intercurrent infection than are patients receiving adequate therapy. Extra exertion, or mental stress, in the absence of an increase in the dose of cortisone, may precipitate a crisis. A surgical operation, even a minor one, will lead to an adrenal crisis if therapy is not appropriately augmented. In some patients the diagnosis of Addison's disease has only been made after a minor operation or accident produced an adrenal crisis and drew attention to features of the disease preexisting but not appreciated. Diarrhoea is one of the features of adrenal insufficiency, especially in more severe phases, but it is also true that excessive diarrhoea produced by a drastic purge in a constipated patient may prove intractable and be the immediate precursor of a crisis.

Frequently there is no warning of a crisis, which comes on with great suddenness, but there are significant symptoms of a relatively minor character, which in the writer's experience can be regarded as of a warning nature. Thus the patient may yawn a great deal and stretch himself frequently, perhaps uttering involuntary cries or monosyllabic sighs. The onset and persistence of hiccoughs is an adverse sign. A susceptibility to conjunctivitis should not be disregarded and photophobia may be ominous. An increased sensitivity to cold, or actual shivering, especially if the patient curls himself up well beneath the blankets and refuses to answer question, should quite properly give rise to anxiety as to the state of adrenal insufficiency. In this stage also, there may be an inexplicable tendency to make unpleasant grimaces, and also a degree of irritability and negativeness.

When a crisis has developed, the patient looks dehydrated with sunken orbits and lack of resilience of the tissues; the veins are collapsed and difficult to enter with a needle and blood can only be withdrawn slowly and intermittently, the veins, tending to collapse after each few ml. and refilling only slowly after a pause. These are manifestations of salt depletion and secondary reduced plasma volume with haemoconcentration, increased haematocrit, increased haemoglobin concentration, and increased plasma protein concentration. The plasma sodium is low and the plasma potassium may be raised. Blood urea is usually raised and the alkali reserve is low because of a metabolic acidosis.

The above clinical description of adrenal crisis largely revolves around severe sodium depletion. However, in some patients, hypoglycaemia may cause predominant features, such as convulsions, coma, or peculiarities of behaviour, e.g. irritability, irrationality. The tendency to hypoglycaemia was more frequent in the desoxycortone or pre-cortisone era and in fact was then probably the most frequent cause of death. Even with cortisone, hypoglycaemia is a phenomenon which one must always be on the alert for in Addison's disease. It may happen quite suddenly in a patient who appears reasonably well, as has been described above for the Addisonian crisis, of which it is one important aspect. These clinical observation—emphasizing the importance of hypoglycaemic crises in Addisons's disease—are fully supported by the results of bilateral adrenalectomy in some animals, namely disappearance of hepatic glycogen and fatal hypoglycaemia, as well as by the contrasting diabetogenic action of cortisone.

Associated Endocrine Disorders

Addison's disease may be complicated by the coincident existence of myxoedema, thyrotoxicosis, parathyroid deficiency and tetany, diabetes mellitus, and acromegaly.

Clinical diagnosis

A combination of weakness, loss of weight, anorexia, nausea, or vomiting, low blood pressure, and pigmentation of the skin and mucous membrane, is strongly indicative of Addison's disease, especially when no other organic lesion can be detected. A previous history of abdominal, cervical gland, bone, or pulmonary tuberculosis, or of idiopathic pleurisy, is important supporting evidence, as is also a familial susceptibility to the tubercle bacillus. In patients over 50 years of age metastatic carcinomatosis may rarely involve both adrenal glands, but tuberculous and atrophic adrenal lesions also occur in older people. Some pigmentation of the skin may be present in carcinomatosis without any gross lesion in the adrenals. In a man of 43 under the care of Dr. Cawadias (1946), and seen by the writer, the cause of Addison's disease of three years duration was found to be primary carcinoma of both adrenal glands. The sedimentation rate was 70 mm. in one hour.

When the condition is met with in children it may present initially as arrest of growth with fatigability, or as infantilism, and the diagnosis may be missed for months or even two years. It is very rare for Addison's disease to occur before the age of 11 but Welch (1957) described such a case in a girl of 9, with weakness and anorexia following an attack of mumps, and pigmentation developing six months later. Mumps and other virus infections apparently may affect the adrenals, the gonads, the pancreatic islets, and other endocrine glands.

In Simmonds' disease, the adrenals are atrophied except the zone (probably zona glomerulosa) which secretes aldosterone, so that apart from salt depletion, the clinical pictures have many points in common, e.g. weakness, fatigability, anorexia, loss of weight, hypotension, and a tendency to hypoglycaemic attacks. However, in Simmonds' disease the face is pale, skin pigmentation is absent or slight, and pigmentation of the mucous membranes does not occur. The reason for this is that the destroyed anterior pituitary cannot secrete the ACTH complex, excess of which is an important factor in producing pigmentation. Bradycardia and low basal metabolism are usual with Simmonds' disease. Amenorrhoea or impotence are early manifestations of

Simmonds' disease, but are late or absent features in Addisons' disease. Slight or absent pubic hair is more characteristic of Simmonds' disease but the pubic hair may become thin in chronic Addison's disease. In fact the clinical differentiation of Simmonds' disease and Addison's is sometimes quite difficult and help is required from hormone and biochemical tests, such as the failure to respond to ACTH in Addisons' disease.

Although pigmentation of the skin is almost invariably a feature of Addison's disease, it may be very slight, or even undetected, in fair-skinned people, especially when the onset is acute. Pigmentation of the mucous membrane is usual but not invariable. Rarely the pigmentation of the skin is indistinguishable from that of idiopathic leukodermia. If pigmentation occurs in the mucous membrane of the mouth as well as in the skin, it is almost invariably due to adrenal insufficiency. Such pigmentation has been said to occur in pernicious anaemia, and Addison himself originally confused the two diseases, but neither the writer nor two colleagues, who have examined hundreds of patients with pernicious anaemia, have ever seen pigmentation of the mucous membrane in this disease, although pigmentation of the skin certainly does sometimes occur. In three other conditions pigmentation of the mucous membranes is said to occur: chronic arsenic poisoning, abdonical carcinoma, and haemochromatosis. In the first and-the last of these conditions the adrenal glands are infiltrated with arsenic or with haemosiderin, as are the other organs, and the pigmentation, when present, is therefore an expression of adrenal insufficiency. In abdominal carcinoma, the adrenals themselves may be affected. Pigmentation of the skin only, however, is not uncommon in any form of carcinoma. Apart from disease, pigmentation of the mucous membrane, although perhaps unnoticed until some illness is present, is found in persons in whose ancestry there is a strain of negroid, Indian, Eurasian, or Levantine blood. The racial factor is frequently forgotten, or unknown, but pigmentation may be found in other members of the family. Pregnancy, with vomiting, in such a Addison' disease. Skin pigmentation, however, especially in brunettes, is not uncommon in pregnancy, and is explicable on the basis of the greater demand on adrenal function.

Pigmentation of the skin occurs in pellagra: and a patient with pellagra secondary to gastro-jejunostomy was sent to the writer with the diagnosis of Addison' disease, other features being anorexia, diarrhoea, weakness, anaemia, and wasting. The adrenals, in fatal cases of pellagra, have, however, been observed to be atrophic probably secondary to vitamin deficiency. The skin in pellagra shows sharply demarcated areas of hyperkeratosis. Pigmentation of the skin is a rare complication of exophthalmic goitre, and it is not unlikely that in such cases the adrenals too are affected, if only functionally, by the thyrotoxicosis. It has already been noted that true Addison's disease and thyrotoxicosis can both exist in the same patient. Cutaneous pigmentation, especially of the orbits, may be met with in Cushing's syndrome, the adrenogenital syndrome and in acromegaly, and is attributable to excess of pituitary ACTH.

Hormone tests

The urinary 17-ketosteroids are almost or entirely absent in Addison's disease in women, because the adrenals are the only source of androgens in the female. This is no differentiation from Simmonds' disease in women as the 17-ketosteroids are also very low in this condition. In men with Addison's disease the testes continue to secrete androgens, so that the 17-ketosteroids are low, *e.g.* 6 mg. per twenty-four hours, but not absent. Low values for urinary glucocorticoids in both sexes also favour the diagnosis, but paradoxically with some techniques the values obtained are within normal limits and must be regarded as fallacious and non-specific. The same is true of plasma glucocorticoids. However, since the adrenals are destroyed largely or entirely, there is a failure of an increase in urinary 17-ketosteroids or 11-oxysteroids, or in plasma 11oxysteroids, following ACTH stimulation in Addison's disease. In Simmonds' disease the involuted adrenals are capable of responding to an ACTH stimulus and the above values all rise. In the more chronic forms of Simmonds' disease the adrenals may have lost their ability to respond to ACTH, but this ability to respond is often restored by cortisone therapy over a period. The ACTH may be given intravenously, e.g. 20 Units in 100 mg. of saline in thirty minutes, or as ACTH gel intramuscularly; neither test can be considered as completely free from danger. With the intravenous route severe allergic reactions may occur and with the intramuscular ACTH gel, contaminated in some preparations with vasopressin, water intoxication has led to convulsions. This is less true of more recent preparations and 40 Units daily for three days of the gel preparation given intramuscularly is justifiable, providing the patient's condition is not severe, when it is preferable to postpone such tests, after starting therapy, and only to use them if the diagnosis remains in doubt. Coincidental with hormone assays, the effect of ACTH on the blood eosinophils is of some value, their concentrate decreasing by more than 50 per cent in some normals and disappearing altogether in many other normals, but not being influenced in Addison's disease. The results, however, are not necessarily constant and cannot be regarded as in any sense infallible, but rather as one possible indication.

Therapeutic test

Cortisone, in therapeutic doses of 25 mg. thrice daily by mouth, is rapidly and dramatically effective in both Addison's disease and Simmonds' disease. In these doses such effects are not obtained in other conditions and, if obtained psychogenically, fail to be sustained. The therapeutic test may therefore be regarded as of very great value and is free from danger.

Diagnosis of Pre-addison's disease

Careful clinical history-taking in retrospect, in a large number of classical cases of Addison's disease, in which the diagnosis is beyond doubt, has convinced the writer that adrenal insufficiency may be present for as long as two years, and occasionally longer, before the characteristic clinical picture develops—and in the absence of gross pigmentation. Since the symptoms are often vague, e.g. malaise, lack of strength, poor appetite, the diagnosis of a pre-Addison state may be missed or rejected, particularly if the routine tests described above prove equivocal. Nevertheless, the condition must be borne in mind, particularly if there is a history of personal or familial tuberculosis or if the symptoms developed within months of a virus infection. It has been noted that adrenal insufficiency in children can manifest itself initially as infantilism or failure of the normal puberty evolution.

Where there are good grounds for a provisional clinical diagnosis, the therapeutic test seems justifiable and a positive indication warrants the continuance of appropriate therapy. The lack of a critical objective approach might lead to the treatment of a number of a non-endocrine psychoneurotic states, but nevertheless the physician should not be guilty of missing a true adrenal insufficiency in its early stages.

Treatment

An aqueous adrenal extract, prepared and found effective on adrenalectomized animals by Swingle and Pfiffner (1919), was first used effectively in Addison's disease by Rowntree and Green (1931) in America and by Simpson (1931) in England. The extract was injected intramuscularly or intravenously 5-20 ml. twice daily. In spite of its bulk and inadequacy compared with modern therapy it was a dramatic therapeutic advance at that time. The following year Loeb noted that patients in Addisonian crisis resembled some types of surgical shock and found that sodium chloride by mouth or intravenously was of considerable benefit. His observations were confirmed by Atchley and Stahl (1935) on adrenalectomized dogs.

The next step was the synthesis of desoxycorticosterone by Reichstein (1936) and its first clinical use in Addison's disease by Simpson (1938). The substance was given in oily solution intramuscularly, e.g. 5 mg. daily, or, after some weeks of stabilization, by subcutaneous implantation of sterile tablets, one tablet of 100 mg. for each 1 mg. of desoxycorticosterone (DOCA) injected daily. DOCA corrected the abnormal renal salt loss of adrenal insufficiency and proved more effective than the aqueous extracts or salt; and particularly so when supplemented by aqueous extracts, whose main action was on carbohydrate metabolism. Nevertheless, many patients, often apparently well, died suddenly from hypoglycaemia. More recently, *DOCA* was prepared as 'crystules' of its trimethyl acetate in buffered isotonic aqueous solution, 50-100 mg. being given intramuscularly and being effective, by slow absorbtion, for four weeks. All these methods of giving DOCA must be proceeded with cautiously, as excessive retention of salt and water and secondary excessive elimination of potassium may follow insidiously from overdosage. The features of this vary, and may be weakness, paresis, anorexia, vomiting, crepitations in the lungs, frothy sputum, oedema, and hypertension. Overdosage should be treated by removing the cause, by giving 100 ml. of 2 per cent. potassium chloride intravenously, and by the very cautious use of a mercurial diuretic, e.g. mersalyl, to eliminate the excess of sodium. Lipoid adrenal extracts containing a carbohydrate regulating action, and liquorice extracts, containing a salt-retaining factor were only of temporary limited interest.

The next real advance was the extraction and partial synthesis of cortisone and related compounds and their availability for the treatment of Addisons' disease. Cortisone (11-dehydro-17-hydroxycorticosterone or Compound E) is effective by mouth and is supplied as tablets of 25 mg., easily divided into two. When swallowed, action is rapid, beginning within thirty minutes and lasting several hours. In practice it is usual to give half the tablet twice daily but some patients can manage with the total dose once daily. The usual total daily dose is 25-37 mg. but some patients can manage on 12 mg. and a few require 50 mg. This dose may be just above physiological requirements and some patients become adipose if this dose is maintained for long periods. Cortisone, or the closely related hydrocortisone is the most important missing hormone in Addison's disease and replacement therapy with either substance produces a more complete return to normal than any other therapy. Apart from its beneficial effect on physical strength, resistance of fatigue, stress and infection, pigmentation, hypotension, abnormal electrocardiograms, and electroencephalograms, it results in a return to a normal positive personality and well being. Metabolically, it restores body glycogen stores and blood sugar concentration to normal, and thus minimizes the incidence of hypoglycaemic attacks, which are an adverse feature of therapy limited to DOCA. However, the effect of cortisone on a negative salt balance is only slight and, theoretically at least, a patient so treated would also require salt or DOCA. In actual practice this has been found to be essential only for a minority of patients, cortisone alone proving adequate. When a salt retaining hormone is required, we now have available a halogen derivative, 9 α-fluorohydrocortisone, which is effective as a tablet by mouth in small doses, e.g. 0.1 mg., or less, once daily. Perhaps aldosterone, effective in small daily doses of 0.15 mg, will become more generally available as an alternative to fluorohydrocortisone. The latter is not suitable for the treatment of Addison's disease by itself because its cortisone-like action is very weak compared with its powerful salt-retaining action. In limited trials, aldosterone, 0.15 mg. daily, has also corrected hypoglycaemic tendencies.

The best treatment of Addisonian crisis is its prevention and the oral dose of cortisone should be increased during excessive stress, or infection, or at the first sign of adrenal insufficiency from unknown causes, e.g. malaise, anorexia, increase of pulse rate, lowering of blood pressure, vomiting, diarrhoea, unexplained pyrexia. No harm will result from a temporary increase of cortisone, if in doubt, and many a severe crisis will be thus prevented. If the patient is vomiting, and unable to chew or swallow cortisone, this can be given intramuscularly, its action, however, being slower by this route and not developing effectively until some hours after injection, with a duration effect of sixteen to twenty-four hours. Cortisone is prepared as cortisone acetate, a white powder, only slightly soluble in water, and therefore suspended in saline solution, 25 mg. per ml., in bottles of 20 ml. containing 500 mg. (with benzyl alcohol as preservative). The intramuscular dosage is equivalent to that of cortisone given by mouth. In crisis 100 mg. cortisone should be injected intramuscularly at once and it will provide a depot for steady absorption; but there are now available two preparations suitable for intravenous injection with consequent rapidity of response: (1) a concentrated solution of hydrocortisone, 100 mg. in. 20 ml. 50 per cent. alcohol, which is added to 1 litre of physiological saline and 5 per cent glucose, injected at the rate of 25 mg. per hour for the first two hours and 12.5 mg. per hour subsequently; (2) the sodium salt of hydrocortisone21-hemisuccinate, extremely soluble in water, presented as a powder in a vial, in a dose equivalent to 100 mg. hydrocortisone, to which is added 2 ml. sterile water from a separate vial; the solution of 2 ml. can be injected directly into a vein or added to a saline-glucose drip. Treatment may be reinforced with desoxycortone 10 mg. intramuscularly, or fluorohydrocortisone 0.2 mg. intravenously, but the danger of overdosage with salt-retaining hormones in real. Many patients who look dehydrated and have haemoconcentration recover with cortisone only, without even intravenous saline infusions. Much of the fluid is not lost from the body in crisis but passes from the extracellular spaces to the intracellular spaces of organs, e.g., liver; and DOCA, fluorohydrocortisone, and to a lesser degree also cortisone, reverse this and restore the plasma volume even without the addition of fluid from outside the body, except in the more severe phases of crisis. Intravenous glucose is advisable if severe hypoglycaemia is the predominant feature, but here, too, cortisone by itself has a rectifying action on carbohydrate metabolism. Although a clinician's judgement will influence variations of treatment in each case, cortisone, or hydrocortisone, is now the sheet anchor in the treatment of all phases of adrenal insufficiency. Prednisone and prednisolone, dehydrogenation products of cortisone and hydrocortisone, are four or five times as powerful as cortisone, or hydrocortisone, as judged by their glucocorticoid activity and are sometimes used in Addison's disease as alternative therapy. However, their sodium retaining action is very much less than that of cortisone and hydrocortisone, and although this makes them more suitable for massive doses in certain non-endocrine states, it is to their disadvantage in the therapy of most patients with Addison's disease.

When cortisone was first introduced, it was feared by some clinicians that the tendency of large doses to cause a spread of infections, especially tuberculosis, as proved by animal experiments would be a danger in Addison's disease .due to tuberculous adrenals. In practice this has not proved to be the case, because the relatively small substitution doses used are physiological and not comparable to the large doses employed in non-endocrine conditions. In fact such physiological doses are probably beneficial in tuberculous and other infections, since the adrenalectomized animal is much more sensitive to infections than the normal animal. Where a patient has active tuberculosis, e.g. pulmonary, as well as Addison's disease, it is probably wise to combine cortisone with modern chemotherapy directed to the tuberculosis, e.g., streptomycin &c., until the acute tuberculosis is under control. This may also be wise in the case of severe non-tuberculous infection, but it must always be remembered that the onset of infection calls for an increase in the maintenance dose of cortisone, and such an increase should not be withheld.

The advent of cortisone has completely changed the prognosis in Addison's disease, and not only are patients living many more years, but they are doing so with a capacity to follow their work and enjoy their pleasures. Nevertheless undue stress and fatigue, or exposure to infections, should be avoided as far as is practical because a patient with Addison's disease has not the reserve protection or adjustments of a normal individual.

Phaeochromocytoma

Adrenaline was synthesized in 1904 but the existence of the related pressure amine, noradrenaline, was not recognized until comparatively recently, and its release *in vivo* be excitation of adrenergic fibres was demonstrated in 1949 (Peart). Noradrenaline differs from adrenaline in the absence of an N-methyl group. Both adrenaline and noradrenaline are secreted by the normal adrenal medulla and by phaeochromocytomas. Both produce a rise in systolic and diastolic blood pressure but adrenaline does this mostly by increasing cardiac output and noradrenaline by increasing peripheral resistance; adrenaline produces tachycardia and noradrenaline tends to produce bradycardia.

Incidence

The first complete study of a classical case was made in 1922 by Labbe, Tinel, and Doumer. The condition may be met with at all ages, including childhood and infancy, and in both sexes.

Pathology

The tumours are usually adenomatous but may be malignant, with metastases. Of 103 recorded cases, 13 were bilateral (Brunschwig and Humphreys, 1940) and yet no case due to bilateral medullary hyperplasia has been described. Occasionally the neoplasm arises from para-aortic medullary tissue outside the adrenal gland. The tumours vary in size and may be enormous but as with other endocrine tumours a tiny tumour may be extremely active in a secretion. The name phaeochromocytoma is based on the fact that adrenaline, in the cytoplasm of the cells, reduces chromium salts to form an insoluble peroxide of chromium, which appears in the cytoplasm as fine brown granules. Most tumours secrete both adrenaline and noradrenaline in varying proportions and contain high concentrations of both substances. An intrathoracic tumour in an infant of five months was described by Mason *et al.* (1957).

Clinical features

Attacks are due to paroxysmal secretory activity of the tumour and may occur at intervals of weeks, or several times a day, and may last from a few minutes to a few hours. In an attack the patient may experience anxiety, palpitations, tremors, perspiration, severe headache, blurred vision, faintness, weakness, dyspnoea, a feeling of suffocation or painful constriction of the chest, or angina-like pain, nausea, vomiting abdominal colic, cramps of the calf muscles, and tingling of the extremities, which may undergo colour changes comparable to those found in Raynaud's disease. The skin is pallid and cold in an attack, with beads of perspiration sometimes visible. Goose-flesh is sometimes present. The blood pressure may be normal in quiescent periods and suddenly rise in attacks, or it may be appreciably raised all the time and rise still further in paroxysms. The pulse rate is often considerably accelerated during the attacks but bradycardia, sometimes associated with a high proportion of noradrenaline, is met with and is characteristic of many cases. The heart beat in an attack may be so violent as to shake the bed, even while the patient sleeps. The systolic pressure is said to rise to a greater extent than the diastolic, but both may be raised considerably, e.g. 280 systolic and 200 diastolic. The heart may be considerably enlarged, and left ventricular failure may ensue. Pulmonary oedema with frothy blood-tinged sputum is also met with. One woman of 66 was admitted to hospital three times with attacks of dyspnoea, copious expectoration, sweating, and vomiting. Breathlessness, at rest or on excretion, may be a prominent feature and cardiac asthma is met with, sometimes with left ventricular failure. In one case phases of hypertension (240/150) were followed by phases of hypotension (90/70). The clinical picture may be one of severe progressive hypertension, e.g. 200/110, without any superimposed paroxysms, as in the woman of 29 years of age described by Binger and Craig (1938). The fundi may show papilloedema or hypertensive retinitis, and these changes may be reversible if the tumour is removed early enough. The kidneys mayor may not be secondarily involved, but nephrosclerosis has been recorded (Green, 1946). In most cases the urea clearance test of kidney function gives values well above normal, e.g. 140 per cent., and this is attributed to the increased blood flow through the kidneys.

The basal metabolic rate is often raised to the extent found in hyperthyroidism, e.g., plus 60 per cent but this is not due to hyperthyroidism since it returns to normal after removal of the adrenal tumour, but is unaffected by thiouracil or thyroidectomy. The thyroid epithelium does not show hyperplasia. An apparently normal thyroid gland may become enlarged and vascular during an attack. The increased basal metabolic rate is often associated with pyrexia, which may be constant or intermittent.

The pyrexia is often associated with leucocytosis, the neutrophil polymorphs being 62 and 75 per cent. in two cases.

Diabetes mellitus has often been diagnosed as a complication but here, too, the hyperglycaemia is a result of the hyperadrenalism. The blood sugar concentration is variable and unstable and the patients hypersensitive to insulin. McCullagh and Engel (1942) recorded the following blood sugar tolerance curves before and after removal of an adrenal phaeochromocytoma, the values being fasting and 30-minute intervals after 100 g. glucose:

Before: 128, 267, 254, 75, 45, 76 mg. per 100 ml.

After: 88, 168, 168, 108, 49, 72 mg. per 100 ml.

Malignant phaeochromocytoma may occur at any age, even in children. The symptoms, apart from metastases, are the same as with adenomas but symptoms of hyperadrenalism are not invariably present.

Diagnosis

The best modern paper on this subject, clinically and pharmacologically, dealing with five cases, is by Hamilton *et al.* (1953). As with other endocrine tumours, the history is sometimes of many years' duration and the condition not recognized until the late stages. Thus benign hypertension, with attacks of apprehension and palpitation have been allowed to proceed to malignant hypertension, with severe retinopathy. It is probable that some cases presenting as persistent hypertension were originally paroxysmal. One man, in whom a phaeochromocytoma was removed at the age of 31, had attacks of pallor and bradycardia at the age of 11, and after a quiescent period of some years had more characteristic attacks, with hypertension, from the age of 23.

The following tests have been used diagnostically and are evaluated by Pearl and co-workers in the paper referred to above.

1. Vasoconstriction measured by heat elimination, by the hand being immersed in a water calorimeter and temperature changes being measured. The heat eliminated measured an average of 35 calories per 100 ml. per minute before operation compared with an average normal of 90. This test is 'simple and useful but not infallible'.
2. The early test of Roth and Kvale (1945) was the use of 0-0.25 mg. of histamine intravenously to produce a release of adrenaline and precipitate an attack. This test is not invariably reliable and such an induced attack in a severe case may prove dangerous.
3. Adrenolytic agents antagonistic to adrenaline, and to a lesser extent to noradrenaline, are added by injection into the tubing to an intravenous infusion of saline. Thus, piperoxane, also called benzodioxane, in a maximum dose of 20 mg. injected in one minute, characteristically produces a fall of blood pressure within a few minutes, but negative results have been obtained in cases proved at operation. The test is not infrequently complicated by headache, nausea, sternal and abdominal pain, palpitations and even pulmonary oedema and hypertensive encephalopathy, especially in patients with essential hypertension, not due to phaeochromocytoma.

   Rogitine, or phentolamine, also added to intravenous saline, during one minute, in a dose of 0.08 mg. per kilogram body-weight, is probably the most satisfactory test, the blood pressure falling within a few minutes, and remaining depressed for a few minutes or longer. Side effects are rare but induced tachycardia may be troublesome. This is regarded as a relatively safe and reliable test but false positive and negative findings are met with. Phentolamine may be injected directly into a vein, e.g., in a 5 mg. dose, instead of into an intravenous drip. Barbiturates should be omitted for twentyfour hours before the test. Blood pressure readings are taken at minute intervals for a minimum time of ten minutes. Dibenamine is a powerful adrenolytic substance with a more lengthy action but sometimes give false positive tests in essential hypertension.

4. The measurement of pressure substances in urine and blood is the most direct evidence of an adrenaline-secreting phaeochromocytoma. The amines are extracted by selective adsorption on alumina and assayed biologically, using the blood pressure response on the barbitone anaesthetized rat and comparing the results with known solutions of adrenaline and noradrenaline. Dihydroergotamine reverses the action of adrenaline but not of noradrenaline, so that there is a biphasic response, with an initial fall of pressure due to adrenaline and a subsequent rise due to noradrenaline. The total excretion with phaeochromocytomas varies between 0.5 and 5 mg. of pressure amines, as compared with normals of 0.02 to 00.7 mg. The ratio of noradrenaline to adrenaline varies from 0.5 to 4.0. Rarely the values for pressure amines may be normal in the presence of a functioning phaeochromocytoma. Burn and Field (1956) have developed a chemical calorimetric method for the assay of pressure amines. A tumour may be palpable or visible by direct radiography. Episacral air insufflation, combined with tomography, may be necessary but may precipitate a severe attack. Since a right-sided tumour is more frequent, some surgeons prefer an initial right-sided laparotomy when location is not defined.

Treatment

Atropine is not used before operation as it may cause exacerbation of the hypertension. An intravenous drip is set up and some clinicians give *Rogitine*, or the shorter-acting poperoxane, before anaesthesia, in the intravenous drip to minimize the rise of blood pressure that may be considerable and dangerous on handling the tumour, and it should always be in the operating theatre ready for such emergency. When the tumour is removed there is a severe fall of blood pressure and for this noradrenaline should be added to the intravenous drip in a concentration of 10-20 mg. per litre, varying the rate of drip according to the blood pressure and keeping the infusion running for twentyfour to thirty hours. Local aecrosis occasionally follows the intravenous infusion of noradrenaline and is avoided by choosing a large superficial vein or by using a long polythene tube. The results of operations are excellent and there is usually complete remission unless permanent changes have occurred in blood vessels, when the basis hypertensive level may persist. Noradrenaline is also useful for the treatment of the hypotension which may follow a spontaneous hypertensive crisis with shock quite apart from operation.

9

PANCREAS

Because of its function, the *pancreas* can be classified as both an endocrine and an exocrine gland. The pancreas is a flattened organ located posterior and slightly inferior to the stomach. The adult pancreas consists of a head, body, and tail. Its average length is 12 to 15 cm. The endocrine portion

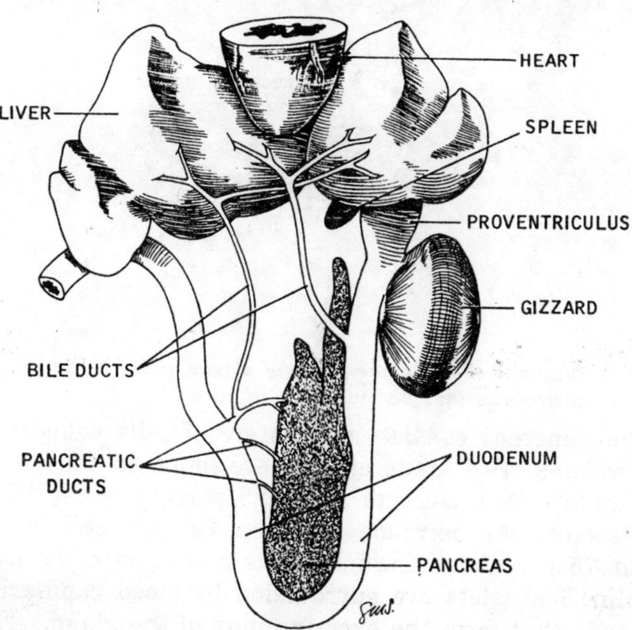

Fig. 9.1. Pancreas of the pigeon (Columba livia).

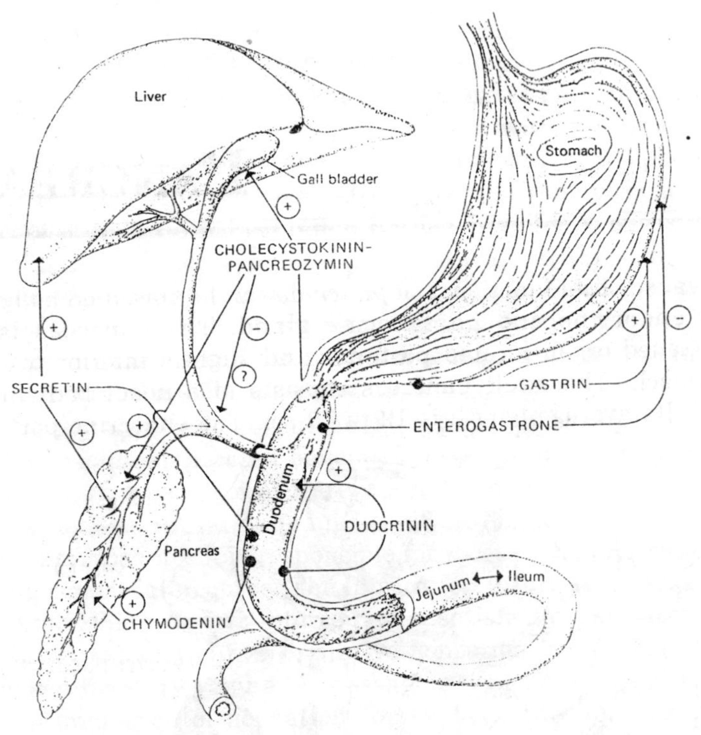

Fig. 9.2. Diagram showing the place of origin and the site of action of gastrointestinal hormones.

of the pancreas consists of clusters of cells called *islets of Langerhans*. Two kinds of cells are found in these clusters: (1) *Alpha cells* constitute about 25 percent of the islet cells and secrete the hormone glucagon; (2) *beta cells* constitute about 75 percent of the islet cells and secrete the hormone insulin. The islets are surrounded by blood capillaries and the cells that form the exocrine part of the gland.

The endocrine secretions of the pancreas—glucagon and insulin—are concerned with control of the blood sugar level. Let us now examine how this regulation takes place.

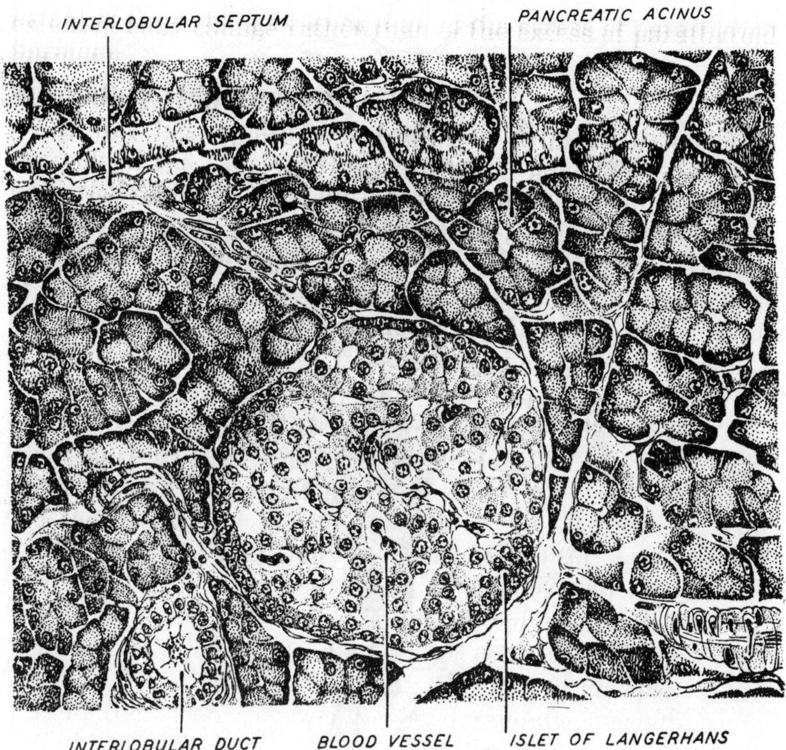

Fig. 9.3. A section through the pancreas of the rat. The islet of Langerhans is a gland of internal secretion, whereas the surrounding acinar tissue forms an exocrine gland.

Effect of Insulin on Carbohydrate Metabolism

The earliest studies of the effect of insulin on carbohydrate metabolism showed three basic effects of the hormone : (1) enhanced rate of glucose metabolism, (2) decreased blood glucose concentration, and (3) increased glycogen stores in the tissues. All of these effects probably result from the following basic mechanisms.

Facilitation of Glucose Transport through the Cell Membrane

The single basic effect of insulin that has been demonstrated many times is its ability to increase the rate of glucose transport through the membranes of most cells in the body. In the complete absence of insulin, the overall rate

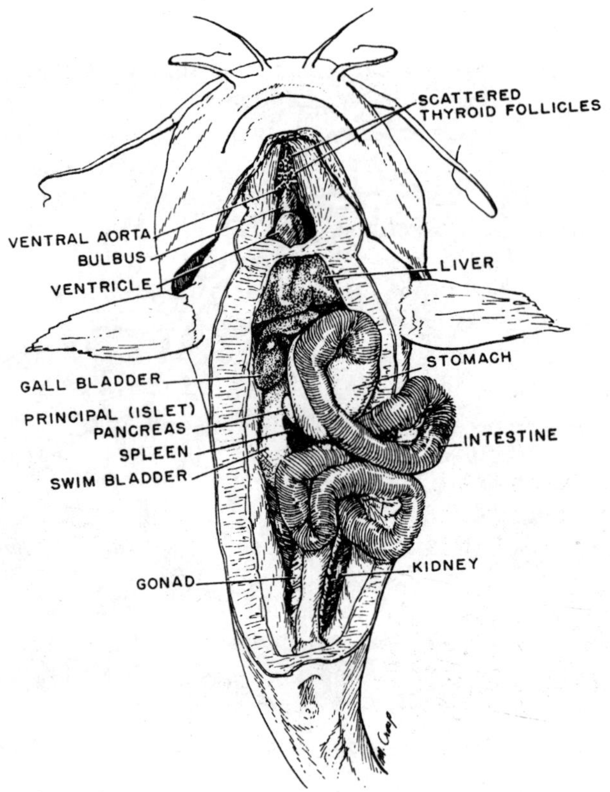

Fig. 9.4. Ventral dissection of the catfish (Ameiurus), showing the location of the principal (islet) pancreas with reference to other organs.

of glucose transport into the cells of the entire body becomes only about one-fourth the normal value. On the other hand, when great excesses of insulin are secreted, the rate of glucose transport into the cells may be as great as five times normal. This means, that, between the limits of no insulin at all and great excesses of insulin, the rate of glucose transport can be altered as much as 20-fold.

Mechanism by which Insulin Accelerates Glucose Transport

As pointed out in the discussion of the cell membrane glucose cannot pass into the cell through the cell pores but instead must enter by some transport mechanism through the membrane matrix. The generally believed method by which

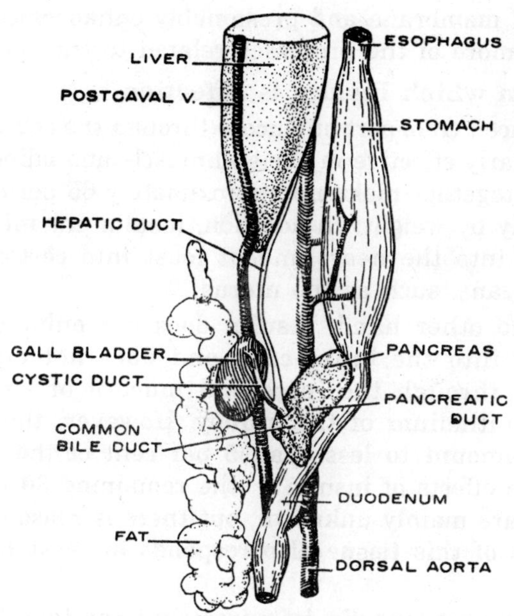

Fig. 9.5. Dissection of the pancreas and adnexa of the snake (Natrix) from ventral view.

glucose enters the cell, showing that glucose probably combines with a carrier subtance in the cell membrane and then is transported to the inside of the membrane where it is released to the interior of the cell. The carrier then returns to the outer surface of the membrane to transport additional quantities of glucose. This process can occur in *either direction*.

Glucose transport through the cell membrane cannot occur against a concentration gradient. That is, once the glucose concentration inside the cell rises to as high as the glucose concentration on the outside, further glucose will not be transported to the interior. Therefore, the glucose transport process is one of *facilitated diffusion*, which means simply that glucose diffuses through the membrane by means of a facilitating carrier mechanism.

In the complete absence of insulin, the carrier mechanism for the transport of glucose to the interior of the cell operates very poorly. Unfortunately, the manner in which insulin affects this mechanism is unknown. The insulin first becomes "fixed" in the cell membrane and presumably enhances the activity of one or more of the processes related to transport.

Tissues in which Insulin is Effective

Enhanced transport of glucose through the cell membrane is particularly effective in skeletal muscle and adipose tissue. They two together make up approximately 65 per cent of the entire body by weight. In addition, insulin enhances glucose transport into the heart and at least into certain smooth muscle organs, such as the uterus.

On the other hand, insulin does not enhance glucose transport into the brain cells and does not enhance its transport through the intestinal mucosa or through the tubular epithelium of the kidney. However, these tissues together amount to less than 5 per cent of the total body mass. The effects of insulin in the remaining 30 per cent of the body are mainly unknown, but there is reason to believe that most of this tissue also responds at least partially to insulin.

In summary, insulin is extremely important for glucose transport into the cells of most tissues of the body. The most important exception to this is the brain; here, glucose transport is probably more dependent on *diffusion* through the blood-brain barrier that through the cell membrane.

Effect of Insulin on Glucose Utilization and Glycogen Storage in Extrahepatic Tissues

Since insulin lack decreases glucose transport into most body cells to about one-fourth normal, without insulin most tissues of the body must depend on other metabolic substrates for energy.

Another major effect of insulin is greatly enhanced storage of glycogen in the skeletal muscle cells throughout the body and moderate enhancement of glycogen storage in the skin and glandular tissues. This effect is undoubtedly caused by the availability of greatly increased intracellular glucose.

Action of Insulin on Carbohydrate Metabolism in the Liver

Knowledge of the action of insulin on carbohydrate metabolism in the liver is still somewhat confused because insulin does not cause immediate increase in glucose transport into liver cells similar to that which occurs in skeletal muscle and most other tissues of the body. Instead, insulin causes initial loss of glycogen from the liver and transfer of this to the skeletal muscle. This effect probably results from the following two factors: (1) The liver cells are naturally highly permeable to glucose, which means that glucose can go through the liver cell membranes in either direction with ease regardless of the presence of insulin. (2) Liver cells contain large quantities of *glucose 6-phosphatase*, an enzyme that continually dephosphorylates glucose that has previously entered the liver cell; therefore, glucose cannot become trapped inside the liver cell as it does in other cells. Because of these two factors, when insulin promotes glucose transport into muscle and other peripheral tissues, the blood glucose concentration falls immediately, which in turn causes glucose mobilization from the liver. Thus, in effect, insulin transfers glycogen from the liver to the peripheral cells.

Long-term effect of Insulin on the Liver

If insulin is administered continuously for days or weeks along with simultaneous availability of adequate carbohydrates, the quantity of glycogen in the liver becomes greater than normal rather than being decreased. This effect is caused by still another effect of insulin in liver cells: it markedly increases the quantity of *glucokinase*, and this accelerates the rate at which glucose can be processed for utilization in liver cells. Therefore, carbohydrate metabolism in the liver becomes greatly enhanced, causing storage of glycogen in the liver cells, which is opposite to the acute effect of insulin.

Regulation of Blood Glucose Concentration, and Control of Insulin Secretion

Glucose concentration in the plasma and other extracellular fluids is controlled by four separate mechanisms: (1) the ability of the liver to "buffer" the blood glucose concentration, (2) an automatic feedback mechanism involving insulin secretion by the pancreas, (3) an automatic feedback mechanism involving sympathetic stimulation, epinephrine, and norepinephrine to cause release of glucose from the liver, and (4) a similar feedback mechanism involving glucagon secreted by the alpha cells of the pancreas.

Glucose-Buffer Function of the Liver

As just discussed, the liver acts as a large storage vault for glucose, and because of this the liver also acts as a blood glucose buffer system. When excess quantities of glucose enter the blood, about two-thirds of this is stored almost immediately in the liver, and this prevents excessive increase in blood glucose concentration. Conversely, when the blood glucose concentration falls below normal, the stored glucose in the liver rapidly replenishes the blood glucose.

Insulin plays an important role in controlling the glucose-buffer function of the liver, because in the absence of insulin, glucokinase is greatly depressed in the liver cells, so that glucose is not readily stored in the liver in times of excess.

In addition to acting as a major store-house for glucose, the liver also helps to regulate blood glucose concentration by increasing its rate of gluconeogenesis when glucose is needed in the blood.

The liver glucose-buffer system reduces the variation in blood glucose concentration to approximately one-third what it is without the liver.

Effect of Insulin on Blood Glucose Concentration

In the absence of insulin, little of the glucose absorbed from the gastrointestinal tract can be transported into the tissue cells. As a consequence, the blood glucose concentration rises very high—from a normal value of 90 mg/100 ml. sometimes to as high as 300 to 1200 mg./100 ml.

On the other hand, in the presence of great amounts of excess insulin, glucose is transported into the cells so rapidly that its concentration in the blood can fall to as low as 20 to 30 mg./100 ml. Therefore, the rate of insulin secretion by the pancreas must be regulated accurately so that the blood glucose concentration also will be regulated accurately at a constant and normal value.

The glucose concentration of the plasma has a direct and immediate effect on the islets of Langerhans to control their rate of insulin secretion. Even the isolated pancreas perfused with a solution containing a high concentration of glucose secretes greatly increased quantities of insulin. Therefore, the secretion of insulin by the pancreas affords an important feedback mechanism for continual regulation of blood glucose concentration.

Besides the immediate increase in insulin secretion that results from increased blood glucose concentration, still and additional gradual increase in insulin secretion occurs if the elevated glucose concentration persists. Over a period of one to three weeks, the islets of Langerhans hypertrophy, and secretion of insulin increases correspondingly. Therefore, if a person suddenly begins to eat a diet containing excessive amounts of carbohydrates, his normal pancreatic insulin secretion may not be capable of taking care of all the excess glucose, and his blood glucose concentration rises above normal. But after several weeks of pancreatic adaptation, and adequate amount of insulin is secreted, and the excess glucose is transported into the cells.

Regulation of Blood Glucose Concentration by the Sympathetic Nervous System and by Glucagon

Decreased blood glucose concentration excites the sympathetic nuclei of the hypothalamus, which then transmit impulses through the sympathetic nervous system to cause epinephrine release by the adrenal medullae and norepinephrine release both by the adrenals and by the sympathetic nerve endings. These hormones, especially epinephrine, exert a direct effect on the liver cells to increase the rate of glycogenolysis. They do this by activating the liver cell enzyme *adenyl cyclase*, which converts much ATP to cyclic AMP. The cyclic AMP in turn activates the normally inactive phosphorylase in the liver cells. This enzyme then causes glycogen to split into glucose phosphate, which is rapidly dephosphorylated by glucose 6-phosphatase and diffuses into the blood to elevate the blood glucose concentration back toward normal.

Later in the chapter we shall see that a decrease in blood glucose concentration also causes glucagon secretion by the pancreas; the glucagon then has the same effect as that of epinephrine to cause glucose release from the liver. In addition, glucagon and glucocorticoids from the adrenal gland enhance gluconeogenesis, which converts amino acids into glucose. These, then, are additional feedback mechanisms for regulation of blood glucose concentration.

Purpose of Blood Glucose Regulation

One might ask the question: Why is it important to maintain a constant blood glucose concentration, particularly since most tissues can shift to utilization of fats and proteins for energy in the absence of glucose? The answer is that glucose is the only nutrient that can be utilized by the brain, retina, and germinal epithelium in sufficient quantities to supply them with their required energy. Therefore, it is important to maintain a blood glucose concentration at a sufficiently high level to provide this necessary nutrition.

OTHER METABOLIC EFFECTS OF INSULIN

Effect on Fat Metabolism

As long as adequate amounts of insulin and glucose are available, glucose is perferentially metabolized by the cells to supply essentially all the energy required. On the other hand, in the absence of adequate glucose or insulin, the major share of the energy required by the body is then supplied by fats.

Mechanism of fat mobilization in insulin lack

The effect of sudden insulin lack on blood glucose and on the fat mobilization products, free fatty acids and acetoacetic acid, showing that all of these increase. The cause of this fat mobilization is two-fold.

First, in the absence of insulin, glucose cannot enter the fat cell with ease. Several products of glucose metabolism are very important for causing, and also for maintaining, fat storage, the most important being a-*glycerophosphate*, which supplies the glycerol nucleus that combines with fatty acids to form the neutral fats in adipose tissue. In the absence of these products of glucose, the process of storing triglycerides in fatty tissue is reversed, causing, instead, release of free fatty acids into the blood.

Second, lack of insulin—in some way not yet understood, but perhaps by acting directly on the fat cell membrane—causes direct enhancement of lipolysis with resultant release of free fatty acids into the circulating body fluids.

Also, additional fat mobilization results from the action of cortisol on the fat cell; this hormone is usually secreted in large quantities in conditions of insulin lack.

Effect of insulin lack on blood lipid concentration

In addition to the increase in free fatty acids in the circulating blood, essentially all the other lipid components of the plasma also become increased in the absence of insulin. This includes an increase in the overall quantity of lipoproteins and of their constituents, *triglycerides*, *cholesterol*, and *phospholipids*. At times the blood lipids increase as much as five-fold, giving a total concentration of plasma lipids of several per cent rather than the normal 0.6 per cent. These high lipid concentrations, especially the high concentration of cholesterol, is the probable cause of extreme atherosclerosis in persons with serious diabetes.

Ketogenic effect of insulin lack

Associated with increased fat mobilization from the tissues is usually also a ketogenic effect of insulin lack. That is, acetoacetic acid, a product of fatty acid metabolism, is released by the liver into the blood in greatly increased quantities, causing the condition called *ketosis*.

Glucose—fatty acid cycle

From this discussion, it is clear that glucose and insulin have a direct inhibitory effect on fatty acid utilization by the body. Conversely, fatty acids have an inhibitory effect on the utilization of glucose, which results primarily from the following effect: When excess fatty acids are available, some of their degradation products inhibit the conversion of fructose 6-phosphate into fructose 1, 6-phosphate, one of the early steps essential to the degradation of glucose before it can be utilized for energy.

This reciprocal inhibitory relationship between glucose and fatty acid utilization is called the *glucose—fatty acid cycle*. It plays a particularly important role in insulin lack, because in addition to the direct effect of insulin lack in depressing glucose utilization, the resultant buildup of fatty acids also indirectly depresses glucose metabolism. This effect almost doubles the glucose metabolic defect in the presence of insulin deficiency.

Effect of Insulin on Protein Metabolism and on Growth

Effect of Insulin on Protein Anabolism—Transport of amino Acids Through the Cell Membranes. The total quantity of protein stored in the tissues of the body in increased by insulin and greatly decreased by insulin lack. Indeed, a person with severe diabetes mellitus can become incapacitated almost as much from protein lack as from failing glucose metabolism.

Most of the effect of insulin on protein metabolism is probably secondary to the action of insulin on carbohydrate metabolism, for utilization of carbohydrates for energy acts as a protein sparer. However, *insulin also increases the transport of amino acids through the cell membranes* in a manner similar to that of its increasing glucose transport. This transport is carrier mediated, as is also true of glucose transport. Insulin enhances the transport several-fold, but the effect is generally far less intense than the effect on glucose transport. Insulin is essential for growth of an animal, principally because of the protein-sparing effect of increased carbohydrate metabolism, but also because of the direct effect of insulin on protein anabolism.

Diabetes Mellitus

Diabetes mellitus is caused by destruction, or a failure to function, of the beta cells in the islets of Langerhans of the pancreas. Though various factors, such as excess growth hormone from the adenohypophysis, can at times cause the beta cells of the islets of Langerhans to "burn out," diabetes is, in general, a hereditary disease, for almost all persons who develop diabetes, especially those who develop it in early years, can trace the disease to one or more forebears.

Pathologic Physiology of Diabetes

Most of the pathology of diabetes mellitus can be attributed to one of the following three major effects of insulin lack: (1) decreased utilization of glucose by the body cells, with a resultant increase in blood glucose concentration to as high as 300 to 1200 mg. per 100 ml.; (2) markedly increased mobilization of fats from the fat storage areas, causing abnormal fat metabolism and especially depositionof lipids in vascular walls to cause atherosclerosis; and (3) diminished deposition of protein in the tissues of the body, caused partly by failure of glucose to be used as a protein sparer and partly by loss of the direct effect of insulin to promote protein anabolism.

However, in addition, some special pathologic physiological problems occur in diabetes mellitus, and these are not so readily apparent. These are as follows:

Loss of Glucose in the Urine of the Diabetic Person

Whenever the quantity of glucose entering the kidney tubules in the glomerular filtrate rises above approximately 225 mg. per minute, a significant proportion of the glucose begins to spill into the urine. If normal quantities of glomerular filtrate are formed per minute, 225 mg. of glucose will enter the tubules each minute when the blood glucose level rises to 180 mg. per cent. Consequently, it is frequently stated that the blood "threshold" for the appearance of glucose in the urine is approximately 180 mg. per cent. When the blood glucose level rises to 300 to 500 mg. per cent—common values in persons with diabetes—several hundred grams of glucose can be lost into the urine each day.

Acidosis in Diabetes

The shift from carbohydrate to fat metabolism in diabetes has already been discussed. When the body depends almost entirely on fats for metabolism, the level of acetoacetic acid and other keto acids in the body fluids may rise from 1 mEq./liter to as high as 30 mEq./liter. This, obviously, is likely to result in acidosis.

All the usual reactions that occur in metabolic acidosis take place in diabetic acidosis. These include *rapid* and *deep breathing* called "Kussmaul respiration," which causes excessive expiration of carbon dioxide and *marked decrease in bicarbonate content of the extracellular fluids.* Likewise, *large quantities of chloride ion are excreted by the kidneys* as an additional compensatory mechanism for correction of the acidosis. These extreme effects, however, occur only in the most severe degrees of untreated diabetes. The overall changes in the electrolytes of the blood as a result of severe diabetic acidosis.

The Glucose Tolerance Test

An important test performed for the diagnosis of diabetes mellitus is the glucose tolerance test. The middle curve is the normal one: when a normal fasting person ingests 50 grams of glucose, his blood glucose level rises from a normal value of approximately 90 mg. per cent to approximately 140 mg. per cent to approximately 140 mg. per cent and falls back to below normal within three hours.

Though an occasional diabetic person has a normal fasting blood glucose concentration, it is usually above 120 mg. per cent, and his glucose tolerance test is almost always abnormal. On ingestion of 50 grams of glucose, these persons exhibit a progressive, slow rise in blood glucose level for two to three hours and the glucose level falls back to the control value only after some five to six hours; it never falls below the control level. This slow fall of the curve and its failure to fall below the control level illustrates that the normal increase in insulin secretion following glucose ingestion does not occur in the diabetic, and a diagnosis of diabetes mellitus can usually be definitely established on the basis of such a curve.

Treatment of Diabetes

The theory of treatment in diabetes mellitus is to adminster enough insulin so that the patient will have normal carbohydrate metabolism. If this is done, most consequence of diabetes can be prevented. Ordinarily, the diabetic patient is given a single dose of one of the long-acting insulins each morning; this increases his overall carbohydrate metabolism throughout the day. Then additional quantities of rapidly acting regular insulin are given at those times of the day when his blood glucose level tends to rise too high, such as at meal times. Each patient must be established on an individualized routine of treatment.

Diet of the Diabetic

The insulin requirements of a diabetic are established with the patient on a standard diet containing normal, well-controlled amounts of carbohydrates; any change in the quantity of carbohydrate intake changes the requirements for insulin. In the normal person, the pancreas has the ability to adjust the quantity of insulin produced to the intake of carbohydrate; but in the completely diabetic person, this control function is totally lost.

Relationship of Treatment of Arteriosclerosis

In the early days of treating diabetes it was the tendency to reduce the carbohydrates in the diet so that the insulin requirements would be minimized. This procedure kept the blood sugar level down to normal values and prevented the loss of glucose in the urine, but it did not prevent the abnormalities of fat metabolism, especially arteriosclerosis with early death from heart disease. Actually, it exacerbated these effects. Consequently, there is a tendency at present to allow the patient a normal carbohydrate diet and then to give large quantities of insulin to metabolize the carbohydrates. This depresses the rate of fat metabolism and also depresses the high level of blood cholesterol that occurs in diabetes as a result of abnormal fat metabolism.

Diabetic Coma

If diabetes is not controlled satisfactorily, severe dehydration and acidosis may result; and sometimes, even when the person is receiving treatment, sporadic changes in metabolic rates of the cells, such as might occur during bouts of fever, can also precipitate dehydration and acidosis.

If the pH of the body fluids falls below approximately 7.0 to 6.9, the diabetic person develops coma. Also, in addition to the acidosis, dehydration is believed to exacerbate the coma. Once the diabetic person reaches this stage, the outcome is usually fatal unless he receives immediate treatment.

Treatment of Diabetic Coma

The patient with diabetic coma is extremely refractory to insulin because acidic plasma has an *insulin antagonist*, an alpha globulin, that opposes the action of the insulin. Also, the very high free fatty acid and acetoacetic acid levels in the blood inhibit the usage of glucose, as was discussed earlier. Therefore, instead of the usual 60 to 80 units of insulin per day, which is the dosage usually necessary for control of severe diabetes, as much as 1500 to 2000 units of insulin must often be given the first day of treatment of coma. Administration of insulin alone may be sufficient to reverse the abnormal physiology and to effect a cure. However, it is usually advantageous to correct also the dehydration and acidosis immediately.

Hyperinsulinism

Though much more rare than diabetes, increased insulin production, which is known as hyperinsulinism, does occasionally occur. This usually results from an adenoma of an islet of Langerhans. To prevent hypoglycemia in some of these patients, more than 1000 grams of glucose have had to be administered each 24 hours.

Insulin Shock and Hypoglycemia

As already emphasized, the central nervous system derives essentially all its energy from glucose metabolism. If insulin causes the level of blood glucose to fall to low values, the metabolism of the central nervous system becomes depressed. Consequently, in patients with hyperinsulinism or in patients who administer too much insulin to themselves, the syndrome called "insulin shock" may occur as follows:

As the blood sugar level falls into the range of 50 to 70 per cent, the central nervous system usually becomes quite excitable, for this degree of hypoglycemia seems to facilitate neuronal activity. Sometimes various forms of hallucinations result, but more often the patient simply experiences extreme nervousness, and he may tremble all over. As the blood glucose level falls to 20 to 50 mg. percent, clonic convulsions and loss of consciousness are likely to occur. As the glucose level falls still lower, the convulsions cease, and only a state of coma remains. Indeed, at times it is difficult to distinguish between diabetic coma as a result of insulin lack and coma due to hypoglycemia caused by excess insulin. If glucose is not administered immediately, permanent damage to the neuronal cells of the central nevous system occurs.

Depressed Glucose Tolerance Curve and Decreased Insulin Sensitivity in Hyperinsulinism

In addition to the usual signs of hypoglycemia that appear in hyperinsulinism, a definite diagnosis of the condition can usually be made by performing a glucose tolerance test. This glucose tolerance curve is usually depressed, and the initial glucose level is low, and the increase after ingestion of glucose is slight.

Functions of Glucagon

Glucagon, like insulin, is a small protein having a molecular weight of 3482. On injecting purified glucagon into an animal, profound *hyperglycemia* occurs, 1 microgram per kg. of glucagon elevating the blood glucose concentration approximately 20 mg./100 ml. of blood.

Glycogenolysis and increased blood glucose concentration caused by glucogon

The most dramatic effect of glucagon is its ability to cause glycogenolysis in the liver, which in turn increases the blood glucose concentration. This effect is similar to but many times more powerful than that of epinephrine in causing glycogenolysis. Like epinephrine, glucagon stimulates the formation of cyclic AMP, which in turn activates the normal inactive phosphorylase in the liver cells. The phosphorylase then causes rapid glycogenolysis followed by release of glucose into the blood.

Infusion of glucagon for about four hours can cause such intensive liver glycogenolysis that all the liver stores of glycogen will be totally depleted.

Prolonged hyperglycemic effect of glucagon

Even after all the glycogen in the liver has been exhausted under the influence of glucagon, continued infusion of this hormone causes continued hyperglycemia and glycosuria. This is due to glucagon's ability to increase the rate of gluconeogenesis.

Regulatory function of glucagon to prevent hypoglycemia

When the blood glucose concentration falls to as low as 70 mg. per cent, the pancreas secretes large quantities of glucagon, which rapidly mobilizes glucose from the liver. Thus, glucagon protects against hypoglycemia.

It has been suggested that glucagon is secreted during muscular activity and that it helps to cause glucose mobilization from the liver for use by the muscles; and in starvation, glucagon is abundantly secreted, perhaps in an attempt to maintain a normal blood glucose concentration.

10

GONADAL HORMONES

Neither cells nor individuals last forever, but new generations of living organisms are produced through the process of *reproduction*. All of the cells in a human body are the mitotic descendants of a single fertilized egg, or *zygote*. A zygote forms through the fusion of two specialized sex cells, or *gametes*, a *sperm* contributed by the father; and an egg, or *ovum* from the mother. The human *reproductive system* produces, stores, nourishes, and transports functional gametes. The reproductive system includes reproductive organs, or *gonads*, that produce gametes and hormones, ducts that receive and transport the gametes, and *accessory glands* that secrete fluids into the ducts. The *external genitalia* are perineal structures associated with the reproductive system, and the term *pudendum* refers to be external genitalia of either sex.

The functional roles of the male and female reproductive systems are quite different. In the adult male the gonads, or *testes* produce large numbers of sperm, perhaps as many as a half a billion per day. While traveling along a lengthy duct system these gametes are mixed with the secretions of accessory glands, creating the mixture known as *semen*. During *ejaculation* the semen is expelled from the body.

The gonads, or *ovaries*, of the sexually mature female usually produce only one mature egg per month. The gamete travels along short *uterine tubes* that terminate in a muscular chamber, the *uterus*. A short passageway, the *vagina*, connects the uterus with the exterior. During the sexual act the male ejaculation introduces semen into the vagina, and the sperm cells ascend the female reproductive tract. If a single sperm contacts the cell membrane of an egg, the two may fuse. This process, called *fertilization*, produces a zygote that subsequently undergoes repeated mitoses. After entering the uterus the resulting cell mass attaches to the uterine wall, developing over the next 9 months into a fetus containing billions of specialized cells. At birth the fetus travels down the vaginal canal and enters the world as a newborn infant.

This chapter will examine the anatomy of the male and female reproductive systems in detail, and consider the physiological and hormonal mechanisms responsible for the homeostatic regulation of reproductive function.

Reproductive System of the Male

The principal components of the male reproductive system proceding from the testes the sperm cells or *spermatozoa* travel along the *epididymis*, the *ductus deferens*, the *ejaculatory duct*,

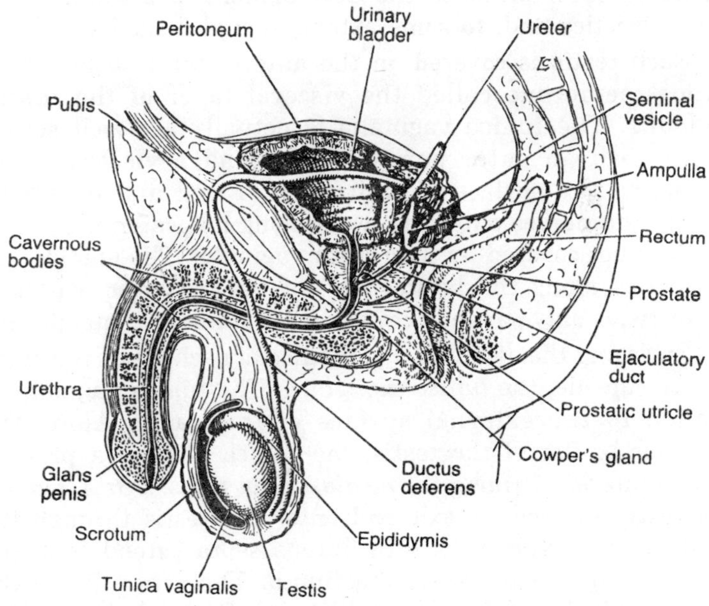

Fig. 10.1. Male reproductive organs in Homo.

and *urethra* before leaving the body. Accessory glands, notably the *seminal vesicles*, the *prostate*, and the *bulbourethral glands* secrete into the ejaculatory ducts and urethra. Externally visible structures include the *scrotum*, which encloses the testes, and the *penis*, an erectile organ that surrounds the distal portions of the urethra. Together the scrotum and penis constitute the external genitalia of the male.

Testis

The testes have two interrelated functions: the production of gametes and the production of steroids. In the male, the production of gametes, the *spermatozoa*, is referred to as *spermatogenesis*. A suspending medium secreted by the testis and other parts of the duct system provides a medium essential for the transport, maintenance, and further maturation of the sperm. Steroid hormones, primarily testosterone, function in the regulation of the development of the spermatozoa and the growth, development, and maintenance of the accessory reproductive glands. These hormones also influence the development of secondary sex characteristics and, to some extent, sexual behaviour.

Each testis is covered on the anterolateral surface by a serous membrane called the visceral layer of the *tunica vaginalis*. The tunica vaginalis is actually a closed serous cavity with an outer parietal layer and a visceral layer contacting the surface of each testis. Blood and lymphatic vessels and nerves enter the testis along its posterior surface, which is not covered by the tunica vaginalis. The epididymis, which lies along the lateral part of the posterior aspect of the testis, is only partially covered with the serous membrane.

Enclosing the seminiferous tubules of each testis is a thick fibrous capsule, the *tunica albuginea*. The tunica vaginalis is applied to the external surface of this layer. Along the posterior border of the testis, the tunica albugniea projects into its interior, forming the *mediastinum testis*. Ducts, nerves, and testicular vessels exit and enter the testis through the mediastinum. Numerous thin fibrous septa extend from the tunica albuginea to the mediastinum. The septa divide the testis into 200 to 300 pyramidal lobules. Each lobule contains one to four high convoluted seminiferous tubules, which produce the spermatozoa. Small blood vessels and lymphatic channels, as well as the *interstitial* or *Leydig cells* lie between the tubules. These are endocrine cells that produce steroid hormones, including testosterone.

Macrophages are also present, but other connective tissue cells including typical fibroblasts are normally absent.

In humans, each testis contains 400 to 600 seminiferous tubules which measure about 150 to 250 μm in diameter and 30 to 80 cm in length. The combined length of the tubules in each testis is about 250 m. The seminiferous tubules usually form an anastomic loop. Near the apex of the lobule the seminiferous tubules open into the *tubuli recti*, which form the first segment of the genital duct system. These ducts are continuous with the *rete testis*, a system of epithelial-spaces in the mediastinum.

Seminiferous Tubule

Boundary tissue or Lamina porpria

The seminiferous tubules are surrounded by an outer layer of adventitial cells and associated connective tissue, the *lamina propria*, and a well-defined basement membrane. The basement membrane lies adjacent to the seminiferous epithelium, separating it from the lamina propria.

The organization of the lamina propria demonstrates considerable species variation. In the common laboratory rodents, it consists of a single layer of flattened epithelioid cells, called *myoid cells* or *peritubular contractile cells*. These contractile cells are in close apposition with one another at their margins and, thus, form a contiguous sheath around the epithelium. In humans and other primates, the contractile cells are not in close apposition, rather they are organized into three to five layers. At the ultrastructural level, the cells demonstrate features associated with smooth muscle cells, including the presence of large numbers of actin filaments. These cells demonstrate rhythmic contractions that assist in the movement of spermatozoa and testicular fluid through the seminiferous tubules and into the genital duct system. Thickening of the lamina propria normally occurs with aging and reduction of the size of the seminiferous tubules, but it has also been associated with many cases of infertility in the human male.

Seminiferous epithelum

The *seminiferous epithelium* is a complex stratified epithelium composed of two basic cells types : *Sertoli cells* and *spermatogenic cells*. The *Sertoli cells* are a nonproliferating population composed of a single cell type. Each Sertoli cell extends from the basement membrane to the lumen of the seminiferous tubule. The *spermatogenic cells* are a proliferating population composed of cells at various phases of a complex differentiative process. For descriptive purposes,

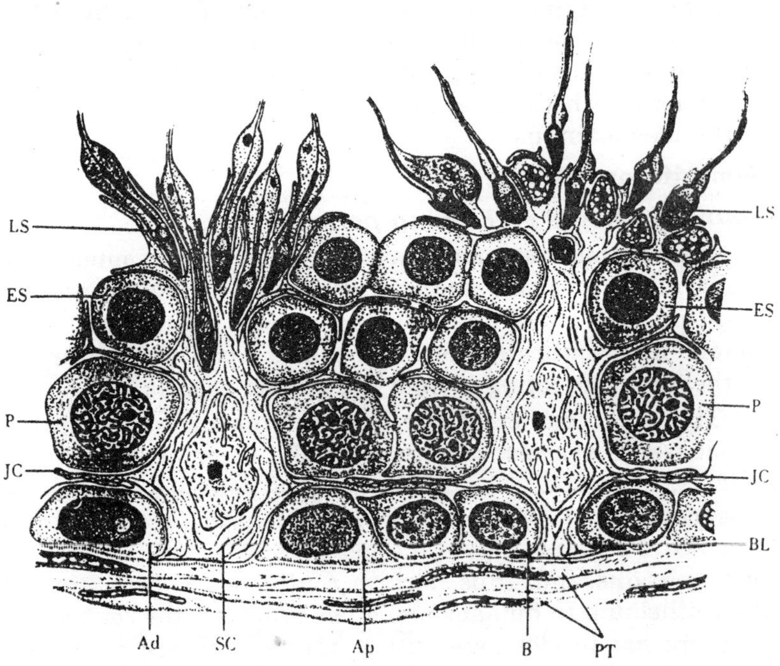

Fig. 10.2. Schematic drawing of human seminiferous epithelium. The relationship of the Sertoli cells (SC) to the spermatogenic cells is indicated. The seminiferous epithelium rests on a basal lamina (BL), and a layer of peritubular cells (PT) surrounds the seminiferous tubule. Pale type A spermatogonium (AP), dark type A spermatogonium (Ad), and type B spermatogonium are located in the basal compartment of the seminiferous epithelium below the junctional complex (JC) between adjacent Sertoli cells; pachytene primary spermatocytes (P), early spermatids (ES), and are late spermatids (LS), with partitioning residual cytoplasm that becomes the residual body, are seen in the adluminal compartment above the junctional complex.

spermatogenesis can be divided into three distinct phases : (1) the spermatogonial phase; (2) the spermatocyte phase or meiosis; and (3) the spermatid phase or spermiogenesis. The most immature cells are located near the basement membrane. As the cells proliferate and undergo differentiation, they move toward the lumen of the seminiferous tubule. Usually four or five concentric layers of cells, representing generations of cells at various phases in their development, can be identified within the seminiferous epithelium. These morphologically distinct spermatogenic cells are at one of the three basic phases in the development and are identified as *spermatogonia*, *spermatocytes*, or *spermatids*. *Spermatogonia*, derived from stem cells, undergo a series of mitotic divisions during which they undergo limited morphological differentiation. The final division of these spermatogonia produces *spermatocytes*, which divide meiotically and produce *spermatids*. These haploid cells undergo dramatic morphological differentiation, referred to as *spermiogenesis*, as they are transformed into spermatozoa.

Sertoli Cell

The *Sertoli cells* are columnar cells with complex apical and lateral processes that surround the adjacent spermatogenic cells and fill the spaces between them. However, the complex, elaborate configuration of the *Sertoli cells* cannot be seen distinctly in routine hematoxylin and eosin preparation. The *Sertoli cells* give structural organization to the tubules as they extend through the full thickness of the seminiferous epithelium.

The *Sertoli cell* nucleus, containing finely dispersed chromatin, is generally ovoid or triangular in shape and may have one or more deep infoldings. Its shape and location vary. The shape may be flattened, lying in the basal portion of the cell near and parallel to the basement membrane, or it may be triangular or ovoid, lying near or some distance from the basement membrane. In some species, the *Sertoli cell* nucleus contains a unique tripartite structure that consists of an RNA-containing nucleolus flanked by a pair of DNA-containing bodies called *karyosomes*.

With the light microscope, the extent and morphology of the apical and lateral processes are poorly defined. At the ultrastructural level, the limits of the cell can be identified

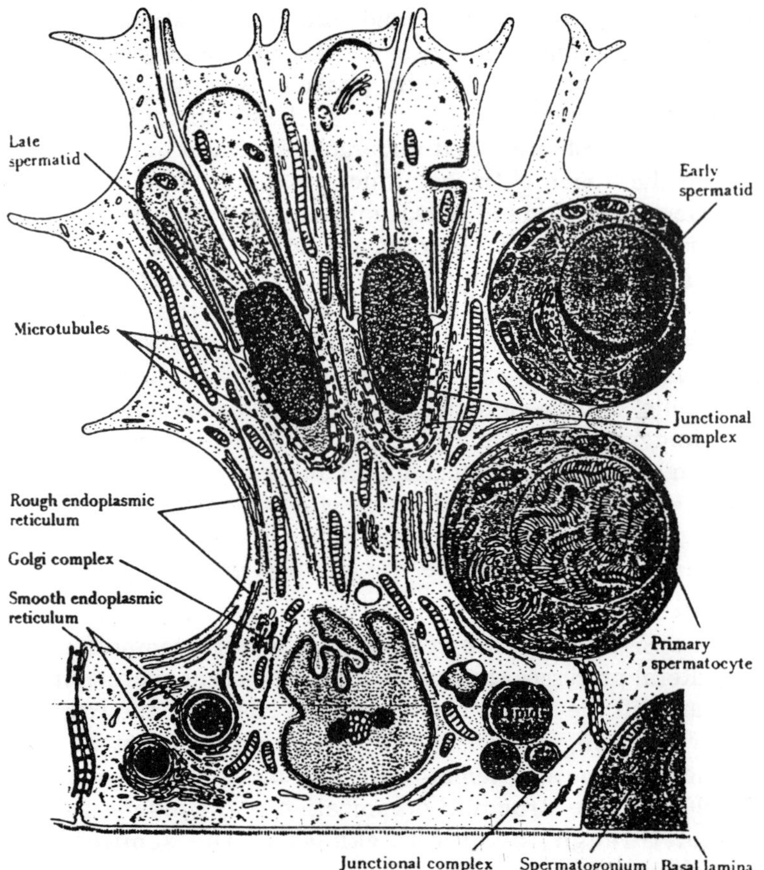

Fig. 10.3. Schematic drawing of a Sertoli cell and adjacent spermatogenic cells.

and the cytoplasm is seen to contain abundant smooth and some granular endoplasmic reticulum, numerous spherical or elongated mitochondria, a well developed Golgi complex, and varying numbers of microtubules, lysosomes, lipid droplets, vesicles, glycogen granules, and fine filaments. A sheath of 7- to 9-nm filaments surrounds the nucleus and separates it from other organelles. In humans, the basal cytoplasm also contains, characteristic *inclusion bodies of Charcot-Bottcher*. These slender, fusiform crystalloid inclusions, measuring 10 to 25 µm in length and 1 µm in width, are often visible with the light microscope. At the ultrastructural level, they appear as rather poorly ordered parallel or converging, straight, dense filaments, 15 nm in diameter. The chemical nature and function of these inclusions is unknown.

Junctional complexes

Adjacent *Sertoli cells* are joined to one another by an unusual type of junction. The *Sertoli cell* junctional complex is morphologically unique, not being found in other kinds of epithelia. A *Sertoli-Sertoli* junctional complex consists of : (1)an exceedingly tight junction composed of up to 50 or more parallel lines of fusion of the apposed membranes; (2) flattened cisternae of endoplasmic reticulum that parallel the plasma membrane in the region of the tight junction; and (3) hexagonally packed bundles of actin filaments that are interposed between the cisternae of endoplasmic reticulum and the plasma membrane. The endoplasmic reticulum associated with the junctional complex has ribosomes only on its cytoplasmic surface, which is directed away from the plasma membrane. A similar junctional complex also develops between *Sertoli* cells and spermatids. The *Sertoli-spermatid* junction is identical to the *Sertoli-Sertoli* junction except that there is no occluding junction between the two cells and there is no complementary specialization in the spermatid at the attachment site. It has been shown that the junctional complex at both sites is a strong adhesion device that holds the adjoining cells together.

Other types of junctions that have been described in *Sertoli cells* include gap junctions; hemidesmosomes, on the basal portion of the cell adjacent to the basal lamina; and desmosome-like complexes, between Sertoli cells and adjacent spermatogenic cells at early phases in their development.

Compartmentalization of the seminiferous epithelium

The *Sertoli-Sertoli* junctional complexes divide the seminiferous epithelium into two compartments: a *basal compartment* containing spermatogonia and young spermatocytes and an *adluminal compartment* containing later spermatocytes and spermatids. The compartmentalization of the seminiferous epithelium enables the Sertoli cells to establish microenvironments that allow the spermatogenic cells to proceed through meiosis and other differentiative events. The young spermatocytes, produced by the final mitotic division of the spermatogonia, must pass through the junctional compartment. This process is not fully understood, but apparently occurs in a manner which maintains the integrity of the two compartments.

Blood-testis barrier

Physiological and morphological studies have established that the *Sertoli-Sertoli* junctional complexes are the site of the *blood-testis barrier*. The concept of physiological barriers that separate tissues from blood-borne substances arose from studies conducted in the early 1900s. Investigators studying the distribution of intravenously injected dyes found that when some dyes were injected, most tissues were stained, but the brain, testes, and a few other tissues were not.

The observation that the seminiferous tubules were not stained received little attention until the late 1960s when studies with a variety of dyes confirmed the earlier observations and further demonstrated that dyes were not excluded from the seminiferous tubules of perpubertal rats. Subsequent physiological studies on the penetration of substances into the seminiferous tubules and the composition of testicular fluids have clearly demonstrated a physiological compartmentalization of the testis. Striking differences have been found in the ionic, amino acid, carbohydrate, and protein composition of fluids collected from the lumens of seminiferous tubules and genital ducts versus blood plasma and testicular lymph. Of particular interest is the absence of many of the serum proteins and immunoglobulins from the fluids of the seminiferous tubules and genital ducts. Morphological studies utilizing electron-opaque markers of various sizes have demonstrated that the junctional complexes between adjacent *Sertoli cells* act as the barrier to the penetration of the markers. Studies in the rat on the development of the testis at puberty have shown that the *Sertoli* junctional complexes are initially formed and establish the permeability barrier when the spermatogenic cells first pass through the early stages of the spermatocyte phase.

In terms of the immunological importance of the blood-testis barrier, two basic facts are well established: (1) spermatozoa and spermatogenic cells possess molecules that are unique to this cell type and are recognized as "foreign" by the immune system; (2) spermatozoa are first produced at puberty long after the individual has become immunocompetent, that is, capable of recognizing foreign molecules and producing antibodies against them. Therefore, the blood testis barrier serves an essential role in isolating the spermatogenic cells from the immune system. Failure of the spermatogenic cells and spermatozoa to remain so isolated results in the production of sperm-specific antibodies. Such an immune response is seen subsequent to vasectomy and in some case of infertility. After vasectomy, sperm-specific antibodies are produced as the cells of the immune system are exposed to the spermatozoa that no longer remain isolated from the immune system within the reproductive tract. It has also been shown that in some cases of infertility, unrelated to vasectomy, sperm-specific antibodies are present in the semen.

Functions

The *Sertoli cells* obviously play an essential role in the development of the spermatogenic cells, however, many of the specific activities remain to be clarified. Many of the *Sertoli cell* functions are directly related to the compartmentalization of the seminiferous epithelium. As the *Sertoli cells* secrete fluids or restrict the movements of molecules, they establish microenvironments that are essential for the spermatogenic cells.

Functions ascribed to the *Sertoli cells* include:

1. Providing physical support for the spermatogenic cells
2. Mediating the movements across the seminiferous epithelium of steroids, metabolites, and nutrients utilized by the spermatogenic cells
3. Restricting the movements of extracellular molecules into the seminiferous epithelium through establishment of occluding junctions between the adjacent *Sertoli cells*
4. Phagocytosing degenerating spermatogenic cells and the excess cytoplasm shed from the differentiating spermatids as the residual bodies.
5. Secreting androgen-binding protein, which serves to concentrate testosterone within the seminiferous epithelium and the proximal part of the genital duct system.

6. Secreting various stimulatory and inhibitory substances involved in the regulation of mitosis, meiosis, steroidogenic functions of the *Leydig cells*, and the release of gonadotropins form the pituitary.
7. Controlling the movements of the spermatogenic cells within the seminiferous epithelium and the release of the spermatozoa into the lumen of the seminiferous tubule.

Spermatogenesis

Spermatogenesis occurs in all the seminiferous tubules during active sexual life, beginning at the age of approximately 13 as the result of stimulation by adenohypophyseal gonadotropic hormones and continuing throughout the remainder of life.

The seminiferous tubules, one of which contain a large number of small to medium-sized cells called *spermatogonia* located in two to three layers along the outer border of the tubular epithelium. These continually proliferate and differentiate through definite stages of development to form sperm. During this process, each sperm loses one member of

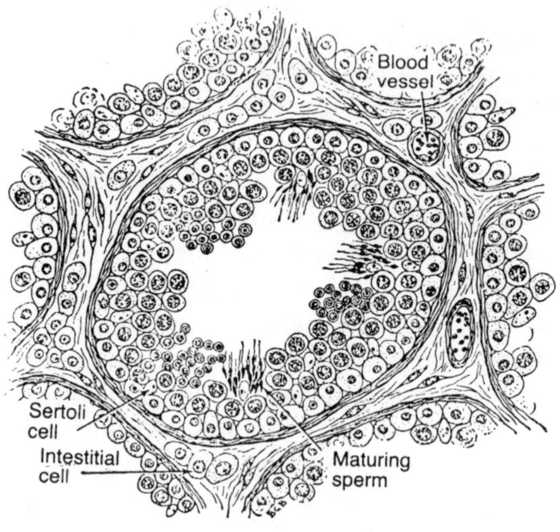

Fig. 10.4. A small area of a mammalian testis, showing one tubule and parts of several others in cross section and the interstitial tissue.

each of its 23 pairs of chromosomes, leaving only 23 chromosomes, none of tem paired.

The sex-determining pair of chromosomes in the spermatogonium is composed of one "X" chromosome, which is called the *female* chromosome, and "Y" chromosome, the male chromosome. During sperm formation, however, the sex-determining chromosomes divide among the sperm, so that half of the sperm become *male sperm* containing the "Y" chromosome and the other half *female sperm* containing the "X" chromosome. The sex of the offspring is determined by which of these two types of sperm fertilizes the ovum.

During formation of sperm form spermatogonia, most of the cytoplasm disappears and each cell begins to elongate into a spermatozoon, composed of a *head*, *neck*, *body*, and *tail*. To form the head, the nuclear material is rearranged into a compact mass, and the cell membrane contracts around the nucleus. It is this nuclear material that fertilizes the ovum. The tail has the same structure as a cilium. Upon release of sperm from the male genital tract into the female tract, the tail begins to wave back and forth, providing snake-like propulsion that moves the sperm forward at a maximum velocity of about 1 foot per hour.

Maturation of Sperm in the Epididymis

Following formation in the seminiferous tubules, the sperm pass through the *vasa recta* into the *epididymis*. After the sperm have been in the epididymis for some 18 hours to 10 days, they mature, develop the power of motility, and become capable of fertilizing the ovum.

Storage of Sperm

A small quantity of sperm can be stored in the epididymis, but probably most sperm are stored in the vas deferens and possibly to some extent in the ampulla of the vas deferens. Though the sperm in these areas become motile if released to the exterior, they are relatively dormant as long as they are stored, probably because of the acidic nature of the fluid in which they are stored. Sperm can maintain their fertility in the genital ducts for as long as 42 days.

Function of the Prostate Gland

The prostate gland secretes a thin, milky, alkaline fluid containing citric acid, calcium, acid phosphate, and fibrinolysin.

During ejaculation, the capsule of the prostate gland contracts simultaneously with the contractions of the vas deferens and seminal vesicles so that the thin, milky fluid of the prostate gland adds to the bulk of the semen. The alkaline characteristic of the prostatic fluid may be quite important for successful fertilization of the ovum, because the fluids of the vas deferens and of the female vagina are relatively acidic and, consequently, inhibit sperm fertility. Sperm do not become optimally motile until the pH of the surrounding fluids rises to approximately 6 to 6.5.

Semen and the Phenomenon of Capacitation

Semen, which is ejaculated during the male sexual act, is composed of the fluids from the vas deferens, from the seminal vesicles, from the prostate gland, and from the mucous glands, especially the bulbo-urethral glands. The average pH of the combined semen is approximately 7.5, the alkaline prostatic fluid have neutralized the mild acidity of the other portions of the semen. The prostatic fluid gives the semen a milky appearance, while fluid from the seminal vesicles and from the mucous glands gives the semen a mucoid consistency. Within approximately one-half hour after ejaculation the mucoid consistency of semen disappears because of proteolytic enzymes in the fluid.

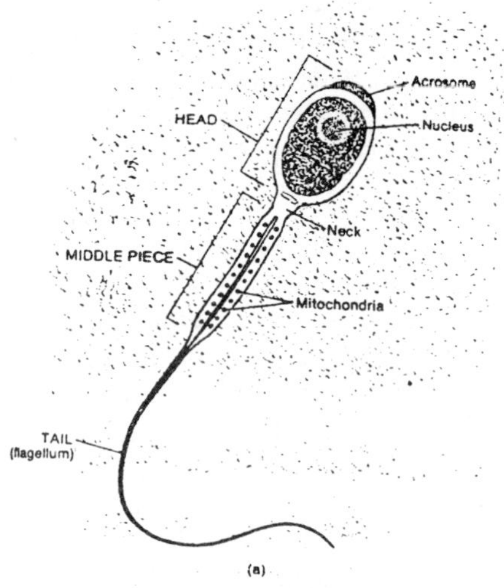

Fig. 10.5. Spermatozoa.

For reasons not yet understood, sperm that are washed clear of the other components of semen become rapidly infertile but regain this fertility in a few minutes to an hour or more when remixed with the other components. This phenomenon is called *capacitation*.

Though sperm can live for many weeks inthe male genital ducts, once they are ejaculated in the semen their maximal life span is only 24 to 72 hours at body temperature. At lowered temperatures, however, semen may be stored for weeks or even years.

Male Fertility

The seminiferous tubular epithelium can be destroyed by a number of different diseases. For instance, bilateral orchiditis resulting from mumps usually causes sufficient degeneration of tubular epithelium that sterility results in a large percentage of males so afflicted. Another disease that frequently localizes in the testes and can cause severe tubular damage is typhus fever. Also, many male infants are born with degenerate tubular epithelium as a result of strictures in the genital ducts or as a result of unknown causes. Finally, a cause of sterility seems to be excessive temperature of the testes.

Effect of Sperm Count on Fertility

An average total of 400,000,000 sperm are usually present in each ejaculate, in a total semen volume of about 3 ml. When the number of sperm falls below approximately 100,000,000, the person is likely to be almost completely infertile. Only a single sperm is necessary to fertilize the ovum; but, for reasons not yet completely understood, the ejaculate must contain a tremendous number of sperm for at least one to fertilize the ovum. It is believed that many sperm must be present to secrete the enzyme hyaluronidase, which removes a barrier of granulosa cells from around the ovum to be fertilized.

Ducts

When the sperm mature, they are moved through the convoluted seminiferous tubules to the *straight tubules*. The straight tubules lead to a network of ducts in the testis called the *rete testis*. Some of the cells lining the rete testis possess cilia that probably push the sperm along. The sperm are next transported out of the testis through a series of coiled *efferent ducts* that empty into a single tube called the *ductus epididymis*. At this point, the sperm are morphologically mature.

Epididymis

The two *epididymides* are comma-shaped organs. Each lies along the posterior border of the testis and consists mostly of a tightly coiled tube: the *ductus epididymis*. This larger, superior portion of the epididymis is known as the *head*. It consists of the efferent ducts that empty into the ductus epididymis. The *body* of the epididymis contains the ductus epididymis. The *tail* is the smaller, inferior portion. Within the tail, the ductus epididymis becomes the ductus deferens.

The ductus epididymis measures about 6m in length and 1mm in diameter. It is tightly packed within the epididymis, which measures only about 3.8 cm. The ductus epithelium is lined with pseudostratified columnar epithelium, and its wall contains smooth muscle. The free surfaces of the columnar cells contain long, branching microvilli called *stereocilia*. Functionally, the epididymis is the site of sperm maturation. It stores spermatozoa and propels them toward the urethra during ejaculation.

Ductus Deferens

Within the tail of the epididymis, the ductus epididymis becomes less convoluted, its diameter increases, and at this point it is referred to as the *ductus deferens* or *seminal duct*. The ductus deferens, about 45 cm long, ascends along the posterior border of the testis, penetrates the inguinal canal, and enters the pelvic cavity, where it loops over the side and down the posterior surface of the urinary bladder. The dilated terminal portion of the ductus deferens is known as the *ampulla*. Histologically, the ductus deferens is lined with pseudostratified epithelium and contains a heavy coat of three layers of muscle. Peristaltic contractions of the muscular coat propel the spermatozoa toward the urethra during ejaculation.

Traveling with the ductus deferens as it ascends in the scrotum are the testicular artery, autonomic nerves, veins that drain the testes, lymphatics, and a small circular band of skeletal muscle called the *cremaster muscle*. These structures constitute the *spermatic cord*, a supporting structure of the male reproductive system. The cremaster muscle elevates the testes during sexual stimulation and exposure to cold. The spermatic cord passes through the *inguinal canal*, a slitlike passageway in the anterior abdominal wall just superior to the medial half of the inguinal ligament. The area of the inguinal canal and spermatic cord represents a weak spot in the abdominal wall, and it is frequently the site of a *hernia*—a rupture or separation of a portion of the abdominal wall resulting in the protrusion of a part of a viscus.

Ejaculatory Duct

Posterior to the urinary bladder, each ductus deferens joins its *ejaculatory duct*. Each duct is about 2 cm long. Both ejaculatory ducts eject spermatozoa into the prostatic urethra. The urethra is the terminal duct of the system, serving as a common passageway for both spermatozoa and urine.

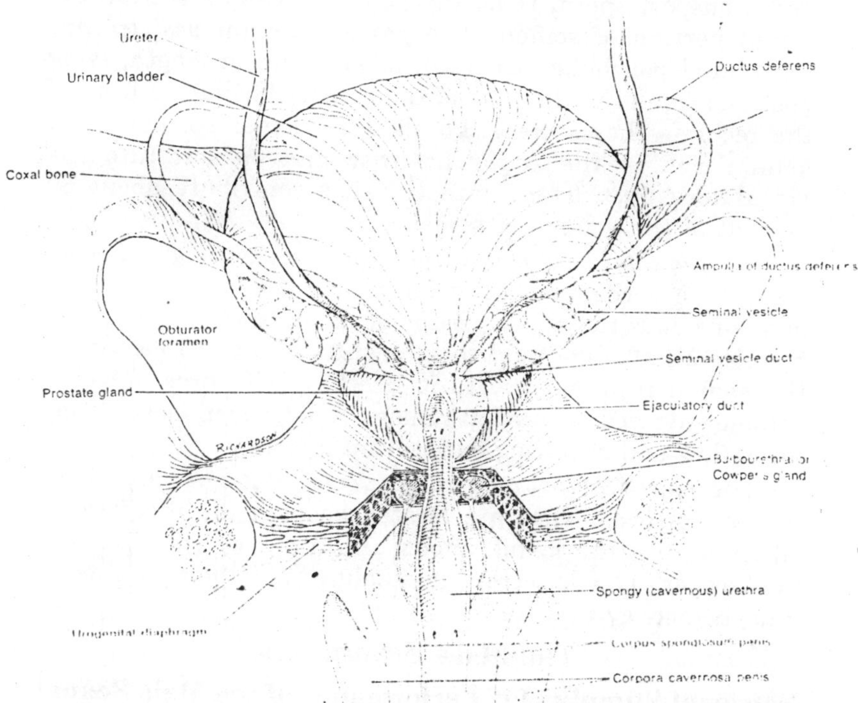

Fig. 10.6. Relationships of some male reproductive organs posterior view of urinary bladder.

Urethra

In the male, the *urethra* passes through the prostate gland, the urogenital diaphragm, and the penis. It measures about 20 cm in length and is subdivided into three parts. The *prostatic portion* is 2 to 3 cm long and passes through the prostate gland. It continues inferiorly as the membranous portion as it passes through the urogenital diaphragm, a muscular partition between the two ischiopubic rami. The *membranous portion* is about 1 cm in length. After passing through the urogenital diaphragm, it is known as the *spongy (cavernous) portion* of the urethra. This portion is about 15 cm long. The spongy urethra enters the bulb of the penis and terminates at the external urethral orific.

Accessory Glands

Whereas the ducts of the male reproductive system store and transport sperm cells, the *accessory glands* secrete the liquid portion of semen. The paired *seminal vesicles* are convoluted pouchlike structure, about 5 cm in length, lying posterior to and at the base of the urinary bladder in front of the rectum. They secrete the alkaline viscous component of semen rich in the sugar fructose and pass it into the ejaculatory duct. The seminal vesicle contribute about 60 percent of the volume of semen.

The *prostate gland* is a single, doughnut-shaped gland about the size of a chestnut. It is inferior to the urinary bladder and surrounds the upper portion of the urethra. The prostate secretes an alkaline fluid that constitues 13 to 33 percent of the semen into the prostatic urethra. The prostate may become enlarged or develop tumors in older men, obstructing urine flow and requiring surgical intervention.

The paired *bulbourethral*, or *Cowper's glands* are about the size of peas. They are located beneath the prostate on either side of the membranous urethra. Like the prostate, the Cowper's glands secrete an alkaline fluid; their ducts open into the spongy urethra.

The Male Sexual Act

Neuronal Stimulus for Performance of the Male Sexual act

The most important source of impulses for initiating the male sexual act is the *glans penis*. The massaging action of intercourse on the glans stimulates the sensory end-organs, and the sexual sensations in turn pass into the sacral portion of the spinal cord, and finally up the cord to undefined areas of the cerebrum. Impulses may also enter the spinal cord form areas adjacent to the penis to aid in stimulating the

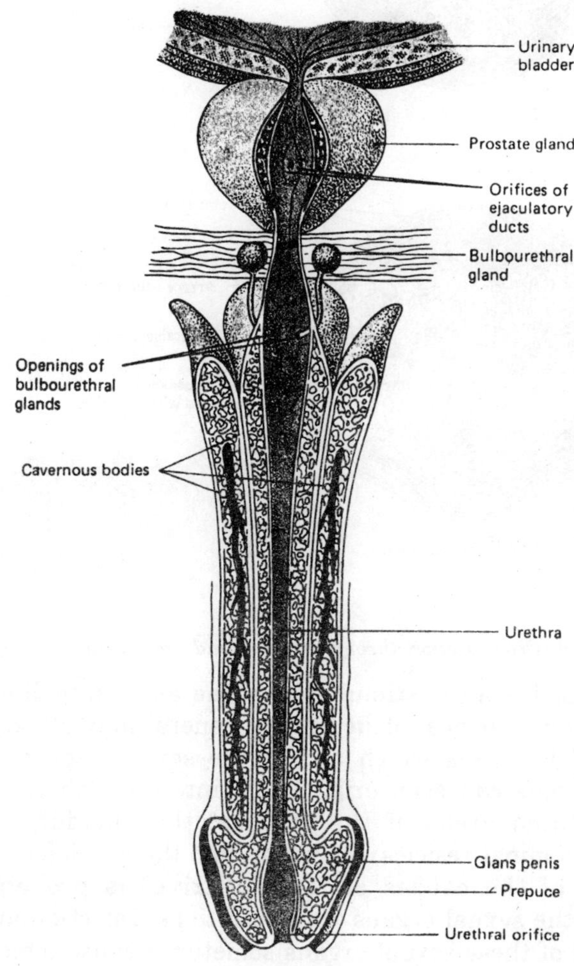

Fig. 10.7. Internal structure of the penis.

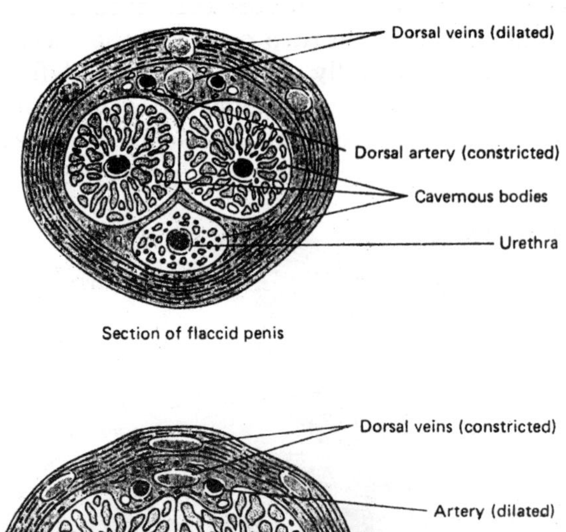

Section of flaccid penis

Section of erect penis

Fig. 10.8. Cross section through flaccid and erect penis.

sexual act. For instance, stimulation of the anal epithelium, the scrotum, and perineal structures in general may all send into the cord impulses which add to the sexual sensation. Sexual sensations can even originate in internal structures, such as irritated areas of the urethra, the bladder, the prostate, the seminal vesicles, the testes, and the vas deferens. Indeed, one of the causes of "sexual drive" is probably overfilling of the sexual organs with secretions. Infection and inflammation of these sexual organs sometimes cause almost continual sexual desire, and "aphrodisiac" drugs, such as cantharides, increase the sexual desire by irritating the bladder and urethral mucosa.

The Psychic Element of Male Sexual Stimulation

Appropriate psychic stimuli can greatly enhance the ability of a person to perform the sexual ct. Simply thinking sexual thoughts or even dreaming that the act of intercourse is being performed can cause the male sexual act to occur and to culminate in ejaculation. Indeed, *nocturnal emissions* during dreams occur in many males during some stages of sexual life, especially during the teens.

Integration of the Male Sexual Act in the Spinal Cord

Though psychic factors usually play an important part in the male sexual act and can actually initiate it, the cerebrum is probably not absolutely necessary for its performance, for appropriate genital stimulation can cause ejaculation in some animals and in an occasional human being after their spinal cords have been cut above the lumbar region. Therefore, the male sexual act results from inherent reflex mechanisms integrated in the sacral and lumbar spinal cord, and these mechanisms can be initiated by either psychic stimulation or actual sexual stimulation.

Stages of the Male Sexual Act

Erection

Erection is the first effect of male sexual stimulation. It is caused by parasympathetic impulses that pass from the sacral portion of the spinal cord to the penis. These parasympathetic impulses dilate the arteries of the penis, thus allowing arterial blood to flow under high pressure into the *erectile tissue* of the penis. This erectile tissue is nothing more than large, cavernous venous sinusoids, which become dilated tremendously when arterial blood flows into them under pressure, since the venous outflow is partially occluded. Also, the erectile bodies are surrounded by strong fibrous coats; therefore, high pressure within the sinusoids causes ballooning of the erectile tissue to such an extent that the penis becomes hard and elongated.

Lubrication

During sexual stimulation, parasympathetic impulses in addition to promoting erection, cause the glands of Littre and the bulbo-urethral glands to secrete mucus. Thus mucus flows through the urethra during intercourse to aid in the lubrication of coitus. However, most of the lubrication of coitus is provided by the female sexual organs rather than by the male. Without satisfactory lubrication, the male sexual act is rarely successful because unlubricated intercourse cause pain impulses which inhibit rather than excite sexual sensations.

Ejaculation

Ejaculation is the culmination of the male sexual act. When the sexual stimulus becomes extremely intense, the reflex centers of the spinal cord begin to emit rhythmic sympathetic impulses that leave the cord at L-1 and L-2 and pass to the genital organs through the hypogastric plexus to initiate emission, which is the forerunner of ejaculation.

Emission begins with peristaltic contractions in the ducts of the testis, the epididymis, and the vas deferens to cause expulsion of sperm into the internal urethra. Simultaneously, rhythmic contractions in the seminal vesicles and the muscular coat of the prostate gland expel seminal fluid and prostatic fluid along with the sperm. These mix with the mucus already secreted by the bulbourethral glands, and all of these different urethral glands, and all of these different types of fluid combine to form the semen. The process to this point is emission.

Then, rhythmic nerve impulses are sent from the cord through S-1 and S-2 and thence to skeletal muscles that encase the base of the erectile tissue, causing rhythmic increases in pressure in this tissue, which expresses the semen from the urethra to the exterior. This process is called *ejaculation* proper.

Reproductive System of the Female

The principal organs include the *ovaries*, the *uterine tubes*, the *uterus*, the *vagina*, and the external genitalia. As in the male a variety of accessory glands secrete into the reproductive tract. Physicians specializing in disorders of the female reproductive tract are called *gynecologists*.

Female Reproductive Anatomy

The ovaries, uterine tubes, and uterus are enclosed within an extensive mesentery known as the *broad ligament*. The reproductive organs and their attendant blood vessels, lymphatics, and nerves lie within this mesentery. The uterine tubes run along the superior border of the mesentery and open into the pelvic cavity lateral to the ovaries. A thickened

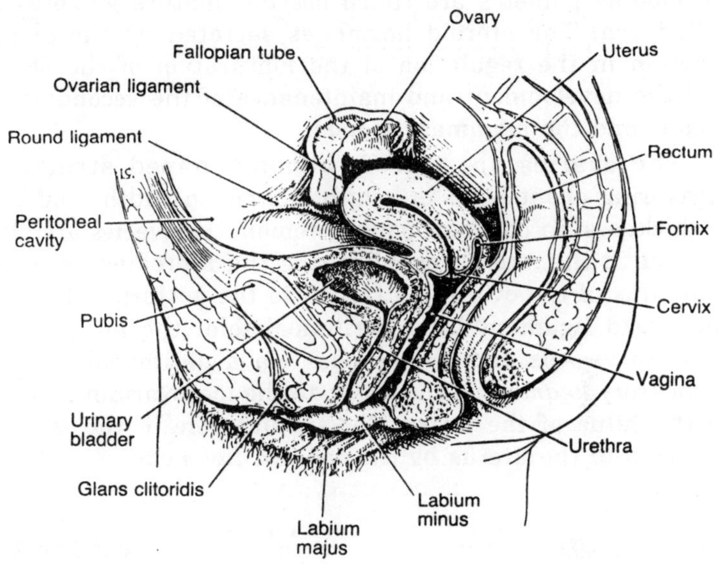

Fig. 10.9. Female reproductive organs in Homo.

fold of mesentery, the *mesovarium*, supports and stabilizes the position of each ovary. The broad ligament attaches to the sides and floor of the pelvic cavity, its epithelium becoming continuous with that of the parietal peritoneum. The broad ligament thus subdivides the pelvic cavity. The pocket formed between the posterior wall of the uterus and the anterior surface of the rectum represents the *rectouterine pouch*, while that between the uterus and the posterior wall of the bladder is the *vesicouterine pouch*.

Several other ligaments assist the broad ligament in supporting and stabilizing the position of the uterus and associated reproductive organs. These ligaments travel within the mesentery sheet of the broad ligament on the way to the ovaries or uterus. The broad ligament limits side to side movement, and the other ligaments provide support in other planes.

Ovary

The ovaries have two interrelated functions : the production of gametes and the production of steriods. In the femalem the production of gametes is called *oogenesis*. Developing gametes are called *oocytes*; mature gametes are called *ova*. The steroid hormones secreted by the ovaries function in the regulation of the maturation of the oocytes and the development and maintenance of the secondary sex organs and the mammary glands.

The ovaries are paired, almond-shaped structures, measuring about 3 cm in length, 1.5 cm in width, and 1 cm in thickness. In postmenopausal women, the ovaries are about one-fourth the size observed during the reproductive period. The *hilus* of the ovary is attached to the posterior surface of the broad ligament by a peritoneal fold, the *mesovarium*. Ovarian vessels and nerves pass via a peritoneal fold, the *suspensory ligament* of the ovary, to the mesovarium and then to the hilus of the ovary. The medial pole of the ovary is attached to the uterus by the *ovarian ligament*.

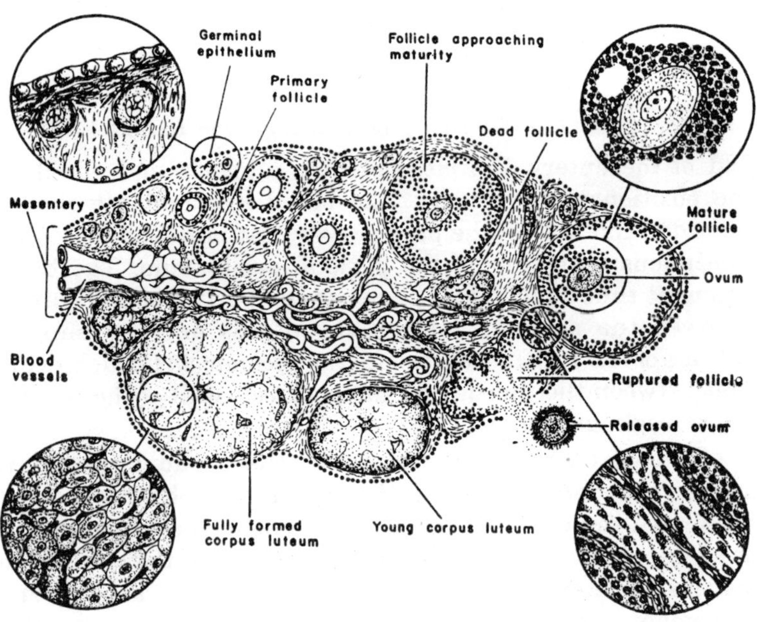

Fig. 10.10. Diagram of the stages in the developmental of an egg, follicle and corpus luteum in a mammalian ovary.

The central portion of the ovary, the *medullary region*, or *medulla*, contains loose connective tissue, a mass of relatively large contorted blood vessels, lymphatic vessels, and nerves. The peripheral portion of the ovary, the *cortical region*, or *cortex*, surrounds the medulla and contains the *ovarian follicles*, embedded in a compact, richly cellular connective tissue. Scattered smooth muscle fibers are present in the stroma around the follicles. The boundary between the medulla and cortex is indistinct. At the hilus, the cortex is interrupted and the mesovarium is continuous with the medulla. Small, irregular ducts, the *rete ovarii*, which are remnants of fetal structures, may be found in this region.

The surface of the ovary is covered by a single layer of cuboidal or squamous cells. This layer, known as the *germinal epithelium*, is continuous with the mesothelium that covers the mesovarium. The germ *germinal epithelium* is carryover from earlier days when it was incorrectly thought to be the site of germ cell formation. It is now known that the primordial germ cells are of extragonadal origin and that they migrate into the gonad where they differentiate into oogonia. A dense connective tissue layer, the *tunica albuginea*, lies between the germinal epithelium and the underlying cortex.

Ovarian follicle

Ovarian follicles, each containing an oocyte, are embedded in the stroma of the ovarian cortex. The size of the follicle is indicative of the developmental state of the oocyte. Early stages of oogenesis occur during early fetal life with mitotic divisions of the oogonia. The oocytes present at birth remain arrested in their development at the first meiotic division. As young girls pass through puberty, the ovaries begin a phase of reproductive activity characterized by the cyclic growth and maturation of small groups of follicles. The first ovulation generally does not take place for a year or more following menarche. A cyclic pattern of follicular maturation and ovulation then continues through a period of 30 to 40 years in parallel with the menstrual cycle. Normally, only one oocyte reaches full maturity and is released from the ovary during each menstrual cycle. Obviously, the maturation and release of more than one egg at ovulation may lead to multiple births. In her reproductive life-span, a woman produces only about 400 mature ova. Most of the estimated 400,000 oocytes present at birth fail to complete maturation and are gradually lost through *atresia*. The few oocytes that remain at menopause degenerate within a few years.

Three basic types of ovarian follicles can be identified : (1) primordial, (2) growing, and (3) mature, or Graafian. The growing follicles can be further categorized as primary and secondary follicles. Some histologists identify additional stages in follicular development. In the cycling ovary, follicles at all stages of development are found. The primordial follicles predominate.

Primordial follicle

The primordial follicles are found in the stroma of the cortrex just beneath the tunica albuginea. A single layer of flattened or squamous *follicular cells* resting on a basement membrane surrounds the oocyte. At this stage, the plasma membranes of the oocyte and the surrounding follicular cells are closely apposed to one another and are relatively smooth. The oocyte in the follicle measures about 30 μm in diameter and has a large, eccentric nucleus containing finely dispersed chromatin and one or more large nucleoli. A well-developed Golgi apparatus surrounded by numerous mitochondria is located near the nucleus. At the ultrastructural level the cytoplasm is seen to contain, in addition to other typical organelles, annulate lamellae and numerous small vesicles that appear to be associated with the Golgi apparatus, but are also seen scattered through the cytoplasm. Pores in the nuclear envelope are numerous.

Growing follicle

As the primordial follicle makes the transition into the growing follicle, changes occur in the follicular cells, oocyte, and adjacent stroma. Initially, the oocyte enlarges and the surrounding flattened follicular cells proliferate and become cuboidal in shape. At this stage, that is, with the follicular cells becoming cuboidal, the follicle is identified as a *primary follicle*. As the oocyte grows, a homogeneous, deeply staining, acidophilic, refractile layer called the *zona pellucida* appears between the oocyte and the adjacent follicular cells. The zona pellucida is first apparent with the light microscope when the oocyte, surrounded by a single layer of cuboidal or columnar

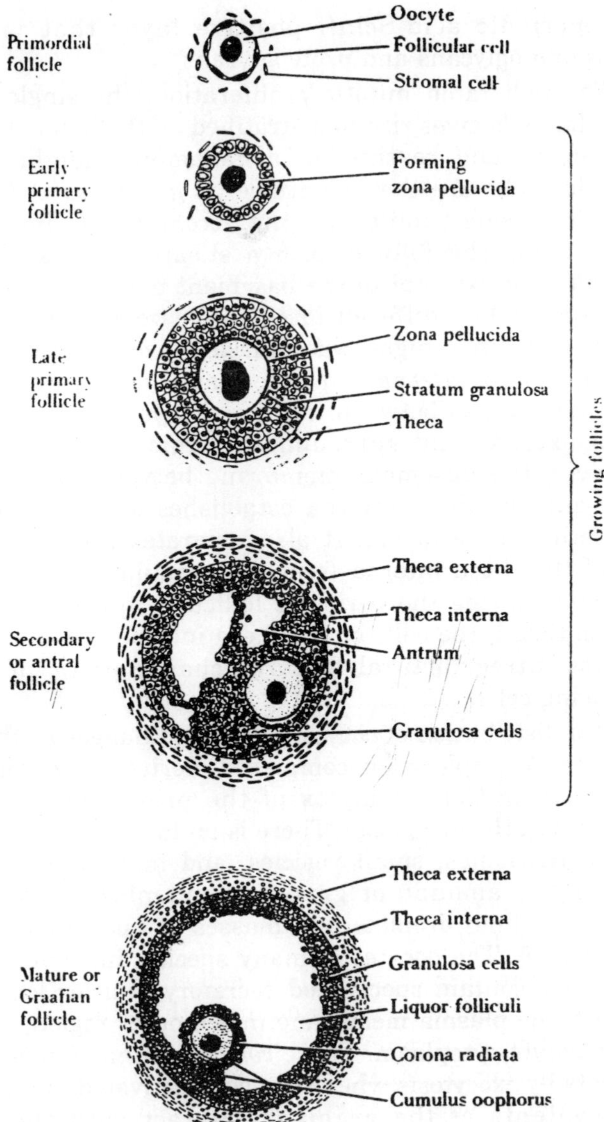

Fig. 10.11. Schematic drawing of the stages in the development of ovarian follicles, beginning with the primordial follicle and ending with the mature, or Graafian follicle.

follicle cells, has grown to a diameter of 50 to 80 μm. The growing oocyte is responsible for the production of this gel-like, periodic acid-Schiff positive layer that is rich in glycosaminoglycans and proteoglycans.

Through rapid mitotic proliferation, the single layer of follicular cells gives rise to a stratified epithelium surrounding the oocyte and resting on a basement membrane. The follicular cells are then identified as *granulosa cells*. As the granulosa cells proliferate, the stromal cells immediately surrounding the follicle form a sheath of cells, the *theca folliculi*, just external to the basement membrane. The theca folliculi further differentiates into two layers: the *theca interna*, an inner, highly vascularized layer of secretory cells, and the *theca externa*, an outer layer of connective tissue cells. Boundaries between the thecal layers and between the theca externa and surrounding stroma are not distinct. However, the basement membrane between the granulosa layer and the theca interna establishes a distinct boundary separating these layers. It also separates the rich capillary bed of the theca interna from the granulosa layer, which is avascular during the period of follicular growth. When fully differentiated, the cells of the theca interna are cuboidal and possess ultrastructural features characteristic of steroid-producing cells.

The distribution of oocyte organelles changes as the oocyte matures. Multiple golgi complexes, derived from the single juxtanuclear Golgi complex of the primordial oocyte, are scattered in the cytoplasm. There is an increase in the number of free ribosomes, small vesicles, and multivesicular bodies and in the amount of granular endoplasmic reticulum. Occasional lipid droplets and masses of lipochrome pigment can be seen. The oocytes of many species, including those of mammals, contain specialized secretory vesicles located just beneath the plasma membrane in the outer region, or *cortex*, of the oocyte cytoplasm. These *cortical granules* release their contents by exocytosis when the egg is activated by the sperm. The contents of the granules interact with the plasma membrane and the coverings around the egg, the egg coat, so as to prevent other sperm from fusing with and entering the egg. Numerous irregular microvilli from the oocyte project into the space between the oocyte and the surrounding granulosa cells as the zona pellucida is deposited. Slender processes from the granulosa cells develop at the same time and project toward the oocyte, intermingling with oocyte microvilli and, occasionally, invaginating into the oocyte plasma membrane. The processes may contact the plasma membrane, but do not establish cytoplasmic continuity between the cells. However, during follicular growth, extensive gap junctions, serving as channels for intercellular communication, develop between adjacent granulosa cells.

The primary follicle initially moves deeper into the cortical stroma as it increase in size, primarily through the proliferation of the granulosa cells. When the stratum granulosa reaches a thickness of 6 to 12 cell layers, fluid-filled cavities appear among the granulosa cells. As the hyaluronic acid-rich fluid, called *liquor folliculi*, continues to accumulate among the granulosa cells, the cavities begin to coalesce, eventually forming a single crescent-shaped cavity, called the *antrum*. The follicle is now identified as a *secondary* or *antral follicle*. The eccentrically positioned oocyte, which has attained a diameter of about 125 μm, undergoes no further growth. The follicle, which was 0.2 mm in diameter when the follicular fluid first appeared, continues to grow and will reach a size of 10 mm or more in diameter.

As the secondary follicle increases in size, the antrum, lined by several layers of granulosa cells, enlarges. The stratum granulosa has a relatively uniform thickness except for the region associated with the oocyte. Here, the granulosa cells form a thickened mound, the *cumulus oophorus*, that projects into the antrum and surrounds the oocyte. Those cells of the cumulus oophorus that immediately surround the oocyte and remain with it at ovulation are also referred to as the *corona radiata*. Extracellular, densely staining material, called *Call-Exner bodies*, may be seen between the granulosa cells. With the PAS reaction, these bodies stain positively. Their role and nature is unknown.

Mature or Graafian follicle

A mature follicle has a diameter of 10 mm or more, extends through the full thickness of the ovarian cortex, and bulges from the surface of the ovary. As a follicle nears its maximum size, there is a decrease in the mitotic activity of the granulosa cells. The stratum granulosa appears to become thinner as the antrum increases in size. As the space between the granulosa cells continues to enlarge, the oocyte and cumulus cells are gradually loosened from the rest of the granulosa cells in preparation for ovulation. The cumulus cells immediately surrounding the oocyte form a single layer of cells, the *corona radiata*, just external to the zona pellucida. These cells and loosely arranged cumulus cells remain associated with the oocyte at the time of ovulation.

During this period of maturation of the follicle, the theca folliculi become more prominent. Lipid droplets appear in the cytoplasm of the theca interna cells, and the cells demonstrate ultrasturctural features associated with steroid-producing cells. The granulosa cells and thecal cells produce the ovarian hormones essential in normal reproductive function. In the human, luteinizing hormone stimulates the cells of the theca interna to secrete testosterone, which serves as the precursor for estrogens. The testosterone diffuses into the follicular stroma and reaches the granulosa cells. In response to follicle-stimulating hormone, the granulosa cells catalyze the conversion of testosterone to estrogen, which, in turn, stimulates the granulosa cells to proliferate and thereby increase the size of the follicle. As estrogen secretion continues and blood estrogen levels increase, a surge in the release of LH is induced in the adenohypophysis. In respose to the LH surge, the granulosa cells are down-regulated and no longer produce estrogens in response to LH. The granulosa and thecal cells then undergo luteinization and produce progesterone.

Ovulation

Shortly before ovulation, the protruding outer wall of the follicle swells rapidly and a small area in the center of the swelling, called the *stigma*, protrudes like a nipple. The pressure in the follicle at this time is about 15 mm. Hg. Within another half hour or so, the stigma ruptures, and fluid begins to ooze from the follicle. About two minutes later, as the follicle becomes smaller because of loss of fluid, a more viscous fluid that has occupied the central portion of the follicle and that contains both the ovum and a mass of granulosa and theca cells is extruded into the abdomen. The cause of ovulation seems to be necrosis or weakening of the cells of the stigma or of the cement substance between them.

Necessity of Luteinizing Hormone (LH) for Ovulation; The Ovulatory Surge of LH. LH is necessary for final follicular growth and ovulation. Without this hormone, even though large quantities of FSH are available, the follicle will not progress to the stage of ovulation. LH acts synergistically with FSH to cause rapid swelling of the follicle shortly before ovulation. The swelling in itself undoubtedly contributes to the rupture, but it is likely that LH also has a direct effect to cause the stigma of break open at the time of ovulation.

It is significant that an especially large amount of LH, called the *ovulatory surge*, is secreted by the adenohypophysis during the day immediately preceding ovulation.

Atretic Follicles

In general only one ovum is expelled from the ovaries during each monthly sexual cycle. All the developing vesicular follicles that to not ovulate begin to become atretic either befor or immediately after the single ovum is expelled. The ovum within each atretic follicle degenerates, and the follicular cells disappear completely, followed by ingrowth of connective tissue. This process of atresia is undoubtedly caused by a hormonal signal from the ovulating follicle, but the nature of this signal is unknown.

The Corpus Luteum

Within a few hours after expulsion of the ovum from the follicle, the granulosa and theca cells of the follicle undergo *luteinization*, and the mass of cells becomes a corpus luteum, which secretes progesterone and also large amounts of estrogens. That is, these cells become greatly enlarged and develop lipid inclusions that give the cells a distinctive yellowish color, whence is derived the term *luteum*.

In the normal female, the corpus luteum grows to approximately 1.5 cm., reaching this stage of development approximately seven or eight days following ovulation. After this, it begins to involute and loses its secretory function, as well as is lipid characteristics, approximately 12 days following ovulation.

Luteinizing function of Luteinizing hormone (LTH) and LH

The change of follicular cells into lutein cells is completely dependent on the secretion of LH by the adenohypophysis. In fact, this function gave LH its name "luteinizing."

Secretion by the corpus luteum

Function of Luleotropic Hormone (LTH) and LH. The corpus luteum is a highly secretory organ, secreting large amounts of both *progesterone* and *estrogens.* Unfortunately, we do not know the precise role of the individual gonodotropic hormones in causing secretion by the corpus luteum. However, total abrogation of gonadotropin secretion prevents further function of the corpus luteum. It is believed, on the basis of animal experiments, that in the human being LH is responsible for initial development of the corpus luteum while LTH is responsible for secretion by the corpus luteum.

Follicular Atresia

As was previously stated, few of the ovarian follicles that begin their differentiation in the ovary are destined to complete their maturation. Most of the follicles degenerate and disappear through a process called *follicular atresia.* Large numbers of follicles undergo atresia during fetal development, early postnatal life, and puberty. After puberty, groups of follicles begin to mature each menstrual cycle, but only one follicle completes is maturation.

A follicle in any stage of its maturation may undergo atresia. The process becomes more complex as the follicle progresses further toward maturation. In the atresia of promordial and small growing follicles, the oocyte becomes smaller and degenerates, and then similar changes occur in the follicular cells. As the cells are resorbed and disappear, the surrounding stromal cells migrate into the space previously occupied by the follicle, leaving no trace of its existence. In the atresia of large growing follicles, the degeneration of the oocyte appears to occur secondarily to degenerative changes in the follicular wall, which includes the following sequential events : (1) invasion of the granulosa layer by strands of vascularized connective tissue, (2) sloughing of the granulosa cells into the antrum of the follicle, (3) hypertrophy of the theca interna cells, (4) collapse of the follicle as the degeneration continues, and (5) invasion of connective tissue into the cavity of the follicle. The oocyte undergoes typical changes associated with degeneration and autolysis and then disappears. The zona pellucida, which is resistant to the autolytic changes occurring in the cells associated with it, becomes folded and collapses as it is slowly broken down within the cavity of the follicle. Macrophages in the connective tissue are involved in the phagocytosis of the zona pellucida and remnants of the degenerating cells. The basement membrane, separating the follicular cells from the theca interna, may separate from the follicular cells and increase in thickness, forming a wavy hyaline layer called the *glassy membrane*. This structure is characteristic of follicles in late stages of atresia.

Enlargement of the cells of the theca interna occurs in some atretic follicles. These cells, which are similar to theca lutein cells, become organized into radially arranged strands, separated by connective tissue. A rich capillary network develops in the connective tissue. These atretic follicles, which resemble and old corpus luteum, are called *corpora lutea atretica*. As these follicles continue to degenerate, a scar with hyaline streaks develops in the center of the cell mass, giving it the appearance of a small corpus albicans. The follicle eventually disappears as the ovarian stroma invades the degenerating follicle. The strands of luteal cells do not degenerate immediately, but become broken up and scattered in the stroma. These cords of cells contribute to the *interstitial gland* of the ovary, producing steroid hormones.

The Female Sexual Act

Stimulation of the Female Sexual Act

As is true in the male sexual act, successful performance of the female sexual act depends on both psychic stimulation and local sexual stimulation.

The sex hormones seem to exert a direct influence on the woman to create such as sex drive, but, on the other hand, the growing female child in modern society is often taught that sex is something to be hidden and that it is immoral. As a result of this training, much of the natural sex drive is inhibited, and whether the woman will have little or no sex drive or will be more highly sexed depends on a balance between natural factors and previous training.

Local sexual stimulation in women occurs in more or less the same manner as in men, for massage or other types of stimulating of the perineal region, sexual organs, and urinary tract create sexual sensations. The *clitoris* is especially sensitive for initiating sexual sensations. Once these sensations have entered the spinal cord, they are transmitted thence to the cerebrum. Local reflexes that are at least partly responsible for the female orgasm are integrated in the sacral and lumbar spinal cord.

Female Erection and Lubrication

Located around the introitus and extending into the clitoris is erectile tissue almost identical with the erectile tissue of the penis. This erectile tissue, like that of the penis, is controlled by the parasympathetic nerves that pass from the cord to the external genitalia. In the early phases of sexual stimulation, the parasympathetics dilate the arteries to the erectile tissues, and this allows rapid inflow of blood into the erectile tissue so that the introitus tightens around the penis; this aids the male greatly in his attainment of sufficient sexual stimulation for promoting ejaculation.

Parasympathetic impulses also pass to the bilateral Bartholin's glands located beneath the labia minora to cause secretion of mucus immediately inside the introitus. This mucus, along with large quantities of mucus secreted by the vaginal mucosa itself, is responsible for most of the lubrication during sexual intercourse. The lubrication inturn is necessary for establishing during intercourse a satisfactory massaging sensation rather than an irritative sensation, which may be provoked by a dry vagina. A massaging sensation constitutes the optimal type of sensation for evoking the appropriate reflexes that columinate in both the male and female climaxes.

The Female Orgasm

When local sexual stimulation reaches maximum intensity, and especially when the local sensations are supported by appropriate conditioning impulses from the cerebrum, reflexes are initiated that cause the female orgasm, also called the *female climax*. The female orgasm is analogous to ejaculation in the male, and it probably is important for fertilization of the ovum. Indeed, the human female is known to be somewhat more fertile when inseminated by normal sexual intercourse rather than by artificial methods, thus indicating an important function of the female orgasm. Possible effects that could result in this are :

First, during the orgasm the perineal muscles of the female contract rhythmically, which presumably results from spinal reflexes similar to those that cause ejaculation in the male. It is possible, also, that these same reflexes increase uterine and fallopian tube motility during the orgasm, thus helping to transport the sperm toward the ovum, but the information on this subject is scanty.

Second, in many lower animals, copulation causes the neurohypophysis to secrete oxytocin; this effect is probably mediated through the amygdaloid nuclei and then through the hypothalamus to the pituitary. The oxytocin in turn causes increased contractility of the uterus, which also is believed to cause rapid transport of the sperm. Sperm have been shown to traverse the entire length of the fallopian tube in the cow in approximately five minutes, a rate at least 10 times as fast as that which the sperm themselves could achieve.

In addition to the effects of the orgasm on fertilization, the intense sexual sensations that develop during the orgasm also pass into the cerebrum and in some manner satisfy the female sex drive.

The Ovarian Cycle

Ovum production, or *oogenesis* occurs on a monthly basis, as part of the *ovarian cycle*. Ovum development occurs in specialized structures called *ovarian follicles*. In the outer portions of the cortex, just beneath the tunica albuginea, there are scattered clusters of immature eggs, or *oocytes*. Each oocyte within one of these *egg nests* has pale cytoplasm and a large, round nucleus with a prominent nucleolus. It is encircled by a simple squamous layer of *follicular cells*, and the combination is known as a *primordial follicle*. At monthly intervals some of those primordial follicles become active, and the developments that follow constitute the ovarian cycle.

The cycle begins as the activated follicles develop into *primary follicles*. In a primary follicle the follicular cells enlarge, and their mitotic divisions create several layers of the cells around the oocyte. As the wall of the follicle thickens further, a space opens up between the developing oocyte and the innermost follicular cells. Within this *zona pellucida* microvilli originating at the surface of the oocyte interdigitate with those of the follicular cells. These microvilli increase the surface area available for absorption by roughly 35 times, and the follicular cells are continually providing the developing oocyte with nutrients. In addition to diffusion, facilitated diffusion, and contransport, pinocytosis occurs at the base of the microvilli.

Although many primordial follicles develop into primary follicles, usually only a few will take the next step. The transformation begins as the wall of the follicle thickens, and the deeper follicular cells begin secreting small amounts of fluid. This fluid accumulates in small pockets that gradually expand and separate the inner and outer layers of the follicle. At this stage the complex is known as a *secondary follicle*. Although the oocyte continues to grow at a slow pace, the follicle as a whole now enlarges rapidly due to this accumulation of fluid.

Eight to ten days after the start of the ovarian cycle, the ovaries usually contain only a single secondary follicle destined for further development. By the tenth of fourteenth days of the cycle it has formed a mature *tertiary* or *Graafian*, follicle roughly 15 mm in diameter. This complex spans the entire width of the cortex and stretches the ovarian capsule, creating a prominent bulge in the surface of the ovary. The oocyte projects into the expanded central chamber or *antrum* surrounded by a mass of follicular cells.

As the time of egg release, or *ovulation*, approaches, the oocyte and its follicular associates lose their connections with the follicular wall and drift free within the antrum. The follicular cells surrounding the oocyte are now known as the *corona radiata*. The distended follicular wall then ruptures, releasing the follicular contents, including the oocyte, into the pelvic cavity. The corona radiata has a sticky surface, so the oocyte usually attaches to the ovarian surface near the ruptured wall of the follicle. Direct contact with the fimbriae or fluid currents established by the ciliated epithelium then transfer the oocyte to the uterine tube that will carry it toward the uterus.

The empty follicle initially collapses, and ruptured vessels leak blood cells into the lumen. The remaining follicular cells then invade the lumen, proliferating to create and endocrine structure known as the *corpus luteum*, named for its yellow color. Unless pregnancy occurs, after about 12 days the corpus luteum begins to degenerate. Fibroblasts then invade the region, producing a knot of pale scar tissue called a *corpus albicans*. The distintegration of the corpus luteum marks the end of the ovarian cycle, but this event is followed by the activation of another cluster of primordial follicles and the start of an ovarian cycle.

Although many primordial follicles may have developed into primary follicles, and several primary follicles converted to secondary follicles, usually only a single oocyte will be released into the pelvic cavity at ovulation. The rest degenerate, a process termed *atresia*. A puberty a woman has approximately 200,000 primordial follicle in each ovary. Forty years later few if any follicles remain, although a relatively small number will have actually completed their development into functional gametes over the interim.

Hormones and the Ovarian Cycle

Follicular development begins under FSH stimulation, and each month some of the primordial follicles begin their development into primary follicles. As the follicular cells enlarge and multiply, they release steroid hormones collectively known as *estrogens*. The hormone *estradiol* is the most important estrogen. Small quantities of estrogen are also contributed by *interstitial cells* scattered within the ovarian cortex. Changes in circulating estrogen concentration are the primary mechanism for coordinating female sexual function.

As follicular development proceeds the concentration of circulating estrogens rises, for the follicular cells are increasing in number and secretory activity. As estrogen concentrations increase, they inhibit both the hypothalamic secretion of GnRF and the pituitary production and release of FSH. Estrogen also has an interesting effect on the rate of LH secretion. Although the synthesis of LH occurs under GnRF stimulation, the rate of release into the bloodstream depends on the circulating concentration of estrogens. Thus as the follicles develop and estrogen concentrations rise, the pituitary output of LH gradually increases. Despite a slow decline in FSH the combination of estrogens, FSH, and LH contains to support follicular development and maturation.

Estrogen concentrations take a sharp upturn in the second week of the ovarian cycle, as this month's tertiary follicle enlarges in preparation for ovulation. At about day 14 estrogen levels peak, accompanying the maturation of that follicle. The high estrogen concentration then triggers a masive outpouring of LH from the anterior pituitary. That sudden surge in LH concentrations coincides with the rupture of the follicular wall at ovulation.

After ovulation LH stimulates the remaining follicular cells to form the corpus luteum, and the yellow color of this mass results from the presence of lipid reserves. These compounds are used to manufacture steroid hormones known as *progestins*, principally the steroid *progesterone*. Although moderate amounts of estrogens are also secreted by the corpus luteum, progesterone is the principal hormone of the postovulatory period.

Luteinizing hormone levels remain elevated for only two days, but that is long enough to sitmulate the formation of the functional corpus luteum. Progesterone secretion continues at relatively high levels for the next week, but unless pregnancy occurs the corpus luteum then begins to degenerate. Roughly 12 days after ovulation, the corpus luteum becomes nonfunctional, and porgesterone and estrogen levels fall markedly.

The decline in progesterone and estrogen levels stimulates the hypothalamic receptors, and GnRF production increases. This leads to an increase in FSH and LH production in the anterior pituitary, and the entire cycle begins again.

The hormonal changes involved with the ovarian cycle in turn affect the activities of other reproductive tissues and organs. At the uterus, the hormonal changes are responsible for the maintenance of the menstrual cycle.

The Menstrual Cycle

Menses

An average *menstrual cycle* lasts 28 days, but it can range from 21 to 35 days in length. The cycle begins with the onset of *menses*, a period marked by the wholesale destruction of the functional zone of the endometrium. The arteries begin constricting, reducing blood flow to this region, and the secretory glands, epithelial cells, and other tissues of the functional zone begin to die of oxygen and nutrient deprivation.

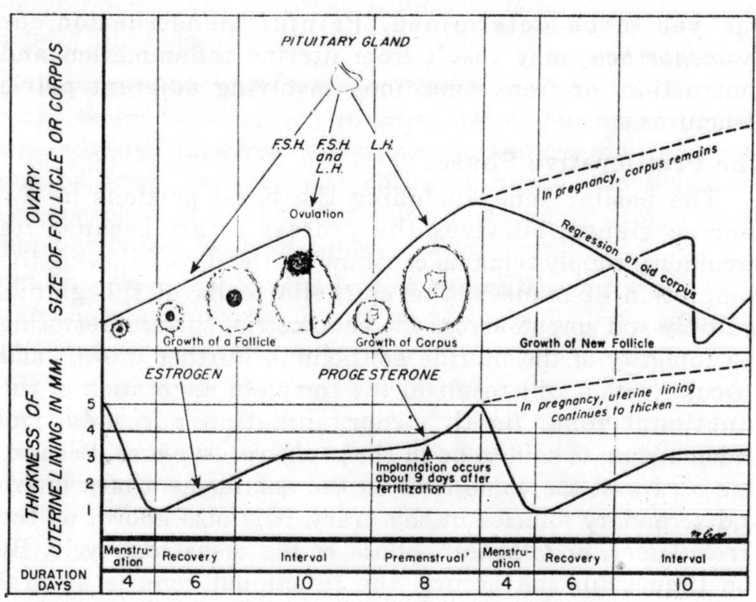

Fig. 10.12. The menstrual cycle in the human female. The solid lines indicate the course of events if the egg is not fertilized; the dotted lines indicate the course of events when pregnancy occurs. The actions of the hormones of the pituitary and ovary in regulating the cycle are indicated by arrows.

Eventually the weakened arterial walls rupture, and blood pours into the connective tissues of the functional zone. Blood cells and degenerating tissues break away and enter the uterine lumen, to be lost by passage through the external orifice and into the vagina. This sloughing of tissue continues until the entire functional zones has been lost. This *menstruation* lasts from 1 to 7 days in normal individuals. Over this period roughly 35 ml of blood is lost.

Despite the magnitude of the changes underway, menstruation is usually relatively painless. However, a number of physical and physiological changes occur in women a few days before the start of menses. Fluid retention, breast enlarement, and an uncomfortable feeling are common complaints. These physical symptoms may be associated with psychological changes producing this *premenstrual syndrome* has yet to be determined. Painful menstruation, or *dysmenorrhea*, may result from uterine inflammation and contraction, or from conditions involving adjacent pelvic structures.

The Proliferative Phase

The basilar zone, including the basal portions of the uterine glands, survives the process intact because its circulatory supply remains constant. In the days following the completion of menses, the epithelial cells of the glands multiply and spread across the endometrial surface, restoring the integrity of the uterine epithelium. Further growth and vascularization will result in the complete restoration of the functional zone. As this reorganization porceeds, the endometrium is said to be in the *proliferative phase*. Because this occurs at the same time as the enlargement of primary and secondary follicles in the ovary, it is also known as the *preovulatory*, or *follicular*, phase of the menstrual cycle. By the time ovulation occurs, the functional zone is several millimeters thick, and prominent mucous glands extend to the border with the basilar zone. At this time the endometrial glands are manufacturing a mucus rich in glycogen. The entire functional zone is highly vascularized, with small arteries traveling toward the inner surface from larger trunks in the myometrium.

The Secretory Phase

During the *secretory phase* of the cycle the glands enlarge, accelerating their rates of secretion, and the arteries elongate and spiral through the tissues of the functional zone. Because this phase begins at the time of ovulation, and it persists as long as the corpus luteum remains intact, it can also be called the *postovulatory*, or *luteal phase* of the menstrual cycle.

Secretory activities peak about 12 days after ovulation, and the uterine lining at this time. Over the next day or two the glandular activity declines, and the menstrual cycle comes to a close. A new cycle then begins with the onset of menses and the disintegration of the functional zone. Unlike the preovulatory phase usually lasts 14 days, and individual variations are uncommon. As a result, the date of ovulation can be determined after the fact by counting backward 14 days from the first day of menses.

This fascinating cycle of events begins with the *menarche*, or first menstrual period at puberty, typically at age 11-12. The cycles continue until age 45-50 when *menopause*, the last menstrual cycle, occurs. Over the interim the regular appearance of menstrual cycle will only be interrupted by unusual circumstances, such as illness, stress, starvation, or pregnancy.

Hormones and the Menstrual Cycle

The precipitous declines in progesterone and estrogen levels that accompany the degeneration of the corpus luteum result in the endometrial breakdown of menstruation. The sloughing of endometrial tissue continues for 3-4 days, until rising FSH, LH, and estrogen levels stimulate the repair and regeneration of the functional zone of endometrium. The preovulatory phase continues until rising progesterone levels heral the arrival of the postovulatory phase. The combination of estrogen and progesterone then causes the enlargement of the endometrial glands and an increase in their secretory activities.

The hormonal fluctuations also cause physiological changes that affect the core body temperature. During the preovulatory period, when estrogen is the dominant hormone, the resting, or "basal," body temperature measured upon awakening in the morning is about one-half a degree Fahrenheit lower than it is during the postovulatory period, when progesterone dominates the endocrine picture. At the time of ovulation, basal temperature declines sharply, and this makes the temperature rise over the following day even more noticeable. As a result, by keeping records of body temperature over a few menstrual cycles a woman can often determine the precise day of ovulation. This information can be very important for individuals wishing to avoid or promote a pregnancy, for such an event can only occur if an ovum becomes fertilized within a day of its ovulation.

Pregnancy, Hormones, and Maternal Systems

If fertilization occurs, the zygote undergoes a series of mitotic divisions, forming a hollow ball of cells known as a *blastocyst*. Upon arrival int he uterine cavity, the blastocyst initially obtains nutrients by absorbing the secretions of the uterine glands. Within a few days it contacts the endometrial wall, erodes the epithelium, and buries itself in the endometrium. This process, known as *implantation,* initiates the chain of events leading to the formation of a special organ, the *placenta,* that will support embryonic and fetal development over the next 9 months.

The placenta provides a medium for the transfer of dissolved gases, nutrients, and waste products between the fetal and maternal bloodstreams. But it also acts as an endocrine organ, producing the hormones. The first of these hormones, called *human chorionic gonadotropin,* appears in the maternal bloodstream soon after implantation has occured. The presence of HCG in blood or urine samples provides a reliable indication of pregnancy, and kits sold for the early detection of pregnancy are sensitive to the presence of this hormone.

In function HCG resembles LH, for in the presence of HCG the corpus luteum does not degenerate. Instead it remains functional and continues to produce progesterone. Because progesterone secretion continues, the endometrial lining remains perfectly functional, and menstruation does not occur. Obviously if it did occur, the pregnancy would end, for the functional zone of the endometrium would disintegrate.

In the presence of HCG, the corpus luteum persists for about 3 months before degenerating. Its departure does not trigger the return of menstrual periods, because by then the placenta is actively secreting both estrogen and progesterone. Over the following months the palcenta also synthesizes two additional hormones, *relaxin* and *human placental lactogen* or HPL. Relaxin increases the flexibility of the symphysis pubis, permitting expansion of the pelvis during delivery. It also causes dilation of the cervix, making it easier for the fetus to leave the uterus.

The resting mammary glands are relatively unaffected by the changes in estrogen and progesterone concentrations that accompany pregnancy. The conversion form resting to active status requires the presence of two additional hormones. One of them is HPL, and the other is the pituitary hormone *prolactin*. The prolactin is produced by cells in the anterior pituitary. But in the nonpregnant woman, the hypothalamus produces an inhibitory factor that keeps prolactin secretion at a minimum. During pregnancy the sustained, elevated levels of estrogens and progesterone depress the production of PIF, and the prolactin concentration in the blood begins to rise. The combination of PRL and HPL then prepares the mammary glands for their secretory functions.

The histological organization of the inactive and active mammary glands. The resting mammary gland is dominated by components of the duct system, rather than glandular cells. The secretory apparatus does not develop until pregnancy occcurs. At that time the ducts become mitotically active, and gland cells begin to appear. By the end of the sixth month of pregnancy the mammary glands are fully developed, and the gland begins producing a secretion known as *colostrum*. Colostrum contains relatively more protein and far less fat than milk, and it will be provided to the infant during the first two or three days of life. Many of the proteins are immunoglobulins that may help the infant ward off infections until its own immune system becomes fully functional.

As colostrum production declines, the mammary glands convert to milk production. Milk consists of a mixture of water, proteins, amino acids, lipids sugars, and salts. It also contains large quantities of *lysozymes*, enzymes with antibiotic properties. Human milk provides roughly 750 Calories per liter. The secretory rate varies, depending on the demand, but a 5-6 kg infant usually requires around 850 ml per day.

The actual secretion of the mammary glands is triggered when the infant begins to suck on the nipple. Stimulation of tactile receptors at that site leads to the release of *oxytocin* at the posterior pituitary, and when oxytocin reaches the mammary gland it causes the contraction of smooth muscles in the walls of the lactiferous ducts and sinuses. The results in the ejection of milk, and this *milk ejection reflex*, or *milk letdown*. This reflex continues to function until *weaning* occurs, typically 1-2 years postdelivery. Milk production ceases soon after, and the mammary glands gradually return to a resting state.

Aging and Menopause

Menopause is usually defined as the time that ovulation and menstruation cease. Menopause typically occurs at around age 50, but in the years preceding the finale the regularity of the ovarian and menstrual cycles gradually fades. A shortage of primordial folicles represents the underlying cause for these developments. It has been estimated that almost 7 million potential oocytes are found in fetal ovaries after 5 months of development, but the number drops to about 1 million at birth, and to a few hundred thousand at puberty. By age 50, there are no primordial follicles left to heed the call of the pituitary; in premature menopause this occurs before age 40. Menopause is accompanied by a sharp and sustained rise in the production of GnRF, FSH and LH, while circulating concentrations of estrogen and progesterone decline.

The decline in estrogen levels leads to reductions in the size of the uterus and breasts, accompanied by a thinning of the urethral and vaginal walls. The reduced estrogen concentrations have also been linked to the development of osteroporosis, presumably because bone deposition proceeds at a slower rate. A variety of neural and cardiovascular effects are also reported, including "hot flashes," anxiety, and depression, but the links to hormonal deficiencies are not clearly established. The symptoms accompanying and following menopause are sufficiently unpleasant that about 40 percent of menopausal women eventually seek medical assistance. Hormone replacement therapies involving a combination of estrogens and progestins can often prevent osteoporosis and the neural and vascular changes associated with menopause.

11

Gonadal Disease

Oophoritis

Primary amenorrhoea or a premature menopause are often described in women with autoimmune disease, particularly 'idiopathic' Addison's disease, myxoedema or hypoparathyroidism. Histologically, the ovaries show lymphocytic infiltration, as do the other target organs in autoimmune endocrinopathies. These women sometimes have steroidal cell antibodies which react with Leydig cells, ovarian granulosa and theca interna cells. The presence of such antibodies predicts ovarian failure, especially in patients who have Addison's or other autoimmune diseases, yet who still have normal menstrual function. The pathogenic significance of ovarian antibodies in autoimmune oophoritis remains to be determined.

Infertility

Case

A 29-year-old builder had been married for 6 years but had no children. His wife had been extensively investigated; she ovulated regularly with a normal menstrual cycle, and had patent fallopian tubes and normal endocrine function. He had normal levels of luteinizing and follicle stimulating hormones and testosterone. He had no past history of orchitis or testicular trauma. On examination, he was a well-virilized, healthy-looking man with normal sized testes. A semen sample showed a low sperm count with sluggishly motile sperm and sperm-associated immunoglobulin (IgA and IgG). The sperm-cervical mucus contact test was abnormal and the use of normal donor sperm and normal cervical mucus confirmed that only the husband had antisperm antibodies. Antibodies to fresh donor sperm were detectable in the serum to a titre of over 1/1000, and in the seminal plasma to a titre of 1/32. The patient was treated with high dose steroids on days 1-10 of his wife's menstrual cycle. His wife became pregnant in the cycle following the fourth course of treatment and subsequently gave birth to a healthy baby girl.

Immunology of Infertility

Human spermatozoa and seminal plasma contain strongly immunogenic material: some of these antigens are unique to sperm or seminal plasma (semen-specific antigens), but others are shared with other fluids, secretions and organs. Five to 14% of infertile couples show evidence of sperm-antibodies. *These antibodies may be produced by the man, the woman,* or *both.*

Experimental male animals can be made sterile by active or passive immunization against testicular or seminal antigens. In man, damage to the seminal tract by surgery, accidental trauma, occlusion or infection may *trigger autoimmunity* to testicular and *seminal antigens.* For example, antisperm antibodies appear in the serum in 50% of vasectomized men within 6-12 months of surgery. Antisperm antibodies seldom appear in seminal plasma following vasectomy as local antibody production occurs proximal to the operation site. High titres of antisperm antibodies may appear in the semen after reversal by vasovasostomy and *modify the success of the reversal.*

Autoantibodies to sperm antigens may cause infertility in otherwise normal men by: (1) immobilization and agglutination of spermatozoa; and (2) inhibition of mucus and/or egg penetration by sperm, possibly by blocking specific receptors on the sperm surface.

Investigations of possible autoimmune infertility include a postcoital test. Poor mobility of sperm in this test suggests the existence of *antisperm antibodies.* Serum from both partners, cervical mucus and seminal plasma are tested for sperm antibodies using normal donor sperm and cervical mucus. When semen is mixed with cervical mucus, spermatozoa normally move rapidly and unidirectionally; IgA antibodies to spermatozoa prevent this type of movement.

Before considering *treatment*, it is important to make sure that there is no additional cause of infertility. Prostatitis has been found in about one-third of men with antisperm antibodies, and prolonged antibiotic treatment may be accompanied by a significant fall in antibody titres and pregnancy in a proportion of the wives. Manipulative techniques such as *in vitro fertilization* (IVF) or *gamete intrafallopian transfer* (GIFT) have been used with limited success but are not widely available. High dose intermittent steroid therapy has many side-effects and its use is therefore debatable but it can be successful.

The harmful effects of *antisperm antibodies* in women is unclear. Since the female genital tract is well endowed with immunocompetent cells, local isoimmunity is probably important in infertility. Where antibodies are found only in the female partner, treatment has been disappointing though controlled studies are lacking. *Immunosuppressive therapy* with steroids is contraindicated as exposure of the zygote and early embryo to high dose steroids may result in congenital abnormalities.

Hypogonadism in Females

Primary Hypogonadism

As in the male this can be either congenital or acquired. Congenital primary hypogonadism is sometimes called ovarian infantilism. Two main groups can be distinguished:

1. developmental inadequacy of the ovaries;
2. ovarian agenesis.

Albright, Smith, and Fraser (1942) were the first to describe patients with sexual infantilism accompanied by an increased excretion of gonadotrophins and not associated with significantly decreased stature. They suggested that the condition could be explained on the basis of a 'premenarchal menopause praecox', the process of follicle atresia which normally commences at birth having proceeded to such a degree that no further responsive follicles were left by the time pituitary gonadotrophin secretion commenced.

Ovarian agenesis is a fascinating form of congenital primary hypogonadism which is usually associated with a

group of other congenital anomalies giving rise to a fairly clear-cut syndrome. Decreased statural growth associated with a rather typical stocky habitus is almost invariably found. Webbing of the neck is another typical feature, but is not uncommonly absent. Cubitus valgus, giving rise to an increased carrying angle of the arms, is seen to some extent in nearly all cases. Other anomalies include cardiovascular abnormalities—in particular, coarctation of the aortadigital deformities, squints, and so on.

The urinary gonadotrophin output is elevated once the age of normal puberty has been passed. Laparotomy reveals the uterus and tubes to be of infantile dimensions, and the ovaries to be represented merely by a fibrous cord, the continuation of the ovarian ligament. Histological examination of this ovarian remnant reveals a normal stroma with rete ovarii and medullary canals, but no trace of follicles in any stage of development. Clumps of hilus cells (large pale staining polygonal cells disposed along the course of nerves and strongly resembling the Leydig cells of the testis) are sometimes prominent.

The growth failure in these patients may conform with one or other of two distinctive patterns. In the first, there is a conspicuous smallness from early infancy, and in the second, growth continues apparently normally until the age of about 9 or 10 years, whereafter it remains at a level appropriate to this age. It is probable that it is in this group of cases alone that the absence of the normal prepuberal and puberal growth spurt is contributory to the growth failure. The fact that some of these patients begin to grow again when treated with oestrogens conforms to this view. Since primary hypogonadism is not always associated with growth failure, it seems clear that the growth failure in this syndrome is an associated defect, either genetically determined, or induced as a result of a mutually injurious influence operating in the 5- to 17-min. phase of embryogenesis. It is at this stage that the cortex of the primitive genital ridge undergoes organization for the process of penetration by the primordial germcells, the sex of which is already determined genetically, which migrate from the endoderm of the yolk-sac. Upon this penetration further sexual differentiation of the gonads would seem to depend, so that with failure of this process the cortical elements specific to the sex of the gonads would not emerge, whereas medullary rudiments would remain relatively unaffected.

This hypothesis of the pathogenesis of ovarian agenesis would seem to receive considerable support from two different sources. The first relates to experimental work in which it has been shown that removal or destruction of the gonads of embryos in mice and rabbits in the sexually indeterminate stage leads to the development of foetuses all of which are apparently female, the males having undergone intersexualization. The second is the demonstration, by means of the skin biopsy technique for differentiating the chromosomal sex, that some, at least, of the patients with Turner's syndrome are of the male chromosomal sex. It is of interest that Polani *et al.* were led to suspect this possibility from the consideration that male examples of Turner's syndrome are very rare; that coarction of the aorta is common in the 'female' cases of Turner's syndrome but is otherwise decidedly commoner in males than in females; and from these facts they suspected that some of the apparent female cases of Turner's syndrome might actually be completely intersexualized genetic males. The three 'female' patients with 'Turner's syndrome associated with coarctation of the aorta on whom they determined the chromosomal sex by skin biopsies all proved to be genetic males. Wilkins et al. reported eight patients with the ovarian agenesis syndrome; six of these had male-type epidermal nuclei and two had the female type.

Acquired primary hypogonadism is rare and due either to surgical trauma or removal, or to local pelvic diseases which is itself rare in prepuberal girls. It is possible that long-standing and debilitating diseases can also lead to intrinsic ovarian failure and so to acquired primary hypogonadism.

Clinical features

Genital organs

The presenting complaint on account of which girls with primary hypogonadism are brought to the physician is usually either failure of the onset of menstruation or failure of sexual development. The external genitalia remain of infantile character and proportions and the vaginal epithelium fails to undergo the changes normally seen in postpuberal girls as a result of endogenous oestrogenic stimulation. The vaginal smear, therefore, consists almost exclusively of small, rounded basal cells with relatively large vesicular nuclei, and it may contain numerous leucocytes. The breasts, including the nipples and areolae, are either completely undeveloped, or at most show no more than trivial enlargement. The uterus remains of the infantile size, and the passage of a sound demonstrates that the cervical canal is considerably longer than the uterine cavity proper (infantile proportions; in the adult uterus these proportions are reversed).

Hair

There is considerable variation in the amount of pubic and axillary hair. This may be quite absent or represented only by scanty hairs on the labia majora, whereas in other patients the growth is more abundant Yet a further pattern is for the pudendal hair to be very scanty, while the axillary hair of normal amount. This latter arrangement depends upon the fact that growth of the axillary hair is more largely due to adrenal androgens which may be produced in normal quantity in these patients, whereas the pudendal hair is more oestrogen dependent.

Skeletal changes

These resemble those of eunuchoidism in the male,' and typically eunuchoidal proportions (span exceeding height and lower measurement exceeding upper) are commonly found even when there is decreased overall stature as in Turner's syndrome. The fingers and toes may be long and slender, though this is less commonly so in Turner's syndrome. The shape of the pelvis approximates to that of the male. Dentition may be delayed. The radiological bone age is less than the chronological age and there is delay in epiphyseal union. Osseous retardation is less marked in Turner's syndrome than in those patients with primary hypogonadism without decreased stature.

Sexual behaviour

These patients are normally completely lacking in libido and have no attraction towards the opposite sex. They usually remain somewhat infantile in outlook and present an immature mentality. They may also feel a sense of inferiority as a result of the knowledge of their physical shortcomings.

Special features in Turner's syndrome

Several of these have already been mentioned. Shortness of stature and associated congenital anomalies distinguish this condition from primary hypogonadism without decreased stature. Generalized osteoporosis of the skeleton has been reported in some patients with this condition. It may give rise to scoliosis and lordosis. Chondrodystrophia of the dorsal vertebrae has also been described.

The commonest congenital deformities are webbing or apparent shortening of the neck, and cubitus valgus, giving rise to an increased carrying angle of the arm. A characteristic shield-shaped chest has also been described, the thorax being prominent anteriorly and broader than normal with an increased antero-posterior diameter. Other defects which have been reported include per cavus, increased pronation of the feet, syndactylism, spina bifida, congenital deafness, and ocular disturbances, such as bilateral ptosis, slight exophthalmos, internal and external squint, cataract, tubular vision, and lack of retinal pigment. Some patients have shown mental deficiency. The occurrence of coarctation of the aorta has already been the subject of comment. Hypertension with a systolic blood pressure between 130 and 150 and a diastolic pressure between 90 and 112 has been noted in many patients with Turner's syndrome, independently of the existence of coarctation of the aorta.

Hormone excretion

Typically, patients with primary hypoovarianism have an increased excretion of pituitary gonadotrophins. The extent, however; of this increase is very variable, and, in not a few patients the excretion has been found to be within the normal range. It seems clear, moreover, that day-to-day fluctuations of considerable magnitude may be encountered. The excretion of 17-ketosteroids is usually somewhat diminished but figures within the normal range have been encountered. Only a few measurements of oestrogen excretion have been made in these patients; the expected low values have been found, but oestrogens are not entirely absent from the urine. The microgram or two presumably arise from the adrenal cortex and not from the gonads.

Diagnosis

The differentiation of primary hypogonadism without decreased stature from secondary hypogonadism due to idiopathic deficiency of pituitary gonadotrophin can be made only the basis of the urinary gonadotrophin excretion, and this is not always reliable. The two forms of primary hypogonadism are distinguished by the shortness of stature and congenital anomalies found in Turner's syndrome but not in the other variety. The differentiation of Turner's syndrome form pituitary infantilism is summarized in the following comparative table, adapted from del Castillo, de la Balze, and Argonz (1947):

Rudimentary Ovaries	*Hypophyseal Dwarfism*
Women of short stature	Dwarfs.
Infantile mammary glands and genital organs.	The same.
Development of pubic and axillary hair.	Lack of pubic and axillary hair.
Well-nourished and strong.	Weak and easily tired.
Bone age some years retarded.	Very marked delay in bone age.
Late closure of the epiphyses.	Lack of closure of the epiphyses.
Very frequently vertebral chondrodystrophia.	The same.
Follicle-stimulating hormones increased in the urine.	Lack of follicle-stimulating hormone.
17-ketosteroids somewhat diminished.	17-ketosteroids considerably diminished.
Normal insulin curve	Persistent hypoglycaemia after intravenous insulin.
Congenital abnormalities.	Not observed.
Diffuse osteoporosis and early senility.	Not observed.
Normal sella turcica.	Pathological modifications may be observed.
Visual fields: some functional alterations.	Abnormalities in the presence of a neoplastic lesion.

Secondary Hypogonadism

Pituitary dwarfism is one cause of primary amenorrhoea, the failure of sexual maturation corresponding with that seen in the same condition in the male. Perhaps the commonest cause of primary amenorrhoea is idiopathic deficiency of pituitary gonadotrophin secretion, without evidence of failure of production of the other pituitary trophic hormones. Girls with this condition are usually of normal or slightly increased stature and may be either thin or adipose. In spite of the rudimentary state of the genital organs and absence of mammary development, pubic and axillary hair may be present in relatively normal amounts. In a few of these patients the clitoris is enlarged, perhaps as result of relative adrenocortical hyperfunction. Secondary hypogonadism may also be the result of cretinism, juvenile myxoedema, and milder forms of hypothyroidism, toxic litre, and diabetes mellitus. When primary amenorrhoea is the result of adrenocortical hyperfunction, virilizing changes are also found. Secondary hypogonadism may result from severe and long-standing disease in other systems, such as anaemia, chronic nephritis, sepsis, and malnutrition.

Diagnosis

The differentiation of secondary from primary hypogonadism can only be made on the basis or urinary gonadotrophin excretion. In secondary hypogonadism the excretory level is too low to be measured by the available clinical tests.

Treatment of Hypogonadism in Females

The treatment of primary hypo-ovarianism can clearly only be substitutive, since it is impossible to replace the functionless or missing ovarian endocrine tissue. Theoretically the ideal treatment of secondary hypo-ovarianism would be the administration of the appropriate pituitary gonadotrophins. In practice, however, this cannot be achieved, and in general it may be said that treatment with the gonadotrophin preparations available for clinical use is mainly disappointing. Consequently the treatment of both primary and secondary hypogonadism in the female resolves itself into the administration of oestrogens so as to bring about growth of the oestrogen-sensitive tissues—principally the genitalia and breasts—and to produce cycles of uterine bleeding. The necessity, or even desirability, of achieving either of these ends is sometimes questioned, but there can be little doubt that most patients afflicted with hypogonadism are very grateful for the changes which can be brought about by oestrogen therapy, and experience great satisfaction in having regular bleeding even though they understand these are essentially artificial and do not indicate their normality from the reproductive standpoint.

Under physiological conditions oestrogens would appear always to be secreted in cyclically fluctuating fashion. It is therefore reasonable to assume that oestrogen substitution therapy ought also to be cyclic. Once a responsive endometrium has been built up, it is imperative that oestrogen administration be discontinuous, since otherwise endometrial hyperplasia with resultant prolonged and excessive irregular bleeding will be the consequence. Some authors consider that continuous oestrogen administration may be employed in the initial stages of treatment so as to produce quicker results. It is doubtful, however, whether there is any advantage or even justification for this.

The simplest form of oestrogen therapy is oral; stilboestrol 2 mg., or ethinyloestradiol 01 mg., daily, is a reasonable average dose and may be given for courses of twenty days. Ten days should elapse between each successive course, unless an oestrogen-withdrawal bleeding occurs in the meantime. When this happens, the succeeding course may begin on the fifth day of the 'cycle' so established, counting the day of commencement of uterine bleeding as day number one of the new cycle. Once begun, these courses of treatment must be continued indefinitely—or at least until the age at which a climacteric might have been expected has been attained. This certainly is true for patients with primary hypogonadism, and is probably true for most of those with secondary hypogonadism also. For some of the latter, however—that is, those patients wherein the defect appears to be an idiopathic deficiency of pituitary gonadotrophin secretion—the hope may be entertained that eventually normal menstrual function might occur, just as in some male eunuchoids of the same type normal testicular function appears to be able to continue after initial treatment. In these cases, therefore, it is reasonable to stop treatment after several months in order to see whether spontaneous menstruation might occur thereafter. On theoretical grounds there is something to be said for combining progesterone of ethisterone with the second half of the oestrogen course in the hope that these hormones will together influence the anterior pituitary in such a way as to evoke gonadotrophin secretion. A method which has sometimes proved successful is as follows:

After several cycles of treatment with oestrogen alone, a daily dose of 40 mg. of ethisterone is given for the last ten days of the twenty-day course of oestrogen (stilboestrol 2 mg., or ethinyloestradiol 0.1 mg., daily). On the fifth day of the next cycle another twenty-day oestrogen course is started, this time at half the previous dose, and on the fifteenth day of that cycle it is combined with 60 mg. daily of ethisterone for ten days. On the fifth day of the next cycle a final oestrogen course is started, the daily dose again being halved (stilboestrol 0.5 mg., or ethinyloestradiol 0.025 mg.), and on the fifteenth day of that cycle a daily dose of 80 mg. of ethisterone is started for ten days. All treatment is then stopped and the patient is observed to see whether spontaneous menstruation will occur.

No useful purpose is served by giving progesterone or ethisterone to patients with primary hypogonadism.

Some girls are unable to tolerate oral oestrogens; complaining of nausea and vomiting; for these, and for these alone, oestrogens must be given by intramuscular injection. Oestradol benzoate or oestradiol dipropionate, 5 mg. twice weekly for three weeks out of every four, is a convenient regime; a further possible alternative is oestradiol monobenzoate in microcrystalline suspension, in a dose of 10 mg. every four weeks. It is claimed that oestradiol valerianate provides prolonged oestrogenic stimulation when administered intramuscularly, in oily solution, 10-20 mg. every three or four weeks being an appropriate dose. Being in solution, this long-acting ester preparation does not suffer from the disadvantages of microcrystalline suspensions.

In addition to stimulating growth of the genitalia and breasts, oestrogen therapy for hypogonadal females may be expected to have import psychological effects. The patient loses her childlike mannerisms and psyche, becoming more adult in mentality and outlook. Whether this is a direct effect of oestrogens on the psyche, or whether it be a psychological response to the morphological maturation, is unknown.

Oestrogens appear to stimulate growth in some patients with primary hypogonadism and decreased stature. This is especially true of those cases where growth was relatively normal until what should have been the age of puberty, but then ceased. It is less likely to be true for those cases where growth was always deficient. As part of the growth-promoting picture, however, oestrogen therapy leads to epiphyseal closure, whereupon, of course, growth ceases. The maximum growth increment is therefore strictly limited arid unlikely to exceed 3 or 4 inches. Even this however, is much welcomed by these patients.

The Climacteric

The climacteric or 'critical period' or 'change of life' covers the phase of waning ovarian function which is a consequence of the continual loss or degeneration of potential ova and the absence of any provision for their replacement. When no more primordial follicles remain in the ovaries reproductive ability obviously comes to an end, but, on the other hand, the endocrine activity of the ovary undergoes a more gradual phase of diminution. This phase begins before the menstrual periods cease and continues for some time afterwards. The menopause, or cessation of menstruation, is merely a single incident of this climateric period.

Physiology of the Climacteric

In contrast to spermatogenesis, which is essentially a function of sexually mature males, oogenesis-the actual production of oocytes—is probably purely a foetal activity. Though the more or less classical view as put forward by Swezy and Evans (1930) was that the production of new oocytes from the germinal epithelium continued throughout life, the idea was opposed by Simpkins (1932) and the accurate observations of Zuckerman and his colleagues (Mandl and Zuckerman, 1951, a, b, c; Mandl, Zuckerman, and Patterson, 1952) have failed to substantiate it. At birth the two ovaries contain some 400,000 (more or less) primordial oocytes, but of these, only a small proportion, perhaps 400, are destined to take part in ovulation. The remainder disappear through the process of atresia. Loss of oocytes by atresia begins at least as early as birth, and is indeed most active before puberty, for the ovary of the new-born has many thousands more primary follicles than that of the adolescent girl. During the proliferative phase of each menstrual cycle several follicles commence to grow but only one reaches the stage of ovulation. The remainder, outstripped by the 'chosen' follicle, regress and become atretic. In this way, during each cycle some 30 or 40 follicles are lost by atresia for each one by ovulation. Since the ovarian hormones are secreted by the follicles or their derivatives, it is clear that when few or no follicles remain, ovarian hormone production must fall to a low level or cease altogether.

The age at which the climacteric commences varies in different women in much the same way as does that at which the menarche occurs; it is in fact more or less normally distributed. Various factors—racial, hereditary (other than racial) general health, sociological and so on—no doubt do determine it, and diseases of various kinds directly affecting ovarian physiology can accelerate it.

It has often been supposed that there is a relationship between the age of the menarche and that the earlier the occurrence of the former, the later is that of the latter. This is probably not true: in an investigation of the menopause of 1,000 women it was found that the average age at the menopause for women whose menarche occurred at 13 years was 47.3 years; while that for women whose menarche occurred at 18 years was 47.5 years. Occasionally, the menopause occurs at a very early age—even before 20 years; in these cases there is nearly always an underlying endocrine abnormality. There have also been reports of the continuation of menstruation until very advanced years—up to the age of 104 in one instance (Novak, 1921). In general, however, prolongation of the menopause after the age of 55 calls for gynaecological examination to exclude the possibility of genital malignancy. There is good reason to suppose that many of the patients with delayed menopause reported in ancient literature had oestrogen-producing tumours of the ovary (such as granulosacell tumours). It is, of course, equally important, even more so, to make a gynaecological examination if, after the menopause has finitely occurred, genital bleeding should reappear. Although this is frequently of benign cause (particularly if injudicious oestrogen treatment for menopausal symptoms has been given) all too commonly malignant disease of the cervix or body of the uterus if found to be responsible.

When the number of ovarian follicles has become significantly reduced, ovulation ceases to occur in every cycle although more or less regular cyclic activity may continue some time. Later, ovulation ceases altogether and by this time some irregularity of the cycles has usually become apparent. Anovular cycles may continue for some time—interspersed with an occasional ovular cycle perhaps—the bleeding becoming more infrequent and scantier, eventually to cease altogether. In other women, the menopause may take the form of an abrupt cessation of previously regular periods. Yet a further variant is that in which the alterations in ovarian hormone production lead to the development of irregular, prolonged and often heavy bleeding—climacteric menorrhagia. It is sometimes considered that in these circumstances a phase of increased oestrogen production precedes the termination of ovarian endocrine activity. It is doubtful, however, this is true and more probable that the menorrhagia is the result of continuous, as posed to discontinuous, oestrogen production, albeit on a decreasing scale.

The gradual lessening of ovarian endocrine activity during the climacteric leads to my secondary changes. Foremost among these are regression of the genital organs—uterus, vagina, vulva, and breasts. But in many women the changes are almost perceptibly slow and long after the menopause, little evidence of genital regression may be seen. In particular, the vaginal smear in post-menopausal women often reveals little evidence of significant oestrogenic deficiency. However, atrophy of the vaginal epithelium may be the cause of post-menopausal bleeding, or pruritus. The tendency to gain weight, which is commonly seen at this stage of life, leading to so-called 'middle_ spread' has been thought to result from failing ovarian function, but this is not likely. As seen in the previous chapter, bilateral ovariectomy by no means invariably leads to the development of adiposity. A more probable explanation is that, as part of the general ageing process, less energy is expended; however, because the appetite remains changed, a previous balance between energy intake (in the food)' and that expended the day's activities becomes converted into an excess of energy intake, thereby leading deposition of fat. It is said that this fat has a peculiar distribution, accumulating mainly around the abdomen, hips, and thighs. The possibility, must not be overlooked It the distribution is the same as would have occurred, given a comparable weight gain, in earlier years. A suggested explanation for the alleged special distribution (assuming that it is in fact special to this time of life) is that the normal ageing process is accompanied by a relatively greater loss of adipose tissue cells in other parts of the body and that, when new fat is laid down, most appears where the adipose cells remain in greater abundance.

Further secondary effects of ovarian failure are changes in the activity of the endocrine glands. These are most important for a complete understanding of the physiology of the climacteric. Because of the progressive failure of ovarian response to pituitary gonadotrophic stimulation, the modifying effect of the ovarian hormones on the activity of the pituitary gland becomes less and less. With the reduced inhibitory effect of oestrogen, the production of pituitary gonadotrophin (mostly of follicle stimulating type) increases, so that in post-menopausal women a great excess can usually be detected in the urine. Sometimes this gives rise to false pregnancy-diagnosis tests, thus confusing an already delicate issue if the woman fears (or hopes) that the delayed period caused by the climacteric is due to pregnancy. Along with this increase in pituitary gonadotrophin production, there is probably also an increase in the output of thyrotrophic and adreno-corticotrophic hormones. These find responsive target organs, and so hyperactivity of the thyroid and adrenal cortex may ensue. It is probably the combination of falling oestrogen level and increased output of thyroid and adrenal hormones which is responsible for most of the untoward effects which may be experienced by women at this stage. An alternative view has supposed that the pituitary overactivity itself is responsible for the climacteric symptoms; but since these are often suppressed by very small doses of oestrogen which have no detectable effect on the pituitary hyperactivity, it seems far more probable that the falling oestrogen level is the more important factor.

The increased activity of the adrenal cortex tends to restore the endocrine balance by taking over some of the functions of the ovary. Thus, there is little doubt that most of the oestrogen which circulates in post-menopausal women, when the ovaries have become quite functionless, is secreted by the adrenal cortex. This oestrogen, and the other adrenal, cortical hormones produced in increased amounts; serve to depress the excessive pituitary activity and so reduce the thyroid overactivity. With still further advance in age, it is likely that some degree of refractoriness occurs in both the adrenal and the thyroid gland, so that in these older women the clinical picture may suggest deficiency, rather than excess, of function of both of these glands.

Clinical features

Certain special features relating to the altered endocrine function of the climacteric will now be considered..

Virilism

Because of the increased adrenal output of androgens, the male characteristics, represented to a slight degree in all females, may become accentuated. Thus, the down on the upper lip and chin tends to become thicker and sometimes the hair is sufficiently coarse and luxuriant to form a moustache and beard, causing great mental anguish. Hair may also increase at the sides of the face, and the pubic hair may extend upwards along the linea nigra towards the umbilicus. It is rare, however, that hirsutism assumes the proportions and distribution of pathological virilism. Nevertheless, in one young woman with a pre-existing tendency to virilism, bilateral ovariectomy, for a gynaecological condition, produced severe hirsutism. The pubic hair tends to become uncurled at the climateric and the development of facial hirsutism may be accompanied by loss of scalp hair. The voice may become deeper and more powerful, so that singers sometimes find themselves able to reach notes lower than any previously possible, the upper notes, on the other hand, becoming more difficult to obtain.

A change in mental outlook with an approximation to 'male' characteristics may take place, greater resolution, command, initiative, originality, and administrative capacity being shown with the attainment perhaps of considerable commercial or public success later life. These changes are more likely to develop in the later part of the climacteric, after the menopause itself has occurred. As further evidence of increased adrenocortical activity, it may be noted that patients with Addison's disease may show considerable amelioration, or even cure, at the climacteric, although this is preceded by an earlier case in which the disease appears to be aggravated. The occurrence of pigmentation in me women at the menopause may well be directly due to the increased pituitary production of melanocyte stimulating hormone, along with adrenocorticotrophic hormone.

Hypertension

This is a common accompaniment of the climacteric and may be of a labile type, disappearing spontaneously after a year or two. In such circumstances it is presumably the direct consequence of vasomotor lability. When, however; it progresses a more severe and permanent type, with secondary changes in the vessels, it seems doubtful if the climacteric itself can be held responsible and more probable that the hypertension is a consequence of general ageing processes. This conclusion follows from the fact, already pointed out in the preceding chapter, that ovariectomy or radium castration does not necessarily lead to hypertension.

Impaired carbohydrate tolerance

Although only a small percentage of women develop clinical diabetes mellitus at the climacteric, it can be shown by investigation of the carbohydrate tolerance that an impairment to some degree occurs in a considerable proportion. It is likely that this is due to pituitary overactivity, since suppression by sufficient doses of oestrogen, so as to cause disappearance of urinary gonadotrophin, may lead to a normal carbohydrate tolerance curve, while cessation of oestrogen treatment followed by a return to the original condition. Presumably, both pituitary growth hormone and the increased production of adrenal glucocorticoids must be held responsible for the impaired carbohydrate tolerance shown by many climacteric women.

Thyroid changes

The excessive production of thyrotrophic hormone by the pituitary gland may initiate exophthalmic goitre at the climacteric, or lead to an exacerbation of pre-existing mild or latent hyperthyroidism. Sometimes a symptomless and long-standing goitre or small adenoma is driven into activity, with the development of symptoms of toxic goitre (secondary thyrotoxicosis). Climacteric hyperthyroidism may be progressive there may be a gradual return to normal in mild cases. Sometimes this involution is excessive and goes on to myxoedema, but the latter, however, may appear without an obvious preceding hyperthyroid phase.

Breasts

At the climacteric, the mammary glands tend to undergo atrophy but this may not be evident owing to the deposition of fat which indeed, can produce an actual enlargement. As a result of loss of cyclic oestrogen production during the climacteric, there may be changes in the ducts. These may undergo lengthening and tortuosity, associated occasionally with epithelial proliferation; and various symptoms, such as tenseness, paraesthesiae, hyperaesthesia (sometimes erotic.) and pain may develop. Occasionally, the breast secrete a thin fluid, probably as the result of the superimposed activity of the pituitary lactogenic hormone (prolactin), which in turn is a consequence of the general hyperactivity of the pituitary; as also may be the prolonged lactation which is sometimes observed in women who become pregnant towards the end of the reproductive period.

Acromegaly

Mild, or fugitive acromegaly sometimes occurs at the climacteric and can be explained by an increased output of pituitary growth hormone.

Hypopituitarism

Occasionally, following the climacteric, and particularly when there is a history of multiple pregnancies, an indefinite syndrome, with some features of Simmonds' disease, may be encountered. Presumably, in these patients, the pituitary gland enters upon a phase of exhaustion, following the phase of hyperactivity, in much the same way as hyperthyroidism may be replaced spontaneously by hypothyroidism.

The Climacteric Syndrome

The precise frequency with which climacteric symptoms of more than minor degree are experienced is uncertain. Hamblen (1947) quotes one authority who estimated 75 -

percent of all women as suffering from distressing symptoms at the climacteric, and other authorities who held that 70 to 90 per cent of climacteric women experience no symptoms materially interfering with general health, or with domestic or social activities,

The symptoms of the climacteric syndrome consist mainly of psychological disturbances and vasomotor instability. There is no doubt that the former are the more important, for it is just those women who have, or through force of circumstances develop psychological inadequacy, who are most liable to suffer from climacteric symptoms and who do so most severely. Two classes of women provide the majority of climacteric sufferers. At one end of the social scale is the woman of wealth and social standing, for whom the menopause is a remainder which cannot be ignored of advancing years with their attendant loss of good looks, sex appeal, and consequent dominance in her social circle. At the other end if the woman who has led a frustrated existence, deprived of good looks and the activities open to the more fortunate, who has been unwanted and unloved; for her the expectation and realization of the approaching end of reproductive function mean the abandonment of all hope of fulfilling her natural childbearing destiny. It is not surprising, therefore, that she should show evidence of despair or, instead, a protest reaction. The least susceptible women are those who have had a healthy, happy life, who have made a successful marriage and have raised a healthy family. For them the menopause is merely another milestone in life passed, with every reason to expect the next phase to be no less worth living than those which had gone before.

The psychic symptoms are largely conditioned by the circumstances which have evoked them. In all instances, however, irritability and depression and predominant. The former makes the woman short-tempered and intolerant; any slight deviation from the expected course of affairs provokes an exaggerated protest. There may be associated anxiety with vague forebodings of ill health or economic disaster in the future. Depression may be associated with tearfulness or with sadness, indifference, and apathy. The woman may lose all ambition, ceasing to care for her husband, her home, and her appearance. As part of the climacteric syndrome, changes in libido may occur. There may be an increase as a compensatory reaction to waning reproductive ability, or as a 'last fling' effect; or it may disappear abruptly as in the case, for example, of an unhappily married woman for whom the menopause can serve as an excuse for avoiding distasteful sexual activity. In the well-balanced woman the climacteric need cause no change in libido and many women continue to have normal and satisfying sexual relations long after menstruation has ceased.

The commonest manifestations of vasomotor instability in the climacteric are hot flushes (vasodilatation) often starting in the face and travelling all over the body. They occur spontaneously, or may be induced by emotion. They may be infrequent, or may be repeated many times during the day and, typically, even more frequently at night. The flushes are often followed by a wave of chilliness (vasoconstriction) and profuse perspiration. The blood pressure may fall during the flushes and rise with the chilliness. Paraesthesiae, cold extremities, tremors, palpirations, colonic spasm, cardiospasm, angioneurotic oedema, and pseudoangina are other features of vasomotor instability.

Migraine is sometimes very troublesome at the climacteric. It appears to be influenced by endocrine factors, as witness its association with puberty, menstruation, and the climacteric, together with a tendency to disappear during pregnancy, lactation, and post-climacteric life. Though often aggravated in the earlier phase of the climacteric many women lose their migrainous symptoms after this period. Its exacerbation at the climacteric might be explained by vasomotor spasm of the cerebral vessels, since cervico-thoracic sympathectomy relieves the condition. An alternative explanation is that it is due to enlargement of the pituitary gland, since migraine is a feature of pituitary syndromes, with or without neoplasm. It has been-thought that migraine occurs more commonly in those women whose suprasellar diaphragm is calcified, so preventing pituitary expansion. The pituitary theory is supported by the beneficial effects of doses of oestrogen which are sufficient to suppress pituitary activity to the extent of causing the disappearance of urinary gonadotrophin. Since benefit may also result from injections of gonadotrophin, which would itself depress pituitary hyperfunction, it is clear that the migraine cannot be due to the excessive production of that hormone. About half the cases of idiopathic migraine obtain relief from an artificial menopause but in the remainder the condition is made worse.

General or local pruritus may be very troublesome and various forms of dermatitis and impetigo occur. Pruritus, leucoplakia, kraurosis, and even superimposed carcinoma may affect the vulvar skin and cause much suffering. These abnormalities are aggravated by glycosuria, but also occur in its absence. It is uncertain to what degree the endocrine changes of the climacteric themselves should be held responsible for those various dermatoses, and to what extent they should be attributed to general ageing processes.

Treatment

The first point to be stressed is that the climacteric itself requires no treatment, being a physiological condition. It is only when symptoms supervene, of a degree sufficient to interfere with general health, or with domestic or social activities, that treatment is required. The second point to be stressed is that the successful treatment of the climacteric syndrome demands common sense and delicacy. The routine administration of oestrogen, often for long periods and in excessively high dosage, is strongly to be condemned; just as is the unhelpful, nihilistic approach which taking its stand on the physiological nature of the climacteric, refuses any form of treatment. It is perfectly true that mild cases can be treated effectively merely by reassurance and perhaps moderate sedation, such as is provided by phenobarbitone, $^1/_2$ gr. twice a day; but it is equally certain that the more severe cases will require oestrogen therapy and, in occasional instances, psychotherapy as well.

The basic principles of intelligent oestrogen therapy for the climacteric are as follows: first oestrogen is never produced continuously under physiological conditions, so that oestrogen treatment should always be in interrupted courses; second, the object of the treatment is to convert an abrupt fall in oestrogen level into a more slowly declining one. It follows, therefore, that if comparatively large doses are given, so that the total oestrogen level is restored to the previously normal value, symptoms will be certain to reappear every time treatment is stopped. Ignorance of this situation is responsible for most of the so-called 'difficult cases' who have been on oestrogen therapy for long periods at a time, but who relapse miserably each time the treatment is stopped. The view is held by some that large doses of oestrogen are required in order to suppress the pituitary hyperactivity—doses which, in fact, are sufficient to bring about the disappearance of urinary gonadotrophin. It is, however, a matter of clinical experience that relief of climacteric symptoms may be obtained with oestrogen doses which are a small fraction of those necessary to produce tangible pituitary depression; and since the drawbacks to the use of large oestrogen doses are serious and numerous, their exhibition in the climacteric seems to be indefensible. Chief among these drawbacks are: nausea; gastro-intestinal disturbance; backache and pelvic congestion; uterine bleeding, which may be prolonged, excessive, and extremely difficult to control, except by surgery; mastopathy; and, possibly, carcinogenesis.

In general it is convenient to start treatment with stilboestrol, 0.2 mg. daily. The treatment should be continued for twenty-eight days and then should be stopped for a fortnight, when it can be resumed at the same dose level. With this dose, most, if not all, of the symptoms will be relieved. Indeed, it is best if an occasional hot flush still occurs while the patient is under oestrogen treatment. Generally some return of symptoms will be experienced during the fortnight off treatment. However, as these monthly courses of treatment proceed, the stage will be reached when symptoms no longer return on omitting treatment. This is the time to reduce the dose of stilboestrol to 0.1 mg daily. Two or three further interrupted courses of treatment at this level will often be sufficient to see the end of all significant symptoms. Occasionally, 0.2 mg. will prove inadequate for controlling the symptoms; if this be so, the dose may be increased to 0.5 mg. daily but it is seldom necessary to exceed this figure.

Ethinyloestradiol may be used instead of stilboestrol; the corresponding dose is lout one-twentieth that of stilboestrol so that 0.2 mg. of the latter would be represented ,0.01 mg. of ethinyloestradiol.

It has been claimed that natural oestrogens (oestradiol, oestrone sulphate, oestriol) are superior to synthetic oestrogens in controlling climacteric symptoms since they promote a sense of well-being which the synthetic oestrogens do not. There is no truth whatever in this contention.

In recent rears, a vogue has developed for the use of mixed hormone preparations, usually containing ethinylo-estradiol and methyltestosterone. The rationale for such treatment is a little dubious but the general idea is that the two hormones are synergistic in many of their desirable characteristics, while being mutually antagonistic as far as some of the undesirable side effects are concerned. By the use of such a mixture, one is often able to control climacteric symptoms with far smaller oestrogen doses than would e possible without the admixture of the androgen. Generally speaking, there is little need to use these mixed preparations in the average case. It does appear, however, that for the more difficult patient who is-responding poorly to oestrogen alone, the mixture may be of real value. The general principles mentioned above apply equally to the use of mixed hormone preparations, the starting dose for which should be about two tablets per day. Some patients prove unduly sensitive to the androgen moiety of these mixed reparations, developing troublesome hirsuties.

Recently, a synthetic oestrogen of the allenolic acid series called methallenoestril (*Vallestril*), for which certain special properties have been claimed, has been advocated for the control of climacteric symptoms. It is claimed for this oestrogen that, whereas in a dose of 3-9 mg. per day climacteric symptoms may be fully controlled and an atrophic vaginal mucosa restored to normal, remarkably little effect is produced on the endometrium. The consequent advantage is that the risk of inducing uterine bleeding is minimized. Although this is of no importance in the majority of cases, if treated with stilboestrol in the manner described above, it is on the other hand true that occasionally women are found to have withdrawal bleedings following treatment even with as little as 0.2 mg. of stilboestrol daily. For these patients, *Vallestril* may be the drug of choice. Another recently introduced oestrogenic substance, chlorotrianisene (tri-panisylchloroethylene marketed under the name TACE) has interesting properties by reason of which, it is claimed; it has particular advantages in the treatment of the climacteric syndrome. It is stored in the body fat, from which there is slow prolonged release. TACE itself is a proestrogen, without direct oestrogenic activity, but is converted in the liver into a true oestrogen, the nature of which is unknown. The recommended course of treatment is two capsules (each of 12 mg. of chlorotrianisene dissolved in oil) by mouth daily for thirty or sixty days. This is stated to ensure complete relief in 50 percent of patients. A second course can be given; not more than 1 per cent of patients are said to require a third course.

Some physicians feel the need for elaborate laboratory investigations, such as repeated vaginal smear studies or even urinary gonadotrophin determinations, as indices to progress in the treatment of the climacteric syndrome. Common sense alone shows that these studies are totally unnecessary, since the treatment is purely a symptomatic one, the climacteric itself being a physiological state, as already stressed. The clinical state of the patient is the only guide needed in the treatment of this syndrome. The uselessness of vaginal smear studies is underlined first by the fact that many post-menopausal women, even though suffering from climacteric symptoms, may still show a relatively well-oestrogenized smear and second vaginal that the production of a fully oestrogenized smear by oestrogen therapy is suggestive not of correct treatment but of over dosage.

Where marked psychological aberrations are present, it is obviously necessary to adopt psychiatric measures in addition to such hormonal therapy as may be required. Since the syndrome is a self-limited one, the prognosis is usually fairly good and it is not often that elaborate psychotherapy is needed.

Climacteric Menorrhagia

Except in mild and transient cases, it is a wise rule to subject every patient with climacteric menorrhagia to a diagnostic dilatation and curettage. This procedure appears to be curative in a fair proportion of patients; estimates vary between 30 and 60 per cent. If the curettage fails to reveal any endometrial pathology, but severe haemorrhage nevertheless persists, a complete menopause, with cessation of bleeding, can be produced by external radiation, insertion of radium into the cervix, or by hysterectomy. Both external radiation and radium are relatively simple measures but may precipitate climacteric symptoms and adiposity in some patients. A careinoma of the body of the uterus may also be overlooked. Hysterectomy does not involve the ovaries and does not therefore disturb the endocrine system. It is, however, a major surgical procedure not entirely free from risk, or from anxiety on the part of the patient, but it is probably the method of choice in severe menorrhagia not responding to hormone therapy.

After a diagnostic curettage it is reasonable to try medical treatment before deciding on surgery or radiation. Androgens, progesterone, and oestrogens are used for their local action on the uterine endometrium, as well as for their inhibition of pituitary gonadotrophic activity. Nevertheless, treatment remains empirical. Ethisterone, 15 mg., together with methyl-testosterone, 5 mg; as two tablets or in a combined tablet (*Androgeston*) may be given twice daily sublingually, commencing a week before the period is expected and continuing until the third day of bleeding. This treatment will often restore the heavy period to more normal proportions. When the bleeding is prolonged and irregular, the choice lies mainly between treatment with progesterone to produce 'medical curettages' or with oestrogens.

A daily intramuscular injection of progesterone, 25 mg., and testosterone, 50 mg., for four days will usually terminate a prolonged bout of bleeding. A few days later a self-limited progesterone withdrawal bleeding will occur. Repetition of the progesterone injections (without the testosterone) at approximately monthly intervals will usually produce regular and normal 'periods'. Oestrogen therapy depends upon the fact that a sufficient dose of oestrogen will usually terminate a bout of bleeding. It may be given conveniently as stilboestrol, 2 mg., or ethinyloestradiol, 0.1 mg., daily, continued for a total of twenty days. If there has been no noticeable effect on the extent of the bleeding thin forty-eight hours, the dose should be doubled and continued at this higher level the remainder of the twenty days. Within a week or ten days after stopping the oestrogen treatment, a withdrawal bleeding will occur and on the fifth day another use of oestrogen therapy should be begun. In this way the regular bleeding is converted into regular cycles and the dose can gradually be reduced in succeeding months. Eventually a stage will be reached when the dose will be too small to provoke oestrogen withdrawal bleeding and the menopause will have been established. If during the course oestrogen administration the bleeding stops but begins again before the course is completed, further oestrogen should be withheld for five days and then a new course commenced as before.

Fears have been entertained that administration of oestrogens to women at the menopause may lead to carcinogenesis. There is no acceptable evidence to substantiate such a fear, and in any case, provided the oestrogen is given discontinuously and in animal effective doses as recommended above, the likelihood, even on theoretical grounds, of its exerting a carcinogenic action seems to be remote.

Failure to respond to hormonal treatment is not uncommon in cases of climacteric menorrhagia; it is obviously unwise to persist in such treatment if it is proving ineffective, and recourse should then be had to surgical measures without undue delay.

Abnormalities of Menstruation

In a textbook of major endocrine disorders, only limited aspects of abnormalities of menstruation call for attention. The gynaecological implications have always to be borne mind, and the information on these should be sought in textbooks of gynaecology. Moreover, abnormal uterine bleeding or amenorrhoea may be part of obvious clinical disorders of the endocrine glands (e.g. adrenogenital syndrome, hyperthyroidism, myxoedema) and as such are considered elsewhere in the present work. This chapter will deal only with certain conditions in which the disorder of menstruation appears to the only evidence of endocrine dysfunction.

AMENORRHOEA

Absence of menstruation for long intervals of months or years is termed amenorrhoea. It is said to be primary when the patient has never menstruated, or secondary if it supervenes after some years of more or less normal menstruation. Physiological amenorrhoea occurs ring pregnancy.

Primary Amenorrhoea

Apart from purely gynaecological causes, such as imperforate hymen or uterine aplasia, primary amenorrhoea is of two main types: the first, due to primary hypogonadism, consists in failure of response of the ovaries to adequate gonadotrophic stimulation and has already been discussed; in the second there is secondary hypogonadism, that is, a lack of gonadal function due to absence of gonadotrophic stimulation. This may be the result of a hypothalamic lesion or ill-understood functional disturbance. It may result from a lesion of the anterior pituitary gland, such as cranio-pharyngioma or chromophobe adenoma. It is also a characteristic feature of pituitary infantilism, or may be the result of an apparent selective deficiency of gonadotrophin secretion. In hyperpituitarism due to an acidophil tumour resulting in giantism, primary amenorrhoea is caused by destruction of the basophil cells which are considered to be the source of the pituitary gonadotrophins.

Other endocrine conditions in which primary amenorrhoea may occur are cretinism, juvenile myxoedema, and milder forms of hypothyroidism, toxic goitre, diabetes mellitus, and adrenocortical tumours. It may also result from severe organic disease of other systems, such as anaemias, chronic nephritis, chronic sepsis, and malnutrition.

It may be very difficult clinically to differentiate between delayed puberty and primary amenorrhoea. Consequently the continuation of normal menstruation after the exhibition of therapy for primary amenorrhoea cannot necessarily be regarded as *prima facie* evidence of cure of the primary amenorrhoea, since one may have been dealing merely with delayed puberty.

Reference has, already been made to male pseudoherm-aphroditism as a possible cause of apparent primary amenorrhoea.

Secondary Amenorrhoea

Without doubt, the commonest cause of cessation of the periods more or less normal menstruation has been established, apart from the occurrence of pregnancy, is psychological. Sometimes such a cause is of major proportions and obvious in its implications, but in many instances the precise psychological factors may be obscure and difficult to elicit.

As in the case of primary amenorrhoea, secondary amenorrhoea may be caused by severe general diseases, such as anaemias, advanced tuberculosis, chronic nephritis, malignant disease, chronic infections, and malnutrition. It is uncertain by what mechanism these conditions bring about the cessation of normal pituitary-ovarian activity.

Secondary amenorrhoea also arises in a large variety of endocrine disorders. These include hypothalamic disease affecting anterior pituitary function; Simmonds disease and other forms of hypopituitarism; ecromegaly, as a result of destruction of basophils by the eosinophil adenoma and Cushing's syndrome. Amenorrhoea may occur in both hyper- and hypothyroidism, in diabetes mellitus, and in Addison's disease. It is present in the adrenogenital syndrome and is also a symptom of virilism due to other causes, such as arrhenoblastoma and adrenal rest tumours of the ovary. It arises when destructive lesions, of whatever kind, of both ovaries destroy a sufficient amount of oestrogen-producing tissue.

The normal climacteric is a natural form of secondary amenorrhoea, resulting from the loss by ovulation and atresia of Graafian follicles. If this happens at an unusually early age, and is not accompanied by typical climacteric symptoms, the secondary amenorrhoea which results may be clinically indistinguishable from that which may occur from other causes; in such cases the finding of a raised gonadotrophin excretion would indicate the climacteric nature of the condition. Other possible causes of amenorrhoea are a functional failure of the pituitary gland to secrete adequate amounts of gonadotrophins, or a loss of the ability of the ovaries to respond to stimulation by gonadotrophins (for reasons other than those which apply at the climacteric), or of the endometrium to respond to the action of ovarian hormones.

Treatment of Amenorrhoea

The treatment of amenorrhoea is largely unsatisfactory and a long digression on the methods which have been employed is not warranted. Suffice it to say that the use of gonadotrophin preparations and of steroid hormones, though occasionally apparently successful, is essentially unreliable and good results cannot be predicted. In primary amenorrhoea the principal indication for long-continued cyclic oestrogen therapy is the effect that this has on the development of secondary sex characters and the psychological advantages to the patient of having apparently normal menstrual bleedings. In secondary amenorrhoea the value is again mainly psychological, except in those instances where, on cessation of a fairly prolonged course of such interrupted therapy, normal menstruation is re-established. For many intelligent women for whom sterility is not a complaint, reassurance that secondary amenorrhoea is not due to underlying disease is often sufficient, and for these hormone therapy is apparently quite unnecessary.

Abnormal Uterine Bleeding

Regular menstruation which is nevertheless excessive in the amount of blood loss, and often at the same time in the duration of the bleeding, is usually referred to as menorrhagia or hypermenorrhoea. Irregular uterine bleeding is usually called metrorrhagia, and when it is associated with hyperplastic changes in the endometrium, the condition is described as metropathia haemorrhagica. The too frequent occurrence of menstrual periods is called polymenorrhoea and their too infrequent occurrence, oligomenorrhoea. Unusually scanty menstrual periods constitute hypomenorrhoea.

In, the elucidation of the causes of abnormal menstrual bleeding, proper gynaecological examination, including, in most cases, diagnostic curettage, is an essential preliminary.

Menorrhagia

In the absence of organic cause, such as fibroids, or pelvic inflammatory disease, this condition most commonly has a psychological basis and would seem to consist essentially in an abnormality of neurovascular control in the endometrium. Biopsy of the endometrium usually reveals an entirely normal secretory pattern in the immediately premenstrual phase, and may show no abnormalities if taken during bleeding itself. In such circumstances it is difficult to imagine any possible endocrine basis for the disturbance and it is not surprising that hormone treatment often has little success.

In some cases, however, an endometrial biopsy taken about the fifth day of bleeding shows the pattern of incomplete shedding of the endometrium, in which mixed proliferative and secretory changes are found alongside each other. It is possible that in these cases there is a delay in the involution of the corpus luteum, with a corresponding prolongation of the secretion of progesterone, though in suboptimal amounts; the continued elimination of pregnanediol during part of the bleeding episode (it usually disappears from the urine, except for traces, before the onset of bleeding) supports this supposition. Unfortunately no treatment based on such an hypothesis has succeeded in dealing effectively with the condition.

Polymenorrhoea

This is often combined with menorrhagia, and when not due to organic cause is again most commonly of emotional origin. It may also arise as a result of early ovulation, and in some cases ovulation may occur even before the menstruation has ceased, thus leading to involuntary sterility. Occasionally it may arise as a result of premature degeneration of the corpus luteum with a corresponding reduction in -the post-ovular phase of the cycle. The fundamental disturbance may then be one of anterior pituitary function.

Metrorrhagia

Irregular uterine bleeding, which may also be prolonged and heavy, can result from organic pelvic lesions, but commonly occurs in the absence of any such obvious causes. Diagnostic curettage may reveal a normal proliferative endometrium, a hypoplastic one or a hyperplastic one, showing typical cystic glandular hyperplasia (the 'Swiss cheese' hyperplasia of Novak). Occasionally mixed hyperplastic and secretory patterns have been described.

This abnormality of menstruation may arise at any age between the menarche and menopause being relatively rather common close to both of these epochs. The essential feature appears to be absence of ovulation, with a consequent lack of cyclic production of oestrogen and progesterone by the ovary. There is no certainty that the typical picture of metropathia haemorrhagica, in which cystic ovaries are found, in association with cystic glandular hyperplasia, is due to an excessive oestrogen production, and the precise hormonal derangements are not known:- Some women appear to have an essentially unstable menstrual rhythm, so that they may alternate between phases of regular menstrual function, and of irregular metrorrhagia. A wholly endocrine cause of metrorrhagia is the occurrence of an oestrogen-secreting tumour of the ovary, such as

Treatment of Abnormal Uterine Bleeding

The diagnostic curettage itself provides effective treatment in a proportion of patients with abnormal uterine bleeding. This proportion, however, seems to vary with different observers. Menorrhagia of emotional origin is unlikely to respond to any kind of hormonal treatment and logically should be dealt with on psychiatric lines. Unfortunately this approach too is often fruitless, and in severe cases hysterectomy has seriously to be considered. In other cases of regular menorrhagia not due to organic disease, the results of any kind of treatment are unpredictable. However, in the writer's experience, a regime worth trying consists of methyltestosterone, 5 mg., and ethisterone, 15 mg., twice daily sublinguaily, beginning a week before the period is expected and continuing until the third day of bleeding. A combined tablet containing these steroids in the above proportions is marketed as *Androgeston*. The methyltesterone can be replaced by *Androstalone*, 25 mg., daily, and for some patients this regime is more effective (Swyer, 1956).

Hormone treatment is most likely to succeed in cases of metrorrhagia without organic cause and may take various forms. For the arrest of prolonged and excessive uterine bleeding, the injection of progesterone and testosterone propionate for a few days is often the most effective procedure (Greenblatt, 1947). If the endometrium is known or suspected to be atrophic the injection should also include oestradiol benzoate or di-propionate. The doses of these steroids are not very critical and something like progesterone 50 mg., testosterone propionate 25 mg., and oestradiol benzoate 2 mg., daily for three or four days is usually sufficient to bring about a complete arrest of bleeding or a very marked amelioration. A few days later a progesterone withdrawal bleeding, which may be likened to a normal period, begins and is self-limited, ending after five or six days. It is of course necessary to inform the patient that this will happen, as otherwise she will fear that the treatment has been ineffective.

It is usually a wise plan to follow this treatment with oestrogen therapy, beginning on about the fifth day of the withdrawal bleeding referred to above. Stilboestrol, 2 mg., or ethinyloestradiol, 0.1 mg., daily should be given orally for a total of twenty days, when an oestrogen withdrawal bleeding will begin some days later.

An alternative method for securing haemostasis, where the bleeding is not severe, is to begin with oestrogen. The dose may be similar to that mentioned above and if there is no reduction in the bleeding within forty-eight hours the dose should be doubled and maintained at that level for the rest of the twenty-day course. Difficulties are sometimes encountered because the dose of oestrogen necessary to secure haemostasis leads to nausea or vomiting as a side reaction. A further possibility is that of the bleeding, having ceased, beginning again before the twenty-day course is completed. If this happens treatment should be stopped for five days and then begun again when effective control will usually be obtained.

Haemostasis having been secured, it is then necessary to re-establish cyclic bleeding, and this is conveniently done by repeating the courses of oestrogen, commencing on the fifth day of each withdrawal bleeding and continuing as before for twenty days. After three such courses there seems to be an advantage in combining the oestrogen with progesterone or ethisterone, since by this means the chances of including ovulation seem to be increased. A method which has been found effective in a fair proportion of patients is as follows (Swyer, 1950): after three controlled cycles with oestrogen alone a fourth cycle is started as before with oestrogen on the fifth day of the cycle (stilboestrol, 2 mg., or ethinylo-estradiol, 0.1 mg., daily for twenty days). On the fifteenth day of the cycle, that is, after oestrogen alone has been given for ten days, ethisterone 40 mg. daily is given for ten days. In this way the ethisterone course, overlaps the second half of the oestrogen course. On the fifth day of the next cycle the dose of oestrogen is reduced to half its previous value, and on the fifteenth day of the cycle ethisterone is begun again, this time at a dose of 60 mg., daily for ten days. In the next and final cycle the dose of oestrogens is again halved (stilboestrol, 0.5 mg., or ethinyloestradiol, 0.25 mg., daily) and on the fifteenth day of the cycle ethisterone is again started, this time at a dose of 80 mg., for ten days. Ovulation has been known to occur during the final cycle of treatment and pregnancy to ensue. Satisfactory results can be expected in some 80 or 90 per cent. of patients with irregular, prolonged, and excessive uterine bleeding after treatment in this way, though some of these may later relapse. In about 40 per cent. of the patients ovulation will be re-established following this regime.

An alternative form of treatment for this kind of menstrual disorder is with progesterone, as originally described by Scowen (1944). Once haemostasis has been secured, a few injections of progesterone (the precise dose seems to be relatively unimportant: 20 mg. on alternate days for three injections, or even 50 mg. in a single injection seems to be satisfactory) will be followed by a withdrawal bleeding. These injections are repeated monthly so as to bring about regular progesterone withdrawal bleedings, which, accompanied as they are by endometrial shedding, prevent the building up of a thick vascular proliferative endometrium from which future irregular bleeding could occur. In some patients it is possible, after a few courses of such injections, to substitute ethisterone (e.g., 50 or 60 mg. daily for about five days, repeated monthly) and to obtain equally satisfactory results.

Other forms of treatment, using androgens and gonadotrophins have often been advocated. The use of androgen combined with progesterone has been mentioned already and is very effective for initial haemostasis. Androgen alone, however, is, in the writer's opinion, much less satisfactory, the dose required to produce a satisfactory effect often being such that, when given for even moderately prolonged periods of time, it tends to cause undesirable virilizing effects. The use of chorionic gonadotrophin has little to recommend it, but it is possible that *Synapoidin*, which is a combination of chorionic and pituitary gonadotrophin, may sometimes prove effective where other measures have failed.

STEIN-LEVENTHAL SYNDROME

Although it was as long ago as 1929 when Stein performed his first wedge resection of bilateral 'polycystic ovaries' for the relief of the accompanying amenorrhoea, it is only within comparatively recent years that general interest has been aroused in what has now come to be known as the 'Stein-Leventhal syndrome', the literature on which, up to 1954, has been summarized by Bishop (1954). The fact that some abnormality of menstruation is usually (though not invariably) a part of the syndrome is the justification for considering it in this chapter.

Bilateral ovarian enlargement is the only invariably finding in this syndrome; however, there is usually amenorrhoea or oligomenorrhoea, and there is commonly also hirutism of varying degree. Though some patients are obese, many are not and there is no justification for including obesity as part of the syndrome.

Ovaries

As mentioned above, bilateral enlargement is invariable and is usually palpable on pelvic examination, if necessary under an anaesthetic. However, cases have been reported in which the ovaries, though found to be enlarged on laparotomy, were not palpably enlarged on pelvic examination. Indeed, Stein, Cohen, and Elson (1949) stated that in only about half of their series of 75 patients could the ovarian enlargement be detected on pelvic examination and they therefore felt that radiological investigation, following the induction of a pneumoperitoneum, which outlines the shape and size of the ovaries, should be an essential procedure in attempting to establish the diagnosis. Other procedures which have been advocated for this purpose are culdoscopy and peritoneoscopy, and combined pneumoperitoneum and hysterosalpingography ('gynaecography').

Although the condition has been referred to as 'polycystic ovaries', the cysts, which are not always present, are in fact no more than slight or moderate enlargements of multiple follicles. Sometimes the ovaries are quite solid, there being no cystic change at all. On microscopical examination, the most striking feature is marked overgrowth of the ovarian stroma. Numerous follicles may be found in all stages of maturation, and in general, there is evidence of increased atresia also. In some cases masses of luteinized cells are found, but these are rather unusual. Hyperplasia of the hilus cells is a feature of some of these ovaries, but again others fail to show it. Although some authors have stressed hyperplasia of the theca interna cells and Fraenkel (1943) coined the term 'hyperthecosis ovarii', such hyperthecosis is by no means invariably and it seems doubtful if the term is at all justifiable.

Menstrual Disturbance

Most commonly this is secondary amenorrhoea, though one case associated with primary amenorrhoea has been described. Some patients have oligomenorrhoea; in which the interval between menstruation may be anything from a matter of several weeks to several months. In yet others, occasional or more frequent irregular heavy bleeding occurs, the clinical picture resembling that of metropathia haemorrhagica. Most often there is absence of ovulation, but some patients certainly do ovulate, and Stokhuyzen (1950) reported a case with normal regular cycles and a secretory endometrium at menstruation. In one of the author's patients a fresh corpus luteum was found at laparotomy, and the endometrium obtained at the same time showed a normal secretory pattern.

Other Signs and Symptoms

Hirsutism is a very common, though again not invariable, component of the syndrome, In a few cases it has been of a very severe degree and even accompanied by the other signs of virilism, such as enlargement of the clitoris, voice changes, acne, and a muscular physique. These, however, are extreme examples. In some the hirsutism may be so mild as to give rise to no complaint.

Sterility is one of the commoner complaints, though of course would be expressed only by those patients who are married. Obesity occurs in some patients, but many are of normal weight, and some individuals with the syndrome may be markedly underweight. It seems probable in fact that any significant variations from the normal in weight are more likely to be due to psychological causes than to underlying endocrine changes.

Hormone Studies

Very little has been published about hormone studies in these patients, but certain points have been established, In the first place, the 17-ketosteroid output is almost invariably within the normal limits, and attempts to demonstrate the excretion of excessive amounts of androgen in the urine do not appear to have met with success. Occasionally pregnanediol may be found in possibly significant amounts in the urine (Fischer and Riley, 1952), but it is certainly not in excess in other cases. Oestrogen excretion studies have not been reported, but on clinical grounds it may be concluded that, although in some patients the oestrogen level is normal or even high, in others it is relatively low. The output of gonadotrophins is uncertain in these cases; in the author's patients no excessive secretion of urinary gonadotrophin has been revealed by techniques which, though in current use, are known to be of limited sensitivity.

Although there has been a good deal of theorizing about the precise nature of the disorder and its hormonal implications, it is fair to say that all this has been purely speculative and that we really do not know the true nature of the condition or how the signs or symptoms are brought about. It may be that in some cases there is an abnormally high output of progesterone or some related hormonal substance with mildly androgenic properties, but this could scarcely hold true for all. It has been suggested that overproduction of luteinizing hormone by the pituitary might be the cause of the ovarian changes, but again this fails to account for those cases in which there is no luteinization in the ovaries. The remarkable response to treatment (see below) does not help in any way to explain the genesis of the condition, and the conclusion seems inescapable that we really are still very ignorant about the precise nature of this syndrome.

Treatment

Whatever may be doubted about the cause and nature of the Stein-Leventhal syndrome, the value of bilateral wedge resection in treatment is undoubted, though the manner in which this procedure produces the beneficial effects is just as mysterious as the cause of the syndrome itself. In point of fact, Stein did his first wedge resection solely for diagnostic purposes, and was very agreeably surprised to find that normal menstruation was restored following the operation. Such restoration of normal ovular menstrual cycles can be expected in about 80 per cent. of cases, and conception is common in those patients who are complaining of sterility. The effects on the hirsutism, however, are a good deal less spectacular. Nearly always there is some reduction in the rate of hair growth, but it is rare for the excessive hair to disappear and the results cannot be compared with those which follow removal of a virilizing adrenal tumour.

The precise duration of the benefits of bilateral wedge resection is uncertain, but it would seem that in general there is little likelihood of relapse, and it may well be that most patients after treatment remain normal until their natural menopause.

In a recent paper Stein (1955) reports that in the past twenty five years he has done wedge resections on 88 carefully selected patients, in 95 per cent of whom menstrual function was restored. Fifty-four became pregnant with a total of 118 pregnancies, and there were no recurrences of bilateral polycystic ovaries.

INDEX